AF804595
WITHDRAWN

Physics and Evolution of the Earth's Interior 1

CONSTITUTION OF THE EARTH'S INTERIOR

ERRATA

Page, line	For:	Read:
12^4	83/89	83/39
13^9	$1+4\frac{r^3p}{mc^2}$	$1+4\pi\frac{r^3p}{mc^2}$
38_2	$K_0' \geqslant 3$	$K_0' > 2$
55_{12}	$\ln K+17.3-4\ln\frac{2\bar{A}}{\varrho}$	$\ln K = 15.3-4.2\ln\frac{2\bar{A}}{\varrho}$
55_{11}	$\ln K = 15.1-3\ln\frac{2\bar{A}}{\varrho}$	$\ln K = 12.2-2.8\ln\frac{2\bar{A}}{\varrho}$
55_{10}	K is in MPa	K is in GPa
88_{14}	$Q_{\text{Earth}} = R^2q$	$Q_{\text{Earth}} = \pi R^2q$
88_4	absorbed by	incident on
203_9	$\lambda \lesssim 0.15\,\mu\text{m}$	$\lambda \gtrsim 0.15\,\mu\text{m}$
212^6	(σ)	(σ_j)
250^{10}	$v_s = \sqrt{\frac{K}{\varrho}}$	$v_s = \sqrt{\frac{\mu}{\varrho}}$
350^{14}	cover, plates with	cover, and plates — both with
$360_{\text{Diortho-silicates}}$	$(Ca,Na)_2(Al,Mg)\ (Si,Al)_2O_7]$	$(Ca,Na)_2(Al,Mg)\ [(Si,Al)_2O_7]$

J. Leliwa-Kopystyński, R. Teisseyre (eds.): *Constitution of the Earth's Interior.*

Physics and Evolution of the Earth's Interior 1

series editor: Roman Teisseyre

Volume 1 Constitution of the Earth's Interior (edited by J. Leliwa-Kopystyński and R. Teisseyre)

Volume 2 Seismic Wave Propagation in the Earth (edited by A. Hanyga)

Volume 3 Continuum Theories in Solid Earth Physics (edited by R. Teisseyre)

Volume 4 Gravity and Low-Frequency Geodynamics (edited by J. Czyż)

Physics and Evolution of the Earth's Interior 1

Constitution of the Earth's Interior

edited by

J. Leliwa-Kopystyński

Institute of Geophysics of the Polish Academy of Sciences

and

R. Teisseyre

Institute of Geophysics of the Polish Academy of Sciences

contributors

N. Bakun-Czubarow

A. Guterch

J. Leliwa-Kopystyński

S. Maj

R. Teisseyre

Elsevier

Amsterdam — Oxford — New York — Tokyo

PWN—Polish Scientific Publishers

Warszawa

1984

Translated from the Polish by
Sławomir Gibowicz, Institute of Geophysics of the Polish Academy of Sciences

Cover design by *Zygmunt Ziemka*

Distribution of this book is being handled by the following publishers:

for the USA and Canada
ELSEVIER SCIENCE PUBLISHING CO., INC.
52, Vanderbilt Avenue
New York, N. Y. 10017

for Albania, Bulgaria, Cuba, Czechoslovakia, German Democratic Republic, Hungary, Korean People's Democratic Republic, Mongolia, People's Republic of China, Poland, Romania, the USSR, Vietnam and Yugoslavia
ARS POLONA
Krakowskie Przedmieście 7, 00-068 Warszawa 1, Poland

QE
509
.C67
1984

for all remaining areas
ELSEVIER SCIENCE PUBLISHERS
Molenwerf 1
P.O. Box 211, 1000 AE Amsterdam
The Netherlands

Library of Congress Cataloging in Publication Data
Main entry under title:
Constitution of the Earth's interior.
(Physics and evolution of the Earth's interior; 1)
Bibliography: p.
Includes index.
1. Earth—Internal structure. I. Leliwa-Kopystyński, J., 1937- . II. Teisseyre R. (Roman), 1929- . III. Series.
QE509.C67 1983 551.1'1 83-11657
ISBN 0-444-99646-X (vol. 1)
ISBN 0-444-99647-8 (series)

Copyright © by PWN – Polish Scientific Publishers – Warszawa 1984

All rights reserved.
No part of this publication may be reproduced, stored in a retrieval system or transmitted in any form or by any means, electronic, mechanical, photocopying, recording, or otherwise, without the prior written permission of the publisher.

Printed in Poland by D.R.P.

Preface to the Series
Physics and Evolution of the Earth's Interior

The physics and evolution of the Earth's interior is a very wide subject today. It is more and more difficult to write a comprehensive monograph on these problems owing to the fast progress in the Earth sciences and the considerable increase in the amount of information about our Planet. Some limitations are necessary. The existing books on this subject differ in their contents and methods of presentation. Some of them present the problems and methods of applied and exploration geophysics relating to studies of shallow structures and formations in the Earth's crust. These problems, though important for practical application, are not included in the present series. It should be noted, however, that the practical aspects of geophysics cannot be restricted to the methods used in exploration geophysics. They also arise from a knowledge of global processes and evolution and their contribution to the formation of either regional or local structures and fields. The authors of the present new series on the physics and evolution of the Earth's interior have concentrated on the physical principles of our knowledge about the Earth, since a proper theoretical background permits not only the formulation of general rules but also the prediction of various particular cases.

The recent development of cosmic research has supplied new information about the planets and their structure. Comparative data obtained for planets of the Earth's group provide a basis for a broader view of terrestrial evolution. Generally speaking, the relative rate of evolution of a planet is a function of its mass; therefore the data concerning different planets may be a source of information about the progress of evolution in time. Thus the objective of the present series is a search for physical and evolutionary principles connecting various elements of our knowledge about structure and dynamics of the Earth's interior.

In the various volumes of this series contemporary dynamical processes occurring in the Earth's interior are considered. Its physical, chemical and petrological state is described in the first volume, which will be followed by consecutive volumes of the series, devoted to the problems of propagation of seismic waves in the Earth's interior, the gravity field, free oscillations and rotation of the Earth, application of the theory of continuous media to the study of earthquakes and dynamical processes in the Earth's interior, including the processes of thermal convection and magnetohydrodynamic dynamo in the Earth's core. An attempt to synthesize Earth dynamics is made in the volume devoted to the Earth's evolution, which is outlined with reference to the general concepts of the evolution of the solar system. The evolution of our planet is related primarily to its thermal history and to the differentiation processes of its matter. Plate tectonics is considered in connection with systems of thermal convection, which is the primary force governing dynamical processes in the Earth.

Thus the series deals with the most important problems of solid Earth physics and presents the most general theories describing contemporary dynamical processes and the Earth's evolution.

Roman Teisseyre

Preface

This volume, the first of the series, contains general data on the depth distribution of basic parameters within the Earth. It presents a general stationary model which is, generally speaking, described by equations of state combining the basic parameters: pressure, temperature, density, gravity acceleration and mineral composition within the Earth. As a rule the variations of these or other equivalent parameters cannot be measured directly, so it is necessary to introduce a model or set of models that would fit the observations as well as possible. In the present volume we deal mostly with spherical-symmetric models.

The majority of the material presented in this volume concerns the chemical and petrological composition of the matter in the Earth's interior; models containing horizontal inhomogeneities were used only to illustrate processes connected with phase transitions. Phase transitions are taken into account already in this volume, which is motivated by the fact that they determine phase boundaries which, along with chemical boundaries, form a general structural pattern. At present it seems reasonable to assume that the Earth's interior is a complicated thermodynamical system in which the processes of phase transition and gravitational differentiation of matter, connected with thermal convection, form a system that generates the present dynamics or, in general, the whole evolution of the Earth's interior.

The authors present each topic in a fairly comprehensive way, with particular regard to the physics of phenomena and do not avoid mathematical and physical descriptions. This directly determines the set of readers—although some chapters can be studied without any special methematical and physical background, such a background is necessary to understand the whole work. Thus the book is mainly addressed to scientists and students of geophysics. It can also be used by physicists as an introduction to geophysical problems. Some chapters will also be of interest and easy to follow for geologists.

As regards the present volume, its authors are convinced that their effort will be rewarded if the reader gets a general idea about how the global structural models of the Earth's interior are being derived from observational data and the basic laws of physics and chemistry.

Roman Teisseyre

Contents

Basic Parameters of the Solar System

EARTH		
Mass	M	5.974×10^{24} kg
Volume	V	$1.0832 \times 10^{21}\ m^3$
Mean radius	R	6.371×10^6 m
Equatorial radius	R_e	$1.00163\,R$
Polar moment of inertia	I_{33}	$0.33168\,MR^2$
Equatorial moment of inertia	$I_{11} = I_{22}$	$0.33058\,MR^2$
Mean density	ϱ	$5.52 \times 10^3\ kg \cdot m^{-3}$
Mean equatorial gravity	g_e	$9.7803\ m \cdot s^{-2}$
Outer core radius		3.485×10^6 m
Inner core radius		1.250×10^6 m
Mean surface heat flow	J	$0.063\ J \cdot m^{-2} \cdot s^{-1}$
Excess luminosity	L	$3.2 \times 10^{13}\ J \cdot s^{-1}$

PLANETS

	Mass	Radius	Density
Terrestrial group			
Mercury	0.055	0.383	0.984
Venus	0.815	0.950	0.949
Earth	1	1	1
Mars	0.107	0.532	0.714
Major planets			
Jupiter	317.9	10.956	0.241
Saturn	95.2	9.072	0.128
Uranus	14.6	3.955	0.236
Neptun	17.2	3.861	0.299

SUN AND PLANETARY SYSTEM

	Mass	Angular momentum
Sun	1.99×10^{30} kg	$1.6 \times 10^{41}\ kg \cdot m^2 \cdot s^{-1}$
Planetary system	2.7×10^{27} kg	$3.1 \times 10^{43}\ kg \cdot m^2 \cdot s^{-1}$

1. Equations of State and Geophysical Equations of State

1.1 INTRODUCTION

All structure models of the Earth, planets or stars are based on certain principal observation data and on general physical laws common to all those bodies.

The observation data can be roughly divided into the following categories:

(a) Astronomic and geodetic measurements related to the total mass, body shape and moment of inertia of the body.

(b) Surface observations of quantities accessible to direct measurements, such as density, temperature, heat flow, gravity acceleration, chemical and mineral composition, and various material parameters.

(c) Seismic observations realizable on the Earth, the Moon and some planets, providing the depth distribution of the velocities of various elastic waves and the eigenperiods of oscillations of those bodies.

The basic physical laws necessary to construct models of large bodies are the law of gravitation, equations of state and equations related to the energy generation and transport inside and energy emission outside of the body. Only the law of gravitation is of a general character. The other equations, i.e., the relations between the pressure p, density ϱ and temperature T or the equations of state $p = p(\varrho, T)$, and the equations of energy transport, combine numerous physical phenomena, and thus they differ according to which of those phenomena are dominant.

The internal structure of large bodies can be recognized from the law of gravitation and from observation data properly arranged and as complete as possible. Various models differ from one another in many details, but certain general features in the structure of similar bodies (e.g., the Earth and the planets of the Earth's group) can be considered as well-established.

Using the law of gravitation in the classical Newton form we will roughly estimate the relations between the pressure p and the radius R of a spherically symmetric and nonrotating body with mass $M_R = M(R)$. The attraction force $\mathrm{d}F(r)$ between the mass $M(r)$ contained in a sphere with the radius r ($r < R$) and mass $\mathrm{d}M(r)$ of a spherical layer with the thickness dr is

$$dF(r) = -G\frac{M(r)dM(r)}{r^2} \tag{1.1.1}$$

where

$$M(r) = 4\pi\int_0^r \varrho(r')r'^2dr', \quad dM(r) = 4\pi r^2\varrho(r)dr \tag{1.1.2}$$

and $G = 6.67\times10^{-11}\ m^3\cdot kg^{-1}\cdot s^{-2}$ is the gravitation constant. The change of pressure along the layer thickness dr is

$$dp(r) = \frac{dF(r)}{4\pi r^2} = -G\frac{M(r)\varrho(r)dr}{r^2} \tag{1.1.3}$$

Taking into account that

$$g(r) = G\frac{M(r)}{r^2} \tag{1.1.4}$$

is simply the gravity acceleration, we can write formula (1.1.3) in the usual hydrostatic form

$$dp(r) = -g(r)\varrho(r)dr \tag{1.1.5}$$

Let us now estimate the pressure inside the body at a distance r from the centre, under the assumption that the density is constant. Equation (1.1.3) takes the form

$$dp(r) = -\tfrac{4}{3}G\pi r\varrho^2dr \tag{1.1.6}$$

or after integration

$$p(r)-p(0) = -\tfrac{2}{3}\pi G\varrho^2r^2 = -\frac{3}{8\pi}G\frac{M^2}{r^4} \tag{1.1.7}$$

The pressure at the centre of a homogeneous sphere can be found from the condition that at the surface, i.e., when $r = R$, the pressure is $p(R) = 0$, and thus

$$p(0) = G\frac{3}{8\pi}\frac{M_R^2}{R^4} \tag{1.1.8}$$

From formula (1.1.8) it follows that the pressure at the Earth's centre ($M_R = 5.98\times10^{24}$ kg, $R = 6.371\times10^6$ m) is $p(0) = 170$ GPa, while in proper models of the real Earth this pressure is twice as high. Such good agreement justifies the further use of our very rough approximation (see Stevenson, 1978). Let us assume that the following equation of state is fulfilled inside the body:

$$p \propto \varrho^n, \quad n > 0 \tag{1.1.9}$$

therefore $p \propto M_R^n R^{-3n}$. Since from eqn. (1.1.8) it follows simultaneously that $p \propto M_R^2 R^{-4}$, comparing the right sides of these formulas we obtain the relationship between the body radius R and its total mass M_R

$$R \propto M_R^{\frac{2-n}{4-3n}} \tag{1.1.10}$$

For $n < \frac{4}{3}$ or $n > 2$ the body radius increases with increasing mass. This intuitively obvious relationship is fulfilled e.g., for bodies described by the Murnaghan ($n \approx 2.8$) or the Murnaghan–Birch ($n = \frac{7}{3}$) equations of state. From relation (1.1.10) it follows, however, that there is an interval of values of the exponent n ($\frac{4}{3} < n < 2$) in which a mass increase is associated with a decrease of radius. In particular, for $n = \frac{5}{3}$, corresponding to the Thomas–Fermi equation of state, the following inverse proportionality is observed

$$R \propto M_R^{-1/3} \tag{1.1.11}$$

which means that the body radius diminishes as the mass becomes greater. The derivation of this relation, based on a dimensional analysis, was given by Landau and Lifshitz (1958). In particular, from a dimensional analysis it follows that $p \propto GM_R^2/R^4$, which is a relation similar to eqn. (1.1.8). This relation is the only possible combination, having the dimension of pressure, of the universal constant G and the quantities M_R and R characterizing the body.

The above relations indicate how important the proper selection of the equation of state becomes for the correct description of the internal structure of bodies with large masses. It is obvious that as the values of the parameters of state increase, i.e., from the surface towards the body centre, different thermodynamic conditions, described by different equations of state, become predominant.

1.2 GENERAL CLASSIFICATION OF EQUATIONS OF STATE

Equations of state can be classified in several ways. In this work it is convenient to introduce two principal types:

(a) "thermodynamical" equations of state,

(b) "geophysical" equations of state.

In a one-component medium which is in a state of one phase, the parameters of state are the pressure p, temperature T, and density ϱ or the specific volume $V = 1/\varrho$. The equation

$$f(p, \varrho, T) = 0 \tag{1.2.1}$$

connecting the parameters of state is called the thermal equation of state or shortly the *equation of state*. In case where the temperature effect can be neglected, the equation of state (1.2.1), assuming that $T = 0$, is reduced to the *barotropic* (isothermal) *equation*

$$p = p(\varrho) \tag{1.2.2}$$

The barotropic equation is not formally dependent upon the temperature and is applicable at the temperature of absolute zero. In practice material parameters dependent on temperature can be included in the equation, extending its applicability to $T > 0$.

Equations connecting the internal energy U with the parameters of state are called *caloric equations of state*

$$U = U(p, \varrho) \quad \text{or} \quad U = U(p, T) \quad \text{or} \quad U = U(\varrho, T) \tag{1.2.3}$$

The relation derived in Section 1.5

$$p = T\int \frac{1}{T^2}\left(\frac{\partial U}{\partial V}\right)_T \mathrm{d}T \tag{1.2.4}$$

is the relation between the thermal and caloric equations of state.

In a multi-component medium the equation of state (1.2.1) depends upon additional variables, which are the concentrations of particular components. Further difficulties might arise from phase transformations which could possibly occur in a given system, within a given interval of pressure and temperature. In this case the concentrations of particular phases become additional variables.

The Earth, as a heterogeneous body with regard to its chemical and mineral composition, and as a medium with a strong vertical gradient of thermodynamical parameters, is an example of a body for which a thermodynamical equation of state of the type of eqn. (1.2.1) cannot be established. The number of variables would be too large. That is why on the basis of laboratory data and seismic observations and possibly some definite form of the isothermic equation of state (1.2.2) we seek the relations between some of the following quantities: ϱ (density), v_P (longitudinal wave velocity), v_S (shear wave velocity), ϕ (seismic parameter), K (bulk modulus), and $\bar{A}$ (mean atomic mass of the medium). Thus the equations

$$f(\varrho, v_i, K, \bar{A}) = 0 \quad \text{where } v_i = v_P \text{ or } v_S \text{ or } \sqrt{\phi} \tag{1.2.5}$$

are called *geophysical equations of state*. In real situations relation (1.2.5) is usually a function of two or three variables, and depends upon empirically selected material coefficients. Different geophysical equations of state are considered in Section 1.10. Since these equations contain the mean atomic mass $\bar{A}$, they permit the estimation of the Earth's composition at various depths, and this is the most significant role played by equations of state in geophysical studies.

1.3 EQUATIONS OF STATE IN A WIDE RANGE OF CHANGES OF THERMODYNAMIC PARAMETERS

The thermodynamic parameters p, ϱ and T describing real physical systems may change by several orders of magnitude. It is not possible, however, to obtain general equations of state of the form

$$f(p, \varrho, T; \text{ universal constants}) = 0 \tag{1.3.1}$$

which would be valid for any medium and within any range of changes of the parameters of state. In practice we usually introduce equations of state in the form

$$f(p, \varrho, T; \text{ material coefficients}) = 0 \tag{1.3.2}$$

which are valid in intervals of changes of the parameters p and T in which only one phase state of the medium is observed. The material coefficients in eqn. (1.3.2) are as a rule quantities determinable in laboratory measurements. It is often assumed that the "material coefficients" are not dependent on pressure and temperature and are therefore "material constants".

Equations of the form (1.3.2) are used within a range of parameter changes corresponding approximately to those existing in the crust and mantle and possibly in the core of the Earth. Thus they relate primarily to the solid state and liquid state under not too high temperature.

Equations of the form (1.3.1) are usually introduced for gas media in the normal sense and for highly dense matter of high temperature. Such equations are also constructed for relativistic ranges of thermodynamic parameter changes. It is possible, therefore, to consider universal equations of state which are valid within a range of parameter changes corresponding to a given phase state. Equations of state describing highly condensed matter are used in the theory of star evolution. In particular, they are applied to the modelling of supernova stars and to the transition to neutron stars (central density of the order of 10^{13} g · cm^{-3} and more), or black hole formation. Considerations relating to media with a density $\varrho > 10^{14}$ g · cm^{-3}, greater than the density of nuclear matter, are very uncertain, since the laws governing interactions between particles are not known.

Below, we present different approaches to the description of the properties of matter under various physical conditions, namely: (1) the diagram of Brush (1966) relating to a very wide range of changes of density and temperature, (2) properties of matter as functions of its density given by Rees *et al.* (1974), and (3) equations of state in the form $p = p(\varrho)$ compiled by Harrison *et al.* (1965). All these descriptions supplement one another, and it seems expedient to present them even though some of the information is repeated.

1.3.1 The Brush diagram

A division of the temperature–density plane into different physical domains, in which the properties of the medium and the methods of their study are different, is shown in Fig. 1.3.1. The numerical values corresponding to the particular domains are approximate. Also, in reality the domains in the T, ϱ plane are not separated by line divisions but rather by certain areas corresponding to transformations between different states of the system. The following domains can be distinguished in Fig. 1.3.1:

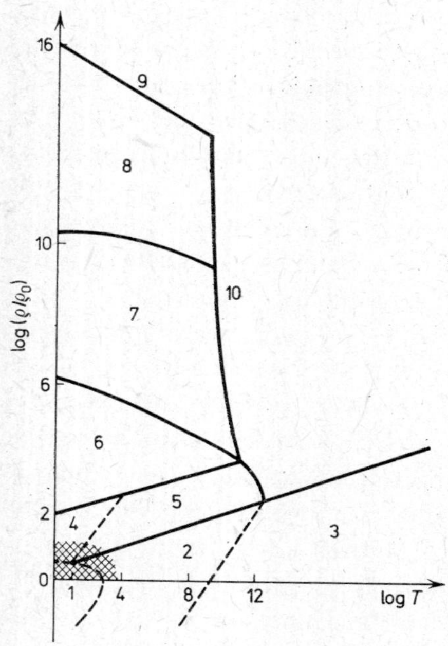

Fig. 1.3.1 Diagram of state according to Brush (1966). Ascribing specific physical properties of matter to various domains in the $\log T$, $\log(\varrho/\varrho_0)$ plane permits to construct equations of state in these areas. The temperature T is in kelvin, and the density ϱ is related to density ϱ_0 under normal conditions. The area of interest for physics of the Earth's and planetary interiors is dashed.

1. The "normal" state of matter, i.e., solid, liquid and gaseous. In this domain two or three states might coexist for a given temperature (three states at a temperature corresponding to the triple point). The line drawn from the critical point of the liquid–gas system towards high pressures but not too high densities and high temperatures, separates domain 2, of highly ionized plasma, from domain 5, of a metallic liquid state. The extension of the curve of melting towards high densities and temperatures separates domain 4 from domain 5, i.e., the areas of a metallic solid state and of a metallic liquid state.

2. Highly ionized gas described by the equation of state

$$\frac{p}{\varrho} = nRT+\delta \quad \text{or approximately} \quad p \propto \varrho \tag{1.3.3}$$

where n is the average number of free particles per one atom, R is the gas constant and δ is a term dependent on internal atomic forces. The state of equilibrium is determined by the Saha equation (Zeldovitch and Rayzer, 1966, p. 160, eqn. (3.4.6)) because of the ionization $\rightleftharpoons$ recombination process.

3. Completely ionized gas described by the equation of state (1.3.3), in which $n = Z+1$, where Z is the atomic number. The pressure of radiation should be taken into account. Electron–positron pairs are created.

4. Extension towards high densities of the solid state in a metallic phase: the nuclei bound in the lattice nodes are responsible for vibration levels; free electrons form the partly degenerated* electron gas of Fermi–Dirac for which

$$p_{\text{electron gas}} \propto \varrho^{5/3} \tag{1.3.4}$$

5. Extension towards high densities and temperatures of the metallic liquid state: the nuclei are either free or weakly bound; the electron gas of Fermi–Dirac. When the temperature increases sufficiently high the Boltzmann distribution for ions can be applied, and then electrostatic interactions between ions can be considered with the use of the Debye–Huckel method. As a result the equation of state is obtained.

6. Free nuclei, the electron gas completely degenerated for which the pressure is $p \propto \varrho^{5/3}$. With increasing density the gas becomes relativistic.

---

* Completely degenerated electron gas corresponds to the temperature of absolute zero and to the situation in which electrons occupy the lowest possible quantum states, and so the energy of the gas is minimal. With increasing temperature and a fixed density the degree of degeneration decreases. The degeneration temperature $T_{\text{deg}} \propto \varrho^{2/3}$ (see Section 1.9).

7. Domain of extremely relativistic electron gas. The beginning of nuclear reactions, especially of neutron synthesis in the reaction p+e = n+ν, where p is the proton, e is the electron, n is the neutron and ν is the neutrino;

$$p_{\text{relativistic electron gas}} \propto \varrho^{4/3} \tag{1.3.5}$$

8. Completely degenerated neutron gas; since neutrons are governed by the Fermi–Dirac statistics, the neutron gas pressure is given by formula (1.3.4), i.e., $p \propto \varrho^{5/3}$.

9. Relativistic neutron gas, i.e., the relativistic gas of Fermi, for which the pressure $p \propto \varrho^{4/3}$, i.e., the same as in formula (1.3.5) describing relativistic electron gas.

10. Domain of unusually high temperatures.

The equations of state in individual domains are derived using various simplifying assumptions, specific for different areas. The most important and most often used assumption is $T = 0$, which means a search for an equation of state independent of temperature. In practice the temperature may be assumed to be equal to absolute zero when it is lower than the temperature of degeneration. Since $T_{\text{deg}} \propto \varrho^{2/3}$, for sufficiently dense media the assumption that $T = 0$ may be valid even at temperatures very high in the normal sense.

For the investigation of the Earth's internal structure the most interesting range of thermodynamic parameter changes is the range corresponding to the borderland of domains 1, 4 and 5 (the dashed area in Fig. 1.3.1). Temperatures up to several thousand degrees, a severalfold relative increase of density and of pressure up to $10^{11} \div 10^{12}$ Pa correspond to this range.

The Brush diagram, dividing the temperature–density plane into various domains, is supplemented by a diagram dividing the pressure–temperature plane, shown in Fig. 1.3.2, according to Gilvarry (1969). From this diagram it can be seen that the thermodynamical conditions in the Earth's interior correspond to a range of state parameter changes difficult to consider, since this range does not belong to one phase, being intersected by a line of solid–liquid phase transitions. For the lower boundary of the domain of applicability of the Thomas–Fermi statistical atom model, Gilvarry assumes the pressure of 10^{13} Pa. This value is considerably higher than the central pressure in the Earth ($\approx 3.5 \times 10^{11}$ Pa). However, the Thomas–Fermi equation of state is often used to describe matter in the cores of terrestrial and giant planets, though a more correct application of the equation is to describe the interiors of stars.

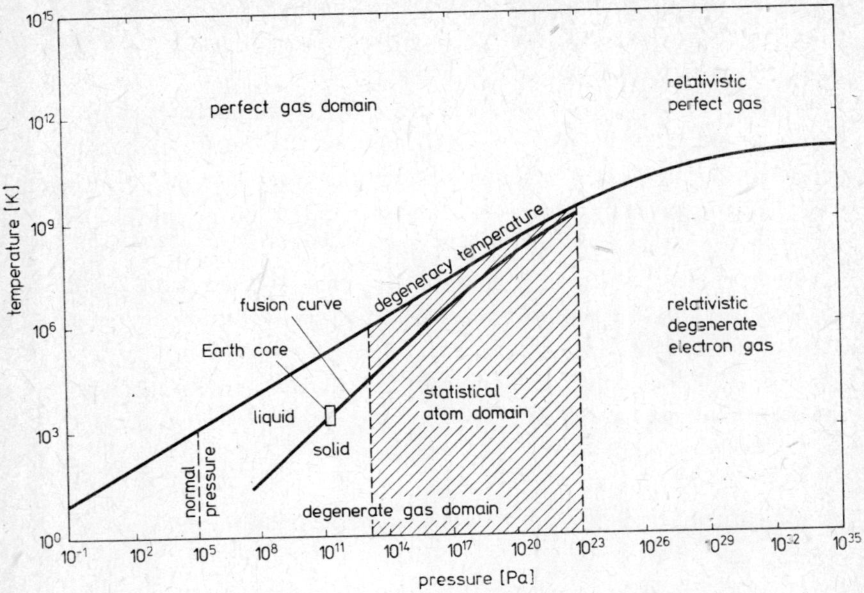

Fig. 1.3.2. Diagram of state according to Gilvarry (1969) in the $\log p$–$\log T$ plane.

1.3.2 The Rees classification

Unlike the Brush diagram, dividing the plane T, ϱ/ϱ_0 into ten domains, the density states considered by Rees *et al.* (1974) relate to the absolute values of the density and not to its dimensionless values, expressed by Brush in imprecisely defined units ϱ_0. Seven ranges of density are distinguished on the Rees diagram. All of them, except the last one, related to extremely high densities, can be compared with Fig. 1.3.1 or, more precisely, with the density intervals marked on the axis $\log(\varrho/\varrho_0)$.

For cold ionized matter the following density ranges are distinguished by Rees:

(1) $7.8 \text{ g} \cdot \text{cm}^{-3} \leqslant \varrho \leqslant 15 \text{ g} \cdot \text{cm}^{-3}$; for this range the equation of state for iron under laboratory conditions was studied.

(2) $15 \text{ g} \cdot \text{cm}^{-3} \leqslant \varrho \leqslant 10^4 \text{ g} \cdot \text{cm}^{-3}$; the range of applicability of the Fermi–Thomas statistical atom model.

(3) $10^4 \text{ g} \cdot \text{cm}^{-3} \leqslant \varrho \leqslant 10^7 \text{ g} \cdot \text{cm}^{-3}$; electron energies are higher than bond energies, and the equation of state describes a gas of free nuclei and electrons, while the main component of pressure comes from electron gas. The electrons become relativistic near the upper boundary.

(4) $10^7 \text{ g} \cdot \text{cm}^{-3} \leqslant \varrho \leqslant 10^{11} \text{ g} \cdot \text{cm}^{-3}$; the most stable nuclei with the atomic number $Z = 28$ and mass number $A = 56$ are transformed into

neutron-rich nuclei as a result of relativistic reactions of electrons with protons (see domain 7 of the Brush diagram).

(5) 10^{11} g · cm^{-3} $\leqslant \varrho \leqslant 4.5\times 10^{12}$ g · cm^{-3}; the nuclei become heavy, $A \approx 122$, and neutron-rich, $(A-Z)/Z \approx 83/89$. The neutrons become disconnected, forming free neutron gas, and at the lower boundary the pressure of the neutron gas is the highest as compared to that of the nuclei and electron gas.

(6) 4.5×10^{12} g · cm^{-3} $\leqslant \varrho \leqslant 10^{14}$ g · cm^{-3}; relativistic gases of electrons, protons and neutrons, which can be considered as the relativistic gases of Fermi.

(7) $\varrho \geqslant 10^{14}$ g · cm^{-3}; a completely relativistic domain, in which the particle abundances are at the ratio

$$n_{\text{electron}} : n_{\text{proton}} : n_{\text{neutron}} = 1:1:8$$

The pressure under such conditions is proportional* to the density

$$p = \tfrac{1}{3}c^2\varrho \tag{1.3.6}$$

1.3.3 The Harrison–Wheeler equations

The equations of state of Harrison and Wheeler (Harrison *et al.*, 1965) are relations of the $p = p(\varrho)$ type relating to four intervals of density, when the lower boundary of density corresponds to the density of iron under room conditions. The equations, expressed by their authors in CGS units, are as follows (ϱ in g · cm^{-3} and p in dyne · cm^{-2}; 1 dyne · cm^{-2} = 10^{-10} GPa):

(1) 7.86 g · cm^{-3} $\leqslant \varrho \leqslant 10^6$ g · cm^{-3}

$$p = 3.0271\times 10^{12}(\varrho^{1/3}-1.4415)^5 - 1.50319\times 10^{11} \tag{1.3.7}$$

approximately $p \propto \varrho^{5/3}$, i.e., just as in domains 4, 5 and 6 of the Brush diagram;

(2) 10^6 g · cm^{-3} $\leqslant \varrho \leqslant 3.22\times 10^{11}$ g · cm^{-3}

$$p = 2.4616\times 10^{15}\varrho^{5/4}(1+2388.8\varrho^{-1/2})^{-5/6} \tag{1.3.8}$$

(3) 3.22×10^{11} g · cm^{-3} $\leqslant \varrho \leqslant 4.63\times 10^{12}$ g · cm^{-3}

$$p = 3.1134\times 10^{21}\varrho^{2/3}(1+0.0031822\varrho^{1/6})^6 \tag{1.3.9}$$

---

* Remembering that the expression $(\partial p/\partial \varrho)_S^{1/2}$ is equal to the hydrodynamic sound velocity, which in turn cannot be higher than the light velocity c, we find that the maximum value of the proportionality constant consistent with the theory of relativity is c^2. Thus the maximum pressure conceivable for highly condensed matter is $p = c^2\varrho$.

(4) $\varrho \geqslant 4.63 \times 10^{12}$ g · cm^{-3}

$$p = 2.9959 \times 10^{20} \varrho^{5/3}(9.1208 \times 10^{8} + \varrho^{5/9})^{-6/5} \tag{1.3.10}$$

For $\varrho \to \infty$ relation (1.3.10) is transformed into the extremely relativistic formula (1.3.6).

The Harrison–Wheeler equation of state was used for analyses of the stability of neutron stars (see e.g., Demiański, 1978). For this purpose the Tolman–Oppenheimer–Volkoff relativistic equation of hydrostatic equilibrium was applied,* namely

$$\frac{dp}{dr} = -G\frac{\varrho m}{r^2}\left(1+\frac{p}{c^2\varrho}\right)\frac{1+4\dfrac{r^3p}{mc^2}}{1-2G\dfrac{m}{c^2r}} \tag{1.3.11}$$

valid for a spherically symmetric nonrotating mass. In eqn. (1.3.11) G is the gravity constant and $m = m(r)$ is the mass contained in a sphere with the radius r

$$\frac{dm}{dr} = 4\pi r^2 \varrho(r) \tag{1.3.12}$$

For nonrelativistic ranges of pressure eqn. (1.3.11) is transformed into the usual Newtonian form

$$\frac{dp}{dr} = -G\frac{\varrho m}{r^2}$$

---

* Equation (1.3.11) can also be written in the form

$$\frac{dp^*}{dr} = -\frac{(\varrho^*+p^*)(m^*+4\pi r^3 p^*)}{r(r-2m^*)}$$

The quantities marked by an asterisk ϱ^*, p^* and m^*, dependent on r, are the density, pressure and mass expressed in units used in the theory of relativity, i.e., under the assumption of $G = 1$ and $c = 1$. The following relations are observed between the quantities with asterisks and the conventional ones (in addition the relations for time and energy are given):

time:	$t^* = ct,$	$[t^*] = \text{cm}$
mass:	$m^* = \dfrac{G}{c^2}m,$	$[m^*] = \text{cm}$
density:	$\varrho^* = \dfrac{G}{c^2}\varrho,$	$[\varrho^*] = \text{cm}^{-2}$
energy:	$E^* = \dfrac{G}{c^4}E,$	$[E^*] = \text{cm}$
pressure:	$p^* = \dfrac{G}{c^4}p,$	$[p^*] = \text{cm}^{-2}$

entirley sufficient for the description of planet interiors (but disregarding their rotation).

Finally, to end this chapter, a hypothetic diagram of state for hydrogen in the pressure–temperature plane is shown in Fig. 1.3.3, according to

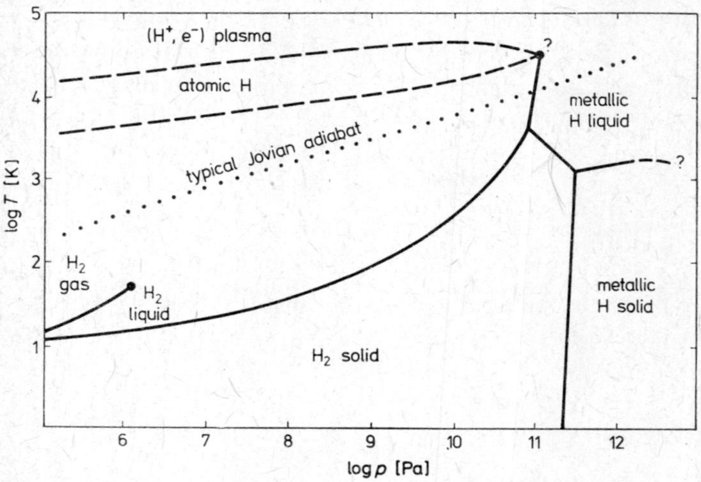

Fig. 1.3.3. Schematic phase diagram for hydrogen. (According to: D. J. Stevenson, "The outer planets and their satellites" in: *The Origin of the Solar System*, S. F. Dermott (ed.), copyright © 1978 by John Wiley and Sons, Ltd. Reprinted by permission of John Wiley and Sons, Ltd.). A typical Jovian adiabat passing through $T = 175$ K at $p = 10^5$ Pa is superimposed. The adiabat does not intercept any phase transition except (possibly) at very high pressures. The phase diagram is very uncertain for $p \geqslant 10^{11}$ Pa. The dashed lines separating liquid H_2, H and H^+ plasma are poorly defined, except at low densities.

Stevenson (1978). In this diagram solid, liquid and gas states of molecular hydrogen H_2 and solid and liquid states of metallic hydrogen H are clearly distinguished. The investigation of the molecular hydrogen → metallic hydrogen transition, occurring most probably under the pressure $p \geqslant 100$ GPa, is still an open experimental problem.

The equation of state for solid hydrogen is interesting for the physics of the planets, since the astrophysical data related to the density of Jupiter and Saturn and their energy balance indicate that hydrogen or hydrogen with an admixture of helium, either in a state of molecular crystal or in metallic state, is most probably their main component. Phase transitions of molecular hydrogen into metallic hydrogen might form a source of energy sufficient to account for the excess of energy radiated by Jupiter over the amount of solar energy reaching Jupiter and then being reemitted (excess luminosity).

1.4 EXPERIMENTAL METHODS IN HIGH PRESSURE PHYSICS

Experimental studies of the physical properties of solids under high pressures and temperatures require various laboratory techniques, which depend on the range of pressures and temperatures and on the type of physical phenomenon being studied.

Two principal methods of investigation can be distinguished: the static and dynamic methods. The main difference between them lies in the fact that in static studies the pressure p and temperature T can be changed independently, while in shock methods p and T are connected by the Hugoniot conditions. Furthermore, a static experiment can be carried out for almost any length of time,* while shock experiments have the nature of an impulse with a duration of the order of microseconds or shorter. The results $A_{stat}(p, T)$ of measurements related to some physical quantity A measured under p and T conditions by static methods are always slightly different from the results $A_{dyn}(p, T)$ of dynamic studies in which the same values of p and T were attained for a very short time. This can be explained by the fact that two identical samples taken from the same material under the initial conditions p_0 and T_0 (e.g., room conditions) form slightly different physical media under the conditions p and T attained by each of the two different methods. A sample undergoing dynamic study has in general no time to reach a thermodynamic equilibrium corresponding to given p and T. Thus for example polymorphic phase transformations of the low-pressure phase → high-pressure phase type would either not occur at all or occur only partially rather than in the whole mass of the sample; an equilibrium degree of ionization could not be reached; the resistance of a sample to mechanical changes (e.g., cracks) is dependent not only on the values of the p and T parameters but on the time of their action as well.

A diagram showing different types of experimental methods and the corresponding pressures is presented in Fig. 1.4.1, where experimental methods and the range of their applicability, the most typical equations of state, and characteristic depths in the Earth's interior and their corresponding pressures are also listed. The pressure axis contains about seven orders of magnitude accessible to experiments. This pressure interval is sufficient to describe pressure changes in the Earth's interior, though the pressure corresponding to that in the core can be obtained only by shock wave

---

* The duration of the experiments is practically sufficient for carrying out any studies except those related to the viscosity (creep) of solids; especially the studies of creep processes at low temperatures require measurements of very long duration.

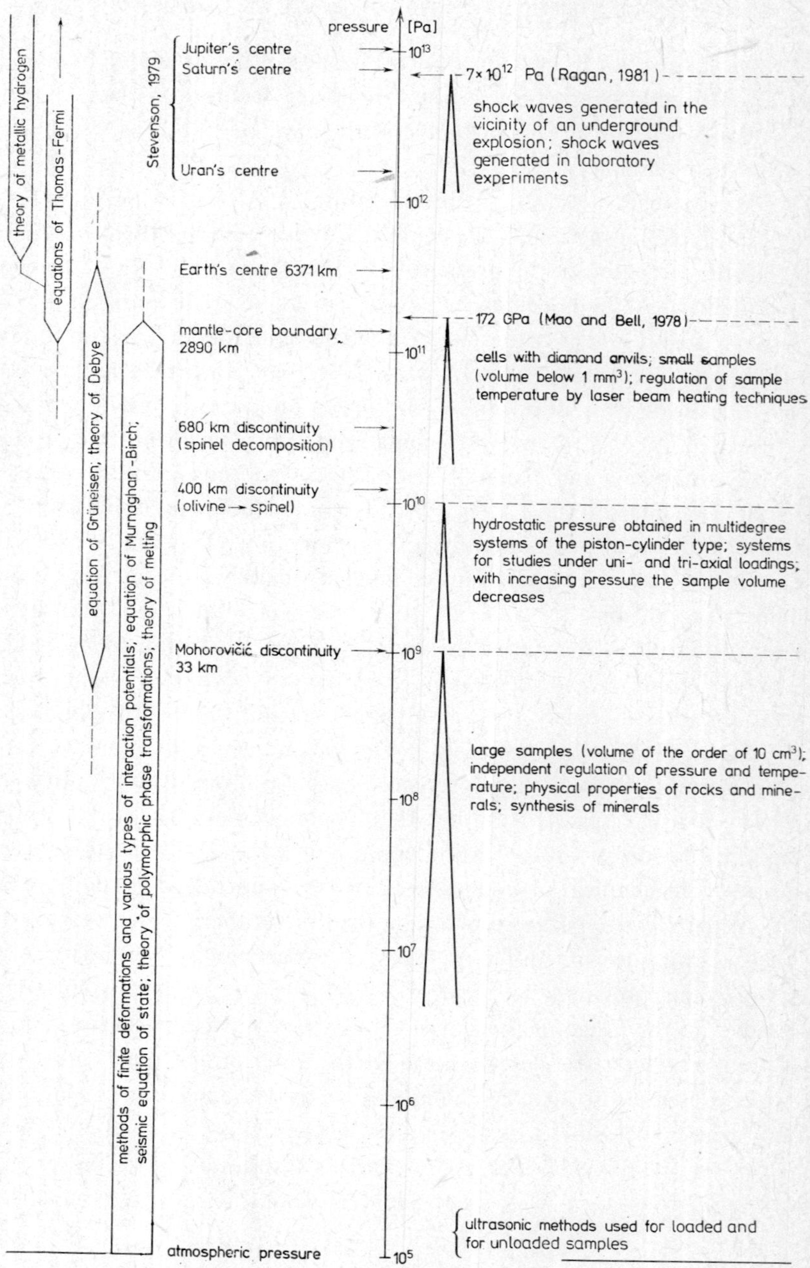

Fig. 1.4.1. Pressure range observed in physics of planetary interiors in comparison to pressure intervals in which different theoretical methods (the left-hand side) can be used and laboratory experiments (the right-hand side) are possible.

methods. The numerical data shown in Fig. 1.4.1 are approximate, new experimental results continuously change the upper boundaries of the pressure intervals corresponding to measurements carried out by various methods. It can be stated as a general rule that the larger the size of a sample, the more difficult it is to obtain high pressure acting on that sample (in static experiments).

Investigations of mechanical properties, especially the velocity measurements of elastic compressional waves v_P and shear waves v_S and of the related bulk modulus K and shear modulus μ, are principally based on ultrasonic measurements, for which relatively large samples at least several millimeters long are needed. Studies of fracture and creep processes also require the use of the largest possible samples, of distinctly larger size than the characteristic sizes of grains or other heterogeneities in the rocks or minerals under consideration. It is quite obvious that some of these measurements, and especially ultrasonic measurements, are carried out beginning from the atmospheric pressure p_0 and room temperature T_0. For $p \lesssim 10$ GPa and $T \lesssim 1000$ K simultaneous changes of pressure and temperature are possible of such a kind that the relation $p = p(T)$ will correspond to the relations $p = p(z)$ and $T = T(z)$ in the Earth's crust at depths $z \lesssim 30$ km, above the Mohorovičić discontinuity. It is also possible to carry out experiments related to the synthesis of minerals under the p and T conditions similar to those existing in the Earth's crust.

Laboratory experiments can be divided into three groups according to the method of applying loads to samples. They are

(1) uniaxial compression (or extension) with the application of confining pressure changing from zero to the value of the axial pressure,

(2) triaxial compression,

(3) hydrostatic compression.

The first two groups of measurements are most interesting for problems related to crack development and fracture mechanics. Measurements under hydrostatic conditions provide data relating to the equation of state, to the synthesis of minerals and to the phase transitions. Hydrostatic pressures are generally obtained by placing a sample in a liquid or gaseous medium which is acted upon by outside pressure. The samples could be cut out from natural rock or mineral, they could be artificially grown monocrystals or constitute an isotropic mixture of initially compressed powder. Powder samples are used in experiments relating to mineral syntheses and phase transformations, but are not suitable for studies of macroscopic mechanical properties.

The experimental methods in which high pressures are created in cells with diamond anvils are of special interest. In such devices static pressures of the order of 10^{11} Pa in flat samples with diameters and thicknesses of the order of 0.1 mm are obtainable. Since the volume of the samples placed between diamond anvils is so small, special methods are needed for pressure measurements (known phase transformations used as reference points or measurements of ruby fluorescence wavelength shift as a function of pressure) as well as for changes of the sample temperatures (sample heating by a beam of laser rays is used). The diamond cell of Mao and Bell (1978), in which a pressure of 172 GPa, i.e., the pressure corresponding to the outmost layers of the Earth's core, was obtained, is shown schematically in Fig. 1.4.2. Thus, diamond cells are suitable for studies of phase transformations which might occur at any depth in the Earth's mantle. They are of particular importance for studies of molecular hydrogen →metallic hydrogen phase transitions.

A separate group of laboratory experiments is formed by those experiments in which shock waves are used. Although shock wave methods provide pressures considerably higher than those obtained by static methods, they have two important shortcomings: the pressures have the nature of impulses of short duration ($\lesssim 10^{-6}$ s), and an increase in pressure is connected with an increase in temperature, which cannot be applied independ-

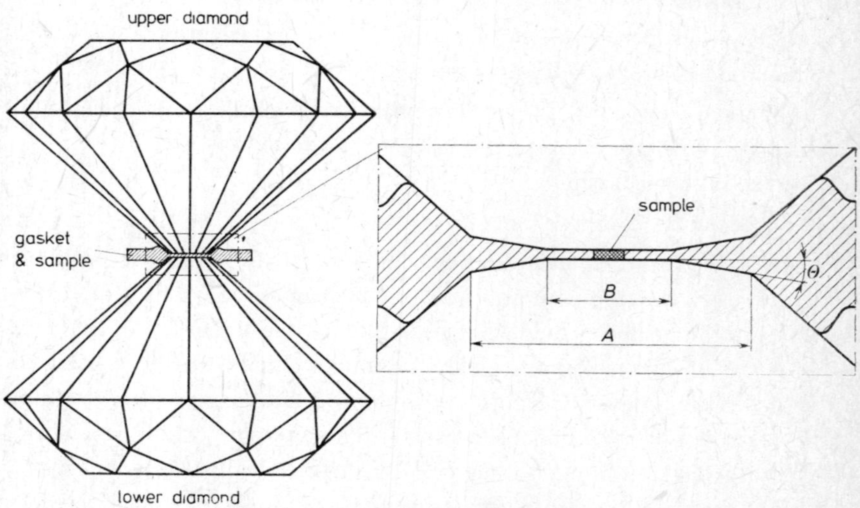

Fig. 1.4.2. Schematic diagram of a high-pressure cell with diamond anvils, according to Mao and Bell (1978). Distance from top surface of upper diamond to bottom surface of lower diamond is approximately 5 mm. The sample width is 250 μm. Copyright 1978 by the American Association for the Advancement of Science.

ently. In spite of these shortcomings shock wave methods are widely used for studies of mineral properties, especially for studies of phase transitions and equations of state. Their most important advantage is the possibility of obtaining pressures one order of magnitude higher than static pressures ($\approx 10^{12}$ Pa), while the sample volumes are relatively large (of the order of a few cubic centimetres). Shock wave studies of equations of state are usually related to the determination of the values of different material coefficients appearing in a chosen equation of state, assumed for a given sample. Ahrens *et al.* (1969), for example, found the material coefficients appearing in the Murnaghan–Birch equation of state applied to a number of minerals, mostly silicates, in low- and high-pressure phases. Shock wave methods are also used for the determination of the Grüneisen parameter.

Mathematical bases of shock wave methods need a more detailed description (see e.g., Zeldovitch and Rayzer, 1966). A shock wave is considered as a surface* of discontinuity of thermodynamic parameters. The wave travels in a medium with velocity D relative to the laboratory system. The equations of mass, momentum and energy conservation (the Rankin–Hugoniot conditions), as the wave passes through this surface, are of the form

$$\varrho_0 D = \varrho(D-v) \tag{1.4.1}$$

$$p_0+\varrho_0 D^2 = p+\varrho(D-v)^2 \tag{1.4.2}$$

$$U_0+\frac{p_0}{\varrho_0}+\frac{1}{2}D^2 = U+\frac{p}{\varrho}+\frac{1}{2}(D-v)^2 \tag{1.4.3}$$

where ϱ_0, p_0 and U_0 are the density, pressure and internal energy for a mass unit in the motionless area before the wave, and ϱ, p and U are the same quantities in the area moving with the particle velocity v ($v < D$) behind the shock wave. Since the medium parameters ϱ_0, p_0 and U_0 in front of the wave are known, and the velocities D and v are measured during the experiments, the parameters ϱ, p and U behind the wave can be found from the equations of conservation (1.4.1)–(1.4.3). When we assume a certain form of the equation of state $U = U(\varrho, T)$, we can determine the temperature T in the medium compressed by a shock wave. Eliminating from eqns. (1.4.1) and (1.4.2) the factor $D-v$, we find the formula for the velocity

* A plane shock wave is considered. In high-pressure experiments systems generating such waves are most often used. The highest pressures, however, can be obtained by generating spherically symmetric shock waves converging towards the centre.

of the shock wave:

$$D^2 = \frac{\varrho}{\varrho_0}\frac{p-p_0}{\varrho-\varrho_0} = V_0^2\frac{p-p_0}{V_0-V}, \quad V = \frac{1}{\varrho} \tag{1.4.4}$$

Eliminating from eqn. (1.4.3) the velocity D and $D-v$, we obtain the equation of shock adiabat, or the Hugoniot adiabat:

$$U-U_0 = \frac{1}{2}(p+p_0)\left(\frac{1}{\varrho_0}-\frac{1}{\varrho}\right) = \frac{1}{2}(p+p_0)(V_0-V) \tag{1.4.5}$$

When formulas (1.4.4) and (1.4.5) are applied to shock waves in solids, the pressure p_0 equal to atmospheric pressure can always be neglected since it is very small in comparison with the pressure p behind the wave.

Any two points on a Hugoniot adiabat can be considered as representing the specific volume and the pressure of the medium in front of and behind the shock wave. The density in front of the wave is always smaller. The diagrams of Hugoniot adiabats for media whose specific volumes reduced to the pressure p_0 are $V_{1,0}$ and $V_{2,0}$ are shown in Fig. 1.4.3. According to formula (1.4.4) the slope of the chords OQ, OAC and OW are connected with the velocity D_1 of the shock wave relative to the motionless medium. The slope of the chord AR or AS is connected with the velocity of the shock

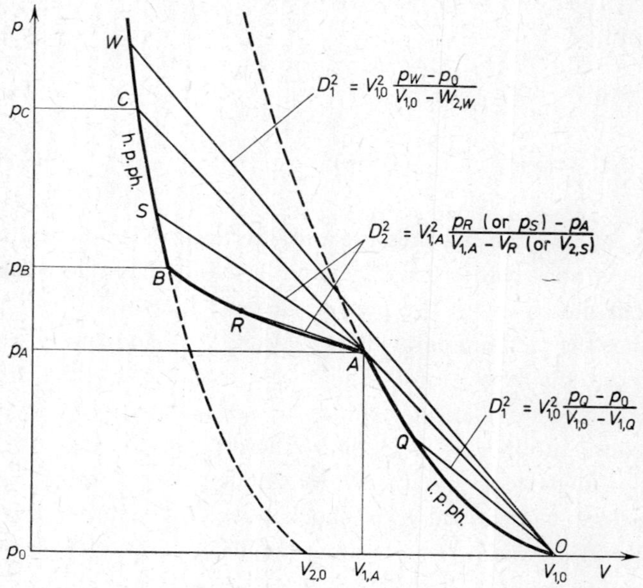

Fig. 1.4.3. Shock adiabat in a medium which can undergo phase transformation. The notation l.p.ph. stands for the low-pressure phase shock adiabat, and h.p.ph. for the high-pressure phase shock adiabat. On the segment ARB in the final state a system of two phases coexists.

wave behind which a phase transition occurs; in this case the velocity D_2 is measured against the medium behind the first shock wave, and thus in relation to the moving medium in the laboratory system. Let us assume that the shock adiabats in Fig. 1.4.3 correspond to the low- and high-pressure phases of the same substance. A pure low-pressure phase is stable at $p < p_A$, while a pure high-pressure phase is stable at $p > p_B$. The shock wave propagating in a low-pressure phase can force a phase transformation of the medium. The transformation could be either complete or only partial, depending on the final pressure. In the case where the final pressure $p_W > p_C$, the increase in pressure on the shock wave occurs at first stepwise from p_0 to p_M and then continuously until the final pressure p_W is reached. The

TABLE 1.4.1
Effects of shock waves of different intensity on the phase transformation of the medium

Final pressure p	Final state	Shock waves in the medium
$p < p_A$	Low-pressure phase	One "weak" shock wave
$p_A < p < p_B$	Mixture of low- and high-pressure phases	Shock wave compressing medium to state A, and the second slower shock wave forming at the same time the surface* of partial phase transformation
$p_B < p < p_C$	High-pressure phase	Shock wave compressing medium to state A, and the second slower shock wave forming at the same time the surface* of complete phase transformation
$p_C < p$	High-pressure phase	One "strong" shock wave forming at the same time the surface* of complete phase transformation

* Since phase transformation processes proceed rather slowly, the "surface" becomes thicker—see Fig. 1.4.4. Because of kinetics of transformation and energy dissipation, it may happen that phase transformation will have no time to occur even when $p > p_C$.

layer in which phase transformations take place corresponds to the interval of continuous increase of pressure from p_M to p_W. Different effects of shock waves on a medium which can undergo phase transformation are presented in Table 1.4.1 and in Figs. 1.4.3 and 1.4.4. The actual course of the shock adiabat of such a medium is shown approximately by the curve $OABW$, in which a distinct bend at point A can be observed.

Certain similarities to shock waves in media which can undergo phase transformations are shown by shock waves in porous media. In this case

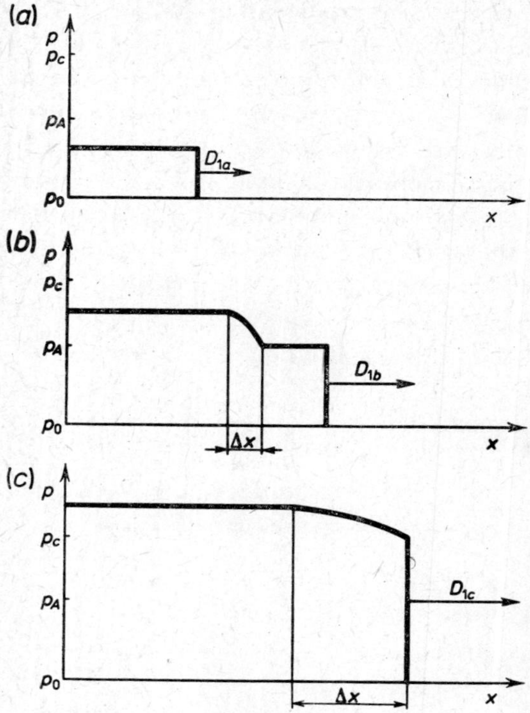

Fig. 1.4.4. Profiles of the pressure p behind the shock wave versus the wave intensity. The symbol x means the shock wave position, D_1 is the primary shock wave velocity ($D_{1a} < D_{1b} < D_{1c}$), and Δx is the phase transition zone; (*b*) partial transformation and (*c*) complete transformation.

the role of the shock adiabat of high-pressure phase is played by the shock adiabat of a porous medium, but the pores are closed by pressure (see e.g., Burke and Weiss, 1971). The rocks forming the Earth's upper mantle are characterized by a certain degree of porosity, and the pores are filled by vapour or other volatiles. Thus the methods and results of high-pressure studies of porous media are also of interest to geology and geophysics.

1.5 THERMODYNAMICAL FOUNDATIONS OF EQUATIONS OF STATE

The total internal energy U of any medium* can be presented as a sum of three components

$$U(V,T) = U_0(V)+U_e(V,T)+U_j(V,T) \tag{1.5.1}$$

* Under conditions close to "normal", i.e., in areas 1–4 of the Brush diagram (Fig. 1.1.1), but not in the areas of pressure and density where nuclear reactions occur or where relativity effects are of significance.

The temperature-independent term U_0 is the elastic energy, the term U_e is the energy of free electron gas, and U_j is the energy of thermal vibrations of ions or atoms forming crystal lattices. Under different thermodynamic conditions different terms in formula (1.5.1) become dominant. In general, the significance of the thermic terms, compared with the elastic term, increases with increasing temperature and pressure.

For several types of media (ion crystals, molecular crystals, metals) general formulas expressing the potential of interaction between the molecules as a function of the distance r between them are well known. This potential in a thermodynamical sense is the elastic component of internal energy per one molecule. The thermic terms can be determined independently and are described by the equation of Grüneisen (the thermal energy of ions) and by the equation of Thomas–Fermi–Dirac (the energy of electron gas). Thus we have to find the equation of state $p = p(V, T)$ assuming that the internal energy $U = U(V, T)$ is known. The appropriate estimations, based on principal thermodynamic relations, are given below.

The free energy F is related to the internal energy U, temperature T and entropy S as in the formula

$$F = U - TS \tag{1.5.2}$$

Using the first law of thermodynamics

$$dU = -p dV + T dS \tag{1.5.3}$$

we have

$$dF = -p dV - S dT \tag{1.5.4}$$

According to formula (1.5.4) the pressure can be expressed by a derivative of free energy

$$p = -\left(\frac{\partial F}{\partial V}\right)_T \tag{1.5.5}$$

and from this, taking into account definition (1.5.2), we obtain

$$p = -\left(\frac{\partial U}{\partial V}\right)_T + T\left(\frac{\partial S}{\partial V}\right)_T \tag{1.5.6}$$

Thus at the temperature of absolute zero the pressure is expressed simply by a derivative of internal energy

$$p(V, T = 0) = -\left(\frac{\partial U}{\partial V}\right)_{T=0} \tag{1.5.7}$$

and depends only on the volume. The same result can also be obtained directly from the first law of thermodynamics. Relation (1.5.7) written in an approximate form

$$p = -\frac{dU}{dV} \quad \text{when } U = U(V) \tag{1.5.8}$$

is often used for the determination of the barotropic equation of state, i.e., an equation of the type $p = p(V)$.

The internal energy $U(V)$ in eqn. (1.5.8) is the sum of the elastic energy $U_0(V)$, the internal energy of electron gas at absolute zero $U_e(V, 0)$, and the energy of zero-point lattice ion vibrations $U_j(V, 0)$. The last term is not important in the geophysical applications of equations of state and is usually neglected. Barotropic equations of state can properly describe reality when the temperature is low enough to make the influence of thermic vibrations negligible in comparison with elastic effects, or when the temperature is considered as a parameter influencing only the material constants appearing in the relation defining the internal energy. An extensive review of the equations of state, especially of the barotropic equations, and numerous examples of their applications in different media (elements, chemical compounds, minerals) were given by Ullmann and Pankov (1976). Since the specific volume V, density ϱ, and characteristic distance r are connected by the relations

$$V = \frac{1}{\varrho} = \frac{V_0}{r_0^3} r^3 \tag{1.5.9}$$

then the barotropic equation of state (1.5.8) can be written in equivalent forms

$$p = \varrho^2 \frac{dU}{d\varrho} \quad \text{or} \quad p = -\tfrac{1}{3} r_0 \varrho_0^{1/3} \varrho^{2/3} \frac{dU}{dr} \tag{1.5.10}$$

The parameters V_0, ϱ_0, and r_0 are the values of the specific volume, density, and characteristic distance (e.g., lattice constant), respectively, under constant pressure p (e.g., $p = 0$).

To find a relation for the pressure valid also for $T \neq 0$ we must go back to formula (1.5.6), which can be transformed by substituting the relation

$$\left(\frac{\partial S}{\partial V}\right)_T = \left(\frac{\partial p}{\partial T}\right)_V \tag{1.5.11}$$

resulting from eqn. (1.5.4). As a result, we obtain the relation between the pressure and the internal energy

$$\left(\frac{\partial U}{\partial V}\right)_T = -p + T\left(\frac{\partial p}{\partial T}\right)_V = T^2 \left[\frac{\partial}{\partial T}\left(\frac{p}{T}\right)\right]_V \tag{1.5.12}$$

and after integration we have

$$p(V,T) = T\int \frac{1}{T^2}\left(\frac{\partial U}{\partial V}\right)_T \mathrm{d}T \tag{1.5.13}$$

Using formula (1.5.1) for the internal energy, we find a similar three-term expression for the pressure

$$p(V,T) = p_0(V)+p_e(V,T)+p_j(V,T) \tag{1.5.14}$$

where p_0 is the elastic pressure, p_e is the pressure of electron gas, and p_j is the pressure induced by thermal vibrations of ions or atoms forming crystal lattices. With such a division of pressure into three physically different components we have $p_0 = -\mathrm{d}U_0/\mathrm{d}V$. It should be noted that the pressure $p_0 = p_0(V)$ defined in this way is different from the pressure $p(V)$ given by formula (1.5.8).

Table 1.5.1 (after Zeldovitch and Rayzer, 1966) gives the values of the components of internal energy and pressure corresponding to formulas

TABLE 1.5.1
Pressure, internal energy, temperature and Grüneisen parameter behind the front of a shock wave in lead as functions of the relative compression ϱ/ϱ_0, after Zeldovitch and Rayzer (1966).

ϱ/ϱ_0	Pressure [GPa			Internal energy [10^4 J · kg^{-1}]			Temperature [K]	Grüneisen parameter γ
	p_0	p_j	p_e	U_0	U_j	U_e		
1.3	21.6	3.35	0.051	15.3	9.6	0.69	1045	1.9
1.5	51.0	13.9	0.63	46.7	42.3	7.4	3580	1.77
1.7	95.3	34.0	3.8	95.8	107.0	39.4	8600	1.60
1.9	156.0	56.0	12.7	163.2	191.0	118.0	15100	1.35
2.1	233.0	73.0	29.0	248.0	284.0	243.0	22300	1.07
2.2	277.0	93.0	41.5	297.0	337.0	332.0	26400	0.98

(1.5.1) and (1.5.14), listed for lead compressed by shock waves up to the density $\varrho = 2.2\varrho_0$, where it can be seen that with increasing density of the medium the significance of the thermic terms also increases in comparison with the elastic component.

A review of the equations of state, taking into account different physical terms of formulas (1.5.1) and (1.5.14), is given in Sections from 1.6 to 1.9. Special attention is paid to those equations which have applications in the physics of the Earth and planetary interiors.

1.6 EQUATIONS OF STATE RESULTING FROM THE THEORY OF A CONTINUOUS MEDIUM

1.6.1 Equation of Murnaghan

Murnaghan (1944) assumed that the isothermal bulk modulus K is a linear function of pressure

$$K(p, T) = \varrho\left(\frac{\partial p}{\partial \varrho}\right)_T = K_0 + K_0' p \tag{1.6.1}$$

where

$$K_0 = K(0, T) \quad \text{and} \quad K_0' = \left(\frac{\partial K}{\partial p}\right)_{T;\, p=0} \tag{1.6.2}$$

After integration we obtain the two-parameter isothermal equation of Murnaghan

$$p = \frac{K_0}{K_0'}\left[\left(\frac{\varrho}{\varrho_0}\right)^{K_0'} - 1\right] \tag{1.6.3}$$

where ϱ_0 is the density under the pressure $p = 0$. The values of the constants appearing in this equation are, according to Bullen (1975), $K_0' \approx 2.8$ and $K_0/K_0' \approx 75$ GPa. For pressures between 40 and 320 GPa Bullen gives a parabolic relation for $K(p)$ instead of a linear relation (1.6.1). The most important advantage of the Murnaghan equation (1.6.3) is its simplicity. The parameters K_0 and K_0' can either be measured in laboratory experiments or determined from a particular seismic observations. The relation $K = K(p)$ is of special importance for the construction of models of the Earth's interior.

1.6.2 The Murnaghan–Birch equation

The Murnaghan–Birch equation (Murnaghan, 1951; Birch, 1952) was given as a semi-empirical conclusion from the theory of finite deformations.

Let us consider an isotropic body under hydrostatic presures. The deformation matrix of Euler* is in this case of a diagonal form

$$\varepsilon_{ij} = \varepsilon\delta_{ij} \tag{1.6.4}$$

---

* Let us denote by a_i the vector of the position of a particle before displacement and by x_i the vector of its position after displacement ($i = 1, 2, 3$; the configuration before and after displacement is described in the same orthogonal rectilinear system of coordinates). The vector

$$u_i = x_i - a_i \tag{1}$$

where ε is the hydrostatic deformation

$$\varepsilon = \frac{1}{2}\left[1-\left(\frac{\varrho}{\varrho_0}\right)^{2/3}\right] \equiv \frac{1}{2}\left[1-\left(\frac{V_0}{V}\right)^{2/3}\right] \qquad (1.6.5)$$

It is usually assumed that the relationship between the free energy F and deformation ε is in the form of a power series

$$F = \sum_{n=1}^{\infty} a_n \varepsilon^n \qquad (1.6.6)$$

where the coefficients a_n are dependent only on the temperature. The pressure can also be found from formula (1.6.6) by differentiating the terms of this series with respect to volume. From the condition that $p(V_0) = 0$ it follows that $a_i = 0$. The coefficients a_2 and a_3 can be expressed by the bulk modulus K_0 and its derivative K_0', defined by formulas (1.6.2).

We thus obtain the Murnaghan–Birch equation which with an accuracy up to the terms of the third order is of the form

$$p = \frac{3}{2} K_0 \left[\left(\frac{\varrho}{\varrho_0}\right)^{7/3} - \left(\frac{\varrho}{\varrho_0}\right)^{5/3}\right]\left\{1-\frac{3}{4}(K_0'-4)\left[\left(\frac{\varrho}{\varrho_0}\right)^{2/3}-1\right]+\ldots\right\} \qquad (1.6.7)$$

---

is the vector of displacement. The tensor of deformation in Euler's sense is defined by the formula (see e.g., Fung, 1965)

$$\varepsilon_{ij} = \frac{1}{2}\left(\delta_{ij}-\delta_{\alpha\beta}\frac{\partial a_\alpha}{\partial x_i}\frac{\partial a_\beta}{\partial x_j}\right) = \frac{1}{2}\left(\frac{\partial u_j}{\partial x_i}+\frac{\partial u_i}{\partial x_j}-\frac{\partial u_\alpha}{\partial x_i}\frac{\partial u_\alpha}{\partial x_j}\right) \qquad (2)$$

in which the displacement is expressed as a function of the configuration after deformation. The tensor of deformation in Lagrange's sens is given by the formula

$$\eta_{ij} = \frac{1}{2}\left(\delta_{\alpha\beta}\frac{\partial x_\alpha}{\partial a_i}\frac{\partial x_\beta}{\partial a_j}-\delta_{ij}\right) = \frac{1}{2}\left(\frac{\partial u_j}{\partial a_i}+\frac{\partial u_i}{\partial a_j}+\frac{\partial u_\alpha}{\partial a_i}\frac{\partial u_\alpha}{\partial a_j}\right) \qquad (3)$$

in which the vector of displacement depends on the coordinates before the deformation. In the theory of infinitely small deformations the products of derivatives are usually neglected and the tensors of Euler and Lagrange are reduced to the tensor of Cauchy.

For isotropic deformations under hydrostatic pressure the following relation is true

$$x_i = (1+A)a_i, \qquad A = \text{const} \qquad (4)$$

Using this relation, we find the ratio of the volume V after deformation to the volume V_0 before deformation

$$\frac{V}{V_0} = \frac{\varrho_0}{\varrho} = \det\frac{\partial x_i}{\partial a_j} = (1+A)^3 \qquad (5)$$

Substituting formula (4) in (2) and (3) and using relation (5), we can obtain eqns. (1.6.5) and (1.6.9).

This equation as well as its simplified form

$$p = \frac{3}{2} K_0 \left[\left(\frac{\varrho}{\varrho_0} \right)^{7/3} - \left(\frac{\varrho}{\varrho_0} \right)^{5/3} \right] \tag{1.6.8}$$

is often used in geophysics. The principal advantage of the equation is its consistence with experimental data, e.g., with the results relating to rock compressibility under high pressures up to at least 10 GPa, i.e., in the region of the upper mantle. From a formal point of view eqn. (1.6.7) has been postulated rather arbitrarily since it is based on the arbitrarily defined deformation matrix ε_{ij}.

Knopoff (1963a) considered an equation of state obtained from the theory of finite deformations and different from eqn. (1.6.6). He assumed a different form for the deformation tensor more complex than that proposed by Birch. The Murnaghan–Birch equation is therefore a specific case of the equation given by Knopoff.

Thomsen (1970) made an interesting comparison of equations of state of the form $p = p(\varrho)$ resulting from the theory of finite deformations. He underlines the fact that this theory can use an unlimited number of different definitions of deformation.

If, instead of the deformation ε in Euler's form (1.6.4), we introduce the deformation in Lagrange's form

$$\eta = \frac{1}{2} \left[\left(\frac{\varrho}{\varrho_0} \right)^{2/3} - 1 \right], \qquad \eta_{ij} = \eta \delta_{ij} \tag{1.6.9}$$

and we assume the free energy F to be given by the series

$$F = \sum_{n=1}^{\infty} A_n \eta^n \tag{1.6.10}$$

similar to series (1.6.6), then with an accuracy up to the terms of the third order we obtain the following equation of state:

$$p = \frac{3}{2} K_0 \left[\left(\frac{\varrho}{\varrho_0} \right)^{1/3} - \left(\frac{\varrho}{\varrho_0} \right)^{-1/3} \right] \left\{ 1 - \frac{3}{4} K_0' \left[\left(\frac{\varrho}{\varrho_0} \right)^{-2/3} - 1 \right] \right\} \tag{1.6.11}$$

The relations $p(\varrho)$ in forms (1.6.7) and (1.6.11) are entirely different. Although both equations are formally of the same value, eqn. (1.6.11) must be rejected as being inconsistent with experimental data.

The ambiguity of the theory of finite deformations cannot be removed on the ground of continuum medium mechanics; the application of models of atomic interactions is needed. The equations of state obtained from this theory provide results which differ from one another more and more substantially as the ratio ϱ/ϱ_0 increases. The order of the approximation used, i.e., the exponent of the term which terminates series (1.6.6), also has a substantial influence on equations of state $p = p(\varrho)$. The usefulness of equations of the fourth order is noted by Thomsen (1970), Ahrens and Thomsen (1972) and other authors.

1.6.3 Fourth-order equation of Thomsen

Cutting off a series of the (1.6.6) or (1.6.10) type at the third term permits us to eliminate the constant a_1 from the condition that $p(\varrho_0) = p_0 = 0$ and to express the constants a_2 and a_3 by the bulk modulus K_0 and its first derivative K_0' referred to the pressure p_0. The use of the fourth term of the expansion of the series requires the introduction of the second derivative K_0'' into the equation describing $p(\varrho)$. Thomsen (1970) found that K_0'' cannot be measured directly and proposed a method for its indirect experimental determination. He gives a general method of constructing the equation of state, assuming that the potential energy can be expressed in the form of an expansion into an exponential series with regard to the deviation from a state of equilibrium. This is, therefore, an approach to the problem connected with the Lagrangian concept of strain. According to Thomsen, the Lagrangian strain, rather than the Eulerian, is more fundamental to the equation of state of solids wherever thermal effects are important. In this case the series for the potential energy can be expressed by the components of the Lagrange strain tensor (1.6.9), and the expansion coefficients are constants dependent only on the initial state. According to Thomsen, a similar series with constant coefficients expressing the potential energy by the components of the Euler strain tensor (1.6.5) cannot be obtained. Some results found by Thomsen and related to the fourth-order anharmonic equation of state elaborated by him are given below.

Thomsen introduces the fourth-order equation of state, using the general form of the Lagrange strain tensor and the Grüneisen equation. He concludes that "It is convenient to restrict the generality of the treatment to those crystal classes whose strain response to the hydrostatic pressure is a diagonal tensor. (This includes all classes but triclinic and monoclinic)." All the constants appearing in the equation of state can be determined from laboratory measurements. Detailed considerations were limited by Thomsen to the case of cubic crystals for which he found that*

---

* In eqns. (1.6.12)–(1.6.16) Thomsen's notation is preserved.

$$p = -3\tilde{K}\left(\frac{a}{\tilde{a}}\right)^{-1}\left(\eta - \frac{3}{2}\eta^2\Gamma + \frac{3}{2}\eta^3\Lambda\right) + \\ + \frac{\tilde{U}_s}{\tilde{V}}\left(\frac{a}{\tilde{a}}\right)^{-1}\left\{\tilde{\gamma} + 3\eta\left[\lambda - \tilde{\gamma}^2\left(1 - T\frac{\tilde{c}_V}{\tilde{U}_s}\right)\right]\right\} \tag{1.6.12}$$

where the strain is

$$\eta = \frac{1}{2}\left(\frac{a^2}{\tilde{a}^2} - 1\right) \tag{1.6.13}$$

and a is the edge of the unit cube. The super-tilde denotes evaluation in the fixed reference state. It is easy to describe the six unknowns $\tilde{a}$, $\tilde{K}$, $\tilde{\gamma}$, Γ, λ, Λ in terms of the six parameters:

$$a_0, \quad K_0^S = -V\left(\frac{\partial p}{\partial V}\right)_S, \quad \alpha_0 = \frac{1}{V}\left(\frac{\partial V}{\partial T}\right)_p, \quad K_0' = \left(\frac{\partial K}{\partial p}\right)_T,$$
$$\left(\frac{\partial K}{\partial T}\right)_p, \quad K_0'' = \left(\frac{\partial^2 K}{\partial p^2}\right)_T$$

measured at the "zero state", $p = 0$, $T = T_0$ = room temperature.

In eqn. (1.6.12) $\tilde{V}$ is the volume and $\tilde{U}_s$ is the vibrational contribution to the internal energy, evaluated at the rest state (summation extends over all wave vectors)

$$\tilde{U}_s = \sum h\tilde{\nu}\left(\frac{1}{2} + \frac{1}{\exp\left(\frac{h\tilde{\nu}}{kT}\right) - 1}\right) \tag{1.6.14}$$

and $\tilde{c}_V$ is the corresponding specific heat

$$\tilde{c}_V = \sum k\left(\frac{h\tilde{\nu}}{kT}\right)^2\left[\exp\left(\frac{h\tilde{\nu}}{kT}\right) - 1\right]^{-2}\exp\left(\frac{h\tilde{\nu}}{kT}\right) \tag{1.6.15}$$

The dependence of the Grüneisen parameter on the volume for a cubic crystal is

$$\gamma(V) = \left(\frac{V}{\tilde{V}}\right)^{2/3}\tilde{\gamma} + 3\lambda\eta \tag{1.6.16}$$

and $\gamma(\tilde{V}) = \tilde{\gamma}$. For a highly compressed medium ($V/\tilde{V} \to 0$) formula (1.6.16) can lead to negative values of the parameter γ.

A further development of the Thomsen theory and its critical review was carried out by Davies (1973). He presented calculations relating to the equation of state, and thus also to the Grüneisen parameter $\gamma = \gamma(V/V_0)$, for MgO based on three different forms of the strain tensor (Lagrange η,

Euler ε, and the tensor $e_{ij} = e\delta_{ij}$, where $e = ((V/V_0)^{1/3}-1)$ and on the Thomsen formula (1.6.16). For the range of compressions V/V_0 from 1 to 0.7 the differences are not larger than about 10%. Davies concluded that, at least for MgO, the expansion according to the exponents of the Euler strain tensor ε leads to results more empirically successful than the expansion according to the exponents of the Lagrange strain tensor η (a conclusion directly opposite to the general considerations of Thomsen).

1.6.4 The Prieto–Renero equation of state

An equation of state that is independent of the material constants when nondimensional coefficients are properly chosen, was proposed by Prieto and Renero (1974, 1976). The only initial assumption required to introduce this "universal" equation of state is the linear relation

$$D = A+Bv \tag{1.6.17}$$

where A and B are constants, between the velocity D of the shock wave in a given medium and the velocity v of the medium behind this wave* (particle velocity).

Using formula (1.6.17) and the Rankin–Hugoniot conditions for the continuity of mass and momentum flow on the surface of the shock wave (1.4.1) and (1.4.2), the nondimensional equation of Hugoniot adiabat is obtained in the following form:

$$p^*-p_0^* = \frac{x}{(1-x)^2}, \qquad x = B\left(1-\frac{\varrho_0}{\varrho}\right) \tag{1.6.18}$$

where

$$p^* = \frac{B}{\varrho_0 A^2}p \tag{1.6.19}$$

---

* McQueen *et al.* (1964) assumed that constant A is equal to the hydrodynamic sound velocity $(\partial p/\partial\varrho)_S^{1/2}$. Constant B determined by them ranges from 1.1 (beryllium, titanium) through 1.6 (zinc, gold) to 1.7 (iron) for metals, and from 1.0 (forsterite: Mg_2SiO_4) through 2.1 (quartz: SiO_2) to 2.2 (rutile: TiO_2) for minerals. The average value of B for 14 metals is 1.4 and for 11 minerals it is 1.3.

Experimental data relating to many different materials (metals, minerals and other investigated in 1969 in Los Alamos Laboratory) indicate that for pressure $p \lesssim 100$ GPa and particle velocity $v \lesssim 3.5$ km·s^{-1} a parabolic relation

$$u = A+Bv+Cv^2 \tag{1}$$

is better than the linear formula (1.6.17). Recently, Prieto and Renero (1980) have used the formula (1) for derivation of the equation of state of NaCl and NaI within the range from 0.5 to 25 GPa.

denotes the nondimensional pressure, and p_0, ϱ_0 and p, ϱ are the pressure and density in front of and behind the shock wave, respectively. The application of eqn. (1.6.17) instead of the third Rankin–Hugoniot condition (the condition of energy flow continuity) led to formula (1.6.18), describing a Hugoniot adiabat which is explicitly independent of the internal energy and thus also of temperature. Using the first principle of thermodynamics, the Grüneisen equation, and formula (1.6.18), Prieto and Renero at first obtained a formula expressing the nondimensional temperature t_h^* on the Hugoniot adiabat

$$t_h^* = t_0^* \mathrm{e}^{mx} + n\mathrm{e}^{mx} \int_0^x \frac{y^2}{(1-y)^3} e^{-my} \mathrm{d}y \tag{1.6.20}$$

and then the nondimensional equation of state

$$p^* = \frac{1}{3}\frac{1}{(1-x)^2} + \frac{7}{9}\frac{1}{1-x} - \frac{10}{9}\mathrm{e}^{\frac{4}{3}x} +$$

$$+\frac{28}{27}\mathrm{e}^{-\frac{4}{3}(1-x)}\left[\int_{-\frac{4}{3}(1-x)}^{\infty}\frac{\mathrm{e}^{-z}}{z}\mathrm{d}z - \int_{-\frac{4}{3}}^{\infty}\frac{\mathrm{e}^{-z}}{z}\mathrm{d}z\right] - t_0^*\mathrm{e}^{\frac{3}{4}x} + t^* \tag{1.6.21}$$

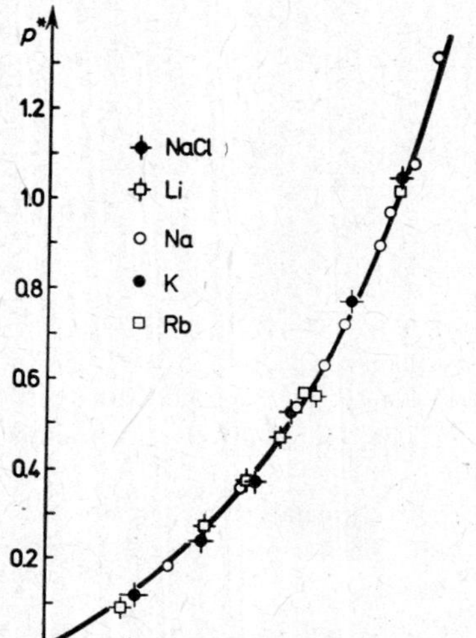

Fig. 1.6.1. Comparison of the relation $p^* = p^*(x)$ resulting from nondimensional equation of state (1.6.21) with experimental data, according to Prieto and Renero (1976).

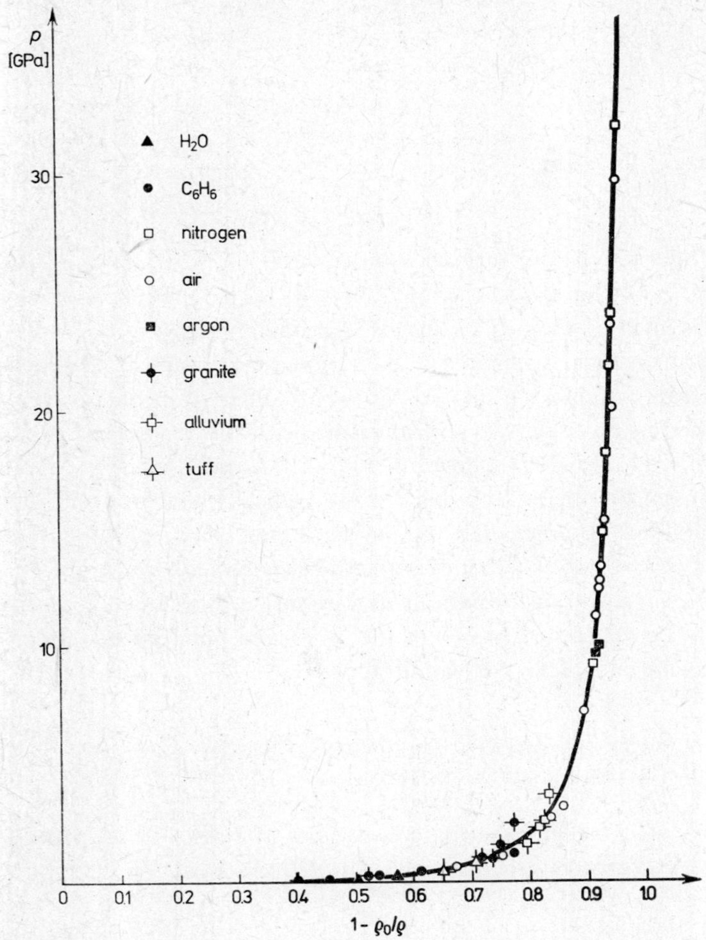

Fig. 1.6.2. Pressure p versus relative density $1-\varrho_0/\varrho$ for eight media, according to Prieto and Renero (1980).

In these equations the nondimensional temperature is defined by the formula

$$t^* = \alpha BT \quad \text{where } \alpha = \frac{1}{V}\left(\frac{\partial V}{\partial T}\right)_p \tag{1.6.22}$$

and t_0^* is the nondimensional temperature corresponding to the nondimensional pressure p_0^*. The numerical coefficients in eqn. (1.6.21) result from the assumption that the quantities

$$m = \frac{\alpha}{\varkappa c_V \varrho_0 B} \quad \text{where } \varkappa = -\frac{1}{V}\left(\frac{\partial V}{\partial p}\right)_T$$

$$n = \frac{\alpha A^2}{B c_V} \tag{1.6.23}$$

$$r = \left(\frac{\partial p^*}{\partial t^*}\right)_x$$

appearing in thermodynamic formulas and in eqn. (1.6.20) are constant and have the following values: $m = \frac{4}{3}$, $n = \frac{4}{3}$, $r = 1$. The coefficient m has the sense of the modified Grüneisen parameter, and the coefficient n^{-1} is the nondimensional specific heat. The numerical values of the coefficients result from a numerical analysis of twenty different media. A comparison of experimental data for several media with the theoretical curve calculated from eqn. (1.6.21) is shown in Figs. 1.6.1 and 1.6.2. Very good agreement between the theory and experiments indicates that eqn. (1.6.21) is really of universal character. It should be stated, however, that this equation will become invalid whenever thermodynamic conditions permit the occurrence of phase transformations in a given medium. In such a case instead of eqn. (1.6.17) two equations of the same type but with two different pairs of coefficients A and B would appear.

1.7 EQUATIONS OF STATE OBTAINED FROM MODELS OF QUANTUM-MECHANICS INTERACTIONS

In this chapter we consider barotropic equations of state $p = p(\varrho)$ based on the internal energy given in the form $U = U(r) \equiv U(\varrho^{-1/3})$, i.e., eqns. (1.5.10):

$$p = \varrho^2 \frac{\mathrm{d}U}{\mathrm{d}\varrho} \quad \text{or} \quad p = -\frac{1}{3} r_0 \varrho_0^{1/3} \varrho^{2/3} \frac{\mathrm{d}U}{\mathrm{d}r} \tag{1.7.1}$$

Various types of interactions between atoms, ions and particles, and their interactions with electron gas are the result of Coulomb electrostatic forces or of quantum effects (exchange energy, zero–point vibration energy, correlation energy). A schematic specification of the most important interactions occurring in solid bodies is given below. It is followed by a classification of different media and the corresponding equations of state based on these interactions and some empirical assumptions. Extensive papers dealing with these problems were published by Born and Huang (1954), Knopoff (1963a), Bernardes and Swenson (1963), Tossi (1964), Brush (1966), Zharkov and Kalinin (1971), Ullmann and Pankov (1976).

The internal energy of interacting particles depends on the distance r between the particles. These relations are as follows (the positive terms correspond to repulsion forces, while the negative terms represent attraction forces):

1. Coulomb attraction of ions with opposite signs

$$U = -\frac{a}{r}, \quad a = \alpha(Ze)^2 \tag{1.7.2}$$

where a is the product of Madelung's constant α, depending on the crystalline structure, and the square of the ion charge Ze.

2. Coulomb interactions between electric multipoles

$$U = -\frac{a}{r^6} - \frac{b}{r^8} - \frac{c}{r^{10}} - \dots \tag{1.7.3}$$

where consecutive terms denote interactions of the following types: dipole–dipole, dipole–quadrupole, quadrupole–quadrupole, ..., i.e., in general interactions of the Van der Waals type.

3. The quantum effect of repulsion, resulting from the energy of electron exchange, and the correlation energy related to electron spin can be expressed by similar relations

$$U = a \exp\left(-\frac{r}{b}\right) \quad \text{or} \quad U = \frac{a}{r^n} \tag{1.7.4}$$

The exponential relation is more proper but the power formula, though formally incorrect, is more convenient and gives reasonable results. The ion overlap energy is also given by a power relation of the $U = ar^{-n}$ type, in which the constant a is related to the coordination number.

4. The kinetic energy of degenerated electron gas

$$U = \frac{a}{r^2} \tag{1.7.5}$$

where $v \propto r^3$ is the volume occupied by one ion and the cloud of electrons neutralizing it.

In formulas (1.7.2)–(1.7.5) the symbols a, b, c, and n denote constant coefficients, or more precisely, they are parameters not dependent on the distance r but possibly on the temperature T. The physical sense of these coefficients is different in different formulas, and this should be remembered since the same symbols are used to denote different quantities.

Using the energies of interactions described above, several authors have proposed a full form of the function $U(r)$ for various media considered. We describe some of these potentials and the corresponding equations of state.

1. The potential of Mie and Borne

$$U(r) = \frac{a}{r^n} - \frac{b}{r^m} \qquad n > m \text{ and } a, b > 0 \tag{1.7.6}$$

A specific case of this potential is the Lennard–Jones potential, otherwise called the 6–12 potential, since $m = 6$ and for n numbers between 9 and 14 were taken and most often $n = 12$. Such values of the exponents m and n adequately describe solidified inert gases except helium, for which the zero-point vibration energy not taken into account is of vital importance. The following equation of state corresponds to the potential (1.7.6)

$$p = \frac{3K_0}{n-m}\left[\left(\frac{\varrho}{\varrho_0}\right)^{\frac{n}{3}+1} - \left(\frac{\varrho}{\varrho_0}\right)^{\frac{m}{3}+1}\right] \tag{1.7.7}$$

where the constants a and b appearing in formula (1.7.6) were eliminated by using the conditions

$$p(r_0) = p_0 \quad \text{and} \quad K(p_0) \equiv K_0 = \left(\varrho\frac{dp}{d\varrho}\right)_{\varrho=\varrho_0} \tag{1.7.8}$$

where $\varrho(r_0) \equiv \varrho_0$ is the density under pressure p_0. In general, it is convenient to take as the reference pressure $p_0 = 0$, i.e., the pressure corresponding to an unstressed state. By using conditions (1.7.8), any two constants appearing in the interaction potential can be replaced by the quantities ϱ_0 and K_0, convenient for direct measurements. The same conditions were used in Section 1.6 for the derivation of the Murnaghan and Murnaghan–Birch equations.

Equation (1.7.7) for $n = 4$ and $m = 2$ becomes the Murnaghan–Birch equation (1.6.8). Such values of n and m, however, have no justification in formulas (1.7.2)–(1.7.5).

2. The potential of Bardeen

$$U(r) = -\frac{a}{r} + \frac{b}{r^2} + \frac{c}{r^3}, \qquad a, b, c > 0 \tag{1.7.9}$$

is used for the description of alkaline metals. The first two terms correspond to the Coulomb attraction between nuclei and electrons and to the energy of electron gas, and the term proportional to r^{-3} includes the shielding, exchange and correlation energies. The equation of state for the potential (1.7.9) is of the form

$$p = 3K_0\left[\left(\frac{\varrho}{\varrho_0}\right)^{5/3} - \left(\frac{\varrho}{\varrho_0}\right)^{4/3}\right]\left\{1 + C\left[\left(\frac{\varrho}{\varrho_0}\right)^{1/3} - 1\right]\right\} \tag{1.7.10}$$

where

$$C = \frac{c\varrho_0}{3K_0 r_0^3}$$

is a coefficient which must be found from experimental data.

3. In general, metals can be described by the potential

$$U(r) = -\frac{a}{r} + \frac{b}{r^2} + c\exp\left(-\frac{r}{d}\right), \quad a, b, c, d > 0 \tag{1.7.11}$$

which differs from the Bardeen formula (1.7.9) only in the form of the last term, describing the energy of exchange. The equation of state in this case is

$$p = A\varrho^{5/3} - B\varrho^{4/3} + C\varrho^{2/3}\exp(-D\varrho^{-1/3}), \quad A, B, C, D > 0 \tag{1.7.12}$$

where two of the constants can be expressed by ϱ_0 and K_0 from relations (1.7.8). In both formulas (1.7.10) and (1.7.12) the term containing $\varrho^{5/3}$ relates to the energy b/r^2 of electron gas. A similar relation $p \propto \varrho^{5/3}$ can be derived by using the Thomas–Fermi method.

4. Ion crystals, e.g., haloids of alkaline metals, zinc sulphide ZnS, periclase MgO, magnetite Fe_3O_4, olivine $(Mg, Fe)_2SiO_4$, form a class of solids for which the theory has been elaborated most fully (especially for haloids). Assuming that the energy is composed of the Coulomb potential and of exchange energy, and that the interactions of Van der Waals can be neglected, the following relations are obtained:

$$U = -\frac{a}{r} + b\exp\left(-\frac{r}{c}\right), \quad a, b, c > 0 \tag{1.7.13}$$

$$p = A\varrho^{2/3}[\exp(-C\varrho^{-1/3}) - B\varrho^{2/3}], \quad A, B, C > 0 \tag{1.7.14}$$

Ion crystals form the most interesting group of bodies in the physics of the Earth's mantle.

5. Molecular crystals or Van der Waals crystals are built of atoms with completely filled electron shells (e.g., solidified inert gases) or of particles with saturated chemical bonds. The energy of interaction is the sum of multipolar interactions and the exchange energy

$$U = -\frac{a}{r^6} + b\exp\left(-\frac{r}{c}\right), \quad a, b, c > 0 \tag{1.7.15}$$

$$p = A\varrho^{2/3}[\exp(-C\varrho^{-1/3}) - B\varrho^{7/3}], \quad A, B, C > 0 \tag{1.7.16}$$

6. Valency crystals (Born and Huang, 1954) differ from molecular crystals by their lack of saturated chemical bonds; their structure is in

general more complex and the potential of interaction is more difficult to establish. Zharkov and Kalinin (1971) gave the following relation for the potential:

$$U = \frac{a}{r}\exp\left(-\frac{r}{c}\right) - b\exp\left(-\frac{r}{c}\right), \quad a, b, c > 0 \tag{1.7.17}$$

where the first term represents the energy of repulsion of positively charged ions screened by the exponentially distributed volume charge of electrons, and the second term describes the covalent attraction forces. Thus the equation of state is

$$p = A(\varrho^{4/3} + B\varrho - C\varrho^{2/3})\exp(-B\varrho^{-1/3}), \quad A, B, C > 0 \tag{1.7.18}$$

Since the structure of valency crystals is very complex, eqn. (1.7.18) does not describe them adequately and therefore is not used in practice.

The equations of state described above are either valid for some precisely specified physical media (e.g., ion crystals, metals) or their applicability for a given medium must be carefully verified against experimental data. Under high pressures and relatively low temperatures, which dominate in the Earth's interior, it proves most reasonable to apply equations of the Born–Mie type in the Earth's mantle, and metallic state equations (of the Thomas–Fermi–Dirac type) in the core.

Let us return once more to an interaction potential of the Born–Mie type (1.7.6), since this potential is connected with two equations of state applied recently to the description of the Earth's and planetary interior.

In the particular case where $n = 2m$, the number of parameters in the potential (1.7.6) is reduced to three and the constants in the equation of state (1.7.7) can be determined from the conditions that $p(\varrho_0) = p_0 = 0$, $K(\varrho_0) = K_0$ and $(\mathrm{d}K/\mathrm{d}p)_{p=0} = K_0'$. As a result, we obtain the equation*

$$p = \frac{3K_0}{K_0' - 2}\left[\left(\frac{\varrho}{\varrho_0}\right)^{(2K_0'-1)/3} - \left(\frac{\varrho}{\varrho_0}\right)^{(K_0'+1)/3}\right] \tag{1.7.19}$$

containing both the Murnaghan–Birch equation, for which $n = 4$ and $m = 2$, and the equation corresponding to the Lennard–Jones potential with $n = 12$ and $m = 6$.

This form of the equation of state was obtained by Ullmann and Pankov (see Model 1 in Ullmann and Pankov (1976) and the work of Ullmann *et al.* (1977) devoted to a numerical analysis of that model). They derived the equation of state not from the Born–Mie potential but from the expan-

* According to experimental data $K_0' \geqslant 3$ always; thus in eqn. (1.7.19) and also in eqn. (1.7.22) no singularity can be expected.

sion into a Taylor series of the free energy $F(\varrho_0/\varrho, T_0)$ or the internal energy $U(\varrho_0/\varrho, S_0)$ around the initial state (ϱ_0, T_0) or (ϱ_0, S_0). Thus the equation in question describes the relation $p = p(\varrho)$ under isothermic or adiabatic conditions. Equation (1.7.19) is an approximate formula in the considerations of Ullmann and Pankov, though it is an exact relation for the pressure corresponding to a given Born–Mie potential. The Ullmann–Pankov relations for the isothermal or adiabatic bulk modulus and for its derivative with respect to the pressure, under (p, ϱ, T_0) or (p, ϱ, S_0) conditions are, respectively,

$$K \approx K_0\left(\frac{\varrho}{\varrho_0}\right)^{(2K_0'-1)/3} + \frac{1}{3}(1+K_0')p \tag{1.7.20}$$

$$K' \approx K_0' + \frac{1}{9}(1+K_0')(1-2K_0')\frac{p}{K} \tag{1.7.21}$$

where $K' \equiv (\partial K/\partial p)_{T=T_0}$ or $K' \equiv (\partial K/\partial p)_{S=S_0}$ and the index zero refers to the initial state, i.e., to the pressure $p_0 = 0$.

O. L. Anderson (1970) presented the following isothermal equation of state:

$$p_{\mathrm{iso}}(\varrho) = \frac{3K_0}{3K_0'-8}\left[\left(\frac{\varrho}{\varrho_0}\right)^{K_0'-4/3} - \left(\frac{\varrho}{\varrho_0}\right)^{4/3}\right] \tag{1.7.22}$$

which corresponds to the Born–Mie potential (1.7.6) with the exponents $m = 1$ and $n = 3K_0'-7$. For the determination of the total pressure in planetary interiors Baumgardner and Anderson (1980) used the two-term equation of state

$$p(\varrho, T) = p_{\mathrm{iso}}(\varrho) + p_{\mathrm{th}}(T) \tag{1.7.23}$$

They found the thermal pressure p_{th} from the condition

$$p_{\mathrm{th}}(T) = \int\left(\frac{\partial p}{\partial T}\right)_\varrho \mathrm{d}T = \int \alpha K \,\mathrm{d}T \tag{1.7.24}$$

where α is the coefficient of thermal expansion and K is the bulk modulus. Since at temperatures $T > 0.8\,\theta$, in which θ is the Debye temperature, the product αK is virtually independent of T, after integration relation (1.7.24) becomes linear

$$p_{\mathrm{th}}(T) = aT - b \tag{1.7.25}$$

where $a = \alpha K$, and b is found from the integral of αK between 0 and $0.8\,\theta$. The final form of the equation of state

$$p(\varrho, T) = aT - b + \frac{3K_0}{3K_0'-8}\left[\left(\frac{\varrho}{\varrho_0}\right)^{K_0'-4/3} - \left(\frac{\varrho}{\varrho_0}\right)^{4/3}\right] \tag{1.7.26}$$

was verified by Baumgardner and Anderson in the modelling of the Earth before its application in the construction of models of the planets. They assumed different values for numerical constants in different layers of the Earth. Thus, for example, in the lower mantle $a = 0.0065$ GPa $\cdot$ K^{-1}, $b = 1.2$ GPa, $K_0 = 263$ GPa, $K_0' = 3.35$, while in the core $a = 0.006$ GPa $\cdot$ K^{-1}, $b = 0$ GPa, $K_0 = 150$ GPa and $K_0' = 4.65$. It was found that such a density model of the Earth is very close (with an accuracy greater than 0.5% for depths $z \geqslant 500$ km) to the PEM model of Dziewonski *et al.* (1975). Such good agreement justifies the application of the above method to the modelling of planetary internal structures (see Section 2.3.5).

1.8 EQUATION OF GRÜNEISEN

The Grüneisen (or Mie–Grüneisen) equation of state is applicable in solids under temperatures too high for the vibrations of the crystalline lattice to be negligible but at the same time so low that the pressure of electron gas can be neglected. The neglecting of the electron gas pressure limits the range of this equation's applicability.

The simplest approximate way to obtain the equation of Grüneisen is to expand the pressure into a Taylor series in relation to the temperature

$$p(V, T) = p(V, 0) + \left[\frac{\partial p(V, T)}{\partial T}\right]_{V, T=0} T + \ldots \tag{1.8.1}$$

and to cut the series after the second term. If it is assumed that $[\partial p(V, T)/\partial T]_{V, T=0} \approx [\partial p(V, T)/\partial T]_V$, then the coefficient at T can be expressed by the isothermic bulk modulus K_T and the coefficient of volume expansion α:

$$K_T = -V\left(\frac{\partial p}{\partial V}\right)_T, \quad \alpha = \frac{1}{V}\left(\frac{\partial V}{\partial T}\right)_p \tag{1.8.2}$$

Hence

$$\left(\frac{\partial p}{\partial T}\right)_V = -\left(\frac{\partial p}{\partial V}\right)_T \left(\frac{\partial V}{\partial T}\right)_p = K_T \alpha \tag{1.8.3}$$

Assuming that the thermal energy of ion vibrations is $U_j = c_V T$, where c_V is the specific heat at constant volume, eqn. (1.8.1) can be transformed into

$$p(V, T) = p_0 + \frac{\gamma U_j}{V} \tag{1.8.4}$$

which is called the *equation of Grüneisen*. The pressure $p_0 = p(V, 0)$

is the elastic component of pressure and the coefficient γ is

$$\gamma = \frac{K_T \alpha V}{c_V} \tag{1.8.5}$$

Formula (1.8.5) can be considered as one of the many equivalent definitions of the parameter of Grüneisen.

A derivation of eqn. (1.8.4) based on quantum mechanics considerations is given below. If it is assumed that each ion of the crystalline lattice of a solid is a harmonic oscillator, the energy of this oscillator in the n-th excited state is $nh\nu$ (where h is the Planck constant, ν is the frequency of vibrations, and $n = 0, 1, 2, \ldots$). Under conditions of thermodynamical equilibrium the probability that the n-th excited state is occupied, is $\exp(-nh\nu/kT)$, where k is the Boltzmann constant and T is the absolute temperature. The mean energy corresponding to one oscillator is

$$u = \frac{\sum_{n=0}^{\infty} nh\nu \exp\left(-\frac{nh\nu}{kT}\right)}{\sum_{n=0}^{\infty} \exp\left(-\frac{nh\nu}{kT}\right)} = \frac{\mathrm{d}}{\mathrm{d}\left(-\frac{1}{kT}\right)} \ln\left[\sum_{n=0}^{\infty} \exp\left(-\frac{nh\nu}{kT}\right)\right]$$

Since the sum appearing in this relation concerns a geometrical series, the mean energy of the oscillator is finally

$$u = \frac{h\nu}{\exp\left(\frac{h\nu}{kT}\right) - 1} \tag{1.8.6}$$

The internal energy of a system containing N ions or $3N$ oscillators can be expressed by the sum of the elastic energy and the vibration energy of those oscillators

$$U = U_0(V) + \sum_{i=1}^{3N} u_i = U_0(V) + \sum_{i=0}^{3N} \frac{h\nu_i}{\exp\left(\frac{h\nu_i}{kT}\right) - 1}$$

$$= U_0(V) + U_j(V, T) \tag{1.8.7}$$

To find the pressure, relation (1.8.7) is substituted in formula (1.5.13)

$$p = T \int \frac{1}{T^2} \left(\frac{\partial U}{\partial V}\right)_T dT$$

$$= T \int \frac{1}{T^2} \frac{\mathrm{d}U_0}{\mathrm{d}V} \,\mathrm{d}T + T \int \frac{1}{T^2} \left[\frac{\partial}{\partial V} \sum_{i=0}^{3N} \frac{h\nu_i}{\exp\left(\frac{h\nu_i}{kT}\right) - 1}\right]_T \mathrm{d}T$$

$$= -\frac{dU_0}{dV} + T\sum_{i=1}^{3N}\int \frac{1}{T^2}\frac{\partial}{\partial V}\left[\frac{h\nu_i}{\exp\left(\frac{h\nu_i}{kT}\right)-1}\right]_T dT$$

The term $-dU_0/dV$ is equal to the elastic pressure p_0. After transforming the second term by substituting $h\nu_i/kT = x_i$, where $\nu_i = \nu_i(V)$, we obtain the following relation:

$$p = p_0 - kT\sum_{i=1}^{3N}\int \frac{d}{dx_i}\left(\frac{x_i}{e^{x_i}-1}\right)\frac{d\nu_i}{dV}\frac{dx_i}{\nu_i}$$

$$= p_0 - \frac{kT}{V}\sum_{i=1}^{3N}\int \frac{d\ln\nu_i}{d\ln V}\frac{d}{dx_i}\left(\frac{x_i}{e^{x_i}-1}\right)dx_i \tag{1.8.8}$$

Grüneisen assumed in 1926 that the derivatives $d\ln\nu_i/d\ln V = -\gamma_i$ are of the same value, equal to $-\gamma$, for all oscillators. This assumption permits us to put the coefficient γ in front of the sum and, therefore, to reduce relation (1.8.8) to the form

$$p = p_0 + \frac{\gamma}{V}\sum_{i=1}^{3N}\frac{h\nu_i}{\exp\left(\frac{h\nu_i}{kT}\right)-1} = p_0 + \frac{\gamma U_j}{V}$$

identical with formula (1.8.4). The parameter of Grüneisen is here defined by

$$\gamma = -\frac{d\ln\nu_i}{d\ln V} = -\frac{d\ln\theta}{d\ln V} \tag{1.8.9}$$

where ν_i is the frequency of vibrations of any oscillator. In particular, it might happen that $\nu_i = \nu_{max}$, where ν_{max} is the highest possible frequency of vibrations. The second part of definition (1.8.9) results from the theory of Debye. According to that theory the highest frequency ν_{max} is related to Debye's temperature θ by the formula

$$\nu_{max} = \frac{k\theta}{h} \tag{1.8.10}$$

and to the velocities v_P and v_S of longitudinal and shear elastic waves by the formula

$$\nu_{max} = \left(\frac{3}{4\pi}\frac{3N}{V}\right)^{1/3}\left(\frac{1}{v_P^3}+\frac{2}{v_S^3}\right)^{-1/3} \tag{1.8.11}$$

where $3N/V$ is the density of the oscillators equal to the number of oscillators (atoms) per unit of body volume.

The velocities v_P and v_S can be expressed by the bulk modulus K and Poisson coefficient σ. After proper substitutions in formula (1.8.11) and using definition (1.8.9), we can represent the parameter of Grüneisen in the following form (Zharkov and Kalinin, 1971; Knopoff and Shapiro, 1969):

$$\gamma = \frac{n}{2} - \frac{2}{3} - \frac{V}{2}\frac{\dfrac{d^2}{dV^2}(pV^n)}{\dfrac{d}{dV}(pV^n)} + \delta$$

$$= \frac{n}{2} - \frac{2}{3} + \frac{1-2n+\dfrac{dK}{dp}+n(n-1)\dfrac{p}{K}}{2-2n\dfrac{p}{K}} + \delta \tag{1.8.12}$$

In particular, for various values of the parameters n and δ we obtain the following formulas (which can also be derived quite independently):

1. The formula of Slater (1939) for $n = 0$

$$\gamma = -\frac{V}{2}\frac{\dfrac{d^2p}{dV^2}}{\dfrac{dp}{dV}} - \frac{2}{3} + \delta \tag{1.8.13}$$

where the correction δ can be expressed by the Poisson coefficient σ.

A simple justification of this relation was given by Zeldovitch and Rayzer (1966). The highest vibration frequency ν_{max} can be estimated as follows:

$$\nu_{max} \propto \frac{\text{velocity of acoustic waves } c_0}{\text{minimum wavelength comparable to the lattice constant } r_0}$$

Since

$$c_0^2 \propto \frac{dp}{d\varrho} = -V^2\frac{dp}{dV}$$

and

$$r_0^3 = \frac{V}{Na} \quad \text{or} \quad r_0 \propto V^{1/3}$$

where N is the number of molecules in the volume V and the coefficient a depends on the lattice structure, we have

$$\nu_{max} = \text{const}\,\frac{c_0}{r_0} = \text{const}\,V^{2/3}\left(-\frac{dp}{dV}\right)^{1/2}$$

Substituting this formula into definition (1.8.9), we obtain the expression

$$\gamma = -\frac{\mathrm{d}\ln \nu_{\max}}{\mathrm{d}\ln V} = -\frac{V}{2}\frac{\dfrac{\mathrm{d}^2 p}{\mathrm{d}V^2}}{\dfrac{\mathrm{d}p}{\mathrm{d}V}} - \frac{2}{3}$$

which differs from the Slater formula (1.8.13) only by the term $\delta(\sigma)$.

2. The formula of Dugdale and MacDonald (1953) for $n = \frac{2}{3}$ and $\delta = 0$ is

$$\gamma = -\frac{V}{2}\frac{\dfrac{\mathrm{d}^2}{\mathrm{d}V^2}(pV^{2/3})}{\dfrac{\mathrm{d}}{\mathrm{d}V}(pV^{2/3})} - \frac{1}{3} \tag{1.8.14}$$

3. The formula of Vashchenko and Zubarev (1963) for $n = \frac{4}{3}$ and $\delta = 0$ results from considerating atom (ion) vibrations in a spherically symmetric field generated by neighbouring atoms (the method of free volume)

$$\gamma = -\frac{V}{2}\frac{\dfrac{\mathrm{d}^2}{\mathrm{d}V^2}(pV^{4/3})}{\dfrac{\mathrm{d}}{\mathrm{d}V}(pV^{4/3})} = \frac{1}{2}\,\frac{-\dfrac{5}{3} + \dfrac{\mathrm{d}K}{\mathrm{d}p} + \dfrac{4}{9}\dfrac{p}{K}}{1 - \dfrac{4}{3}\dfrac{p}{K}} \tag{1.8.15}$$

This formula is often used for the description of the Earth's interior (e.g., Stacey, 1977; Stiller and Frank, 1980).

O.L. Anderson (1979) gives the comparison of the values of the Grüneisen parameter calculated from Vashchenko–Zubarev formula (1.8.15) with those of the "acoustic Grüneisen parameters" defined by the following formulas:

$$\begin{aligned} \gamma_S &= \frac{1}{3} + \frac{\mathrm{d}\ln v_S}{\mathrm{d}\ln \varrho} \\ \gamma_P &= \frac{1}{3} + \frac{\mathrm{d}\ln v_P}{\mathrm{d}\ln \varrho} \\ \gamma_a &= \frac{1}{3}(2\gamma_S + \gamma_P) \end{aligned} \tag{1.8.16}$$

For his calculations he used the velocity distribution of shear waves v_S, longitudinal waves v_P and the density distribution ϱ in the Earth corresponding to the PEM model (Dziewonski *et al.*, 1975). Anderson applied the obtained results for an analysis of thermodynamical properties of the

Earth's interior. He believes that the acoustic Grüneisen parameter γ_a should be regarded as a property of the Earth like the bulk modulus K and density ϱ.

Experimental determinations of the Grüneisen parameter are based on relations (1.8.5), (1.8.11) or (1.8.12), i.e., on thermodynamic or acoustic measurements. The Grüneisen parameter is not explicitly dependent on temperature. Under room conditions $\gamma \approx 2$, e.g., for metals $\gamma_{\text{Al}} = 2.09$, $\gamma_{\text{Cu}} = 1.98$, $\gamma_{\text{Pb}} = 2.46$ (Zeldovitch and Rayzer, 1966). With the increase of temperature the coefficient γ decreases (Table 1.5.1), which can be explained as follows. At very high temperatures the elastic pressure p_0 can be neglected when compared with the thermal term $\Delta U_j/V$, and so the equation of state (1.8.4) is transformed into the equation of state of mono-atomic ideal gas $pV = \gamma U$, where U is the internal energy of the gas and $\gamma = \varkappa - 1$, where $\varkappa = \frac{5}{3}$ is the adiabatic exponent. Thus for $T \to \infty$ we have $\gamma \to \frac{2}{3}$.

1.9 THE THOMAS–FERMI EQUATION

The equations of state presented in the previous sections do not take into account the contribution of the energy U_e and pressure p_e of free electron gas to the total internal energy and total pressure. The Thomas–Fermi method permits us to calculate the pressure $p_e = p_e(\varrho_e)$, where ϱ_e is the density of electron gas in a medium. Then, under certain assumptions related to medium density and temperature, it can be shown that the electron component p_e is considerably larger than the elastic pressure p_0 and the pressure of thermal ion vibrations p_j, and thus it can be assumed that

$$p_e \equiv p_e(\varrho_e) \approx p \tag{1.9.1}$$

is the equation of state of the medium. Since the Thomas–Fermi model does not take into account the interaction forces between individual atoms, it does not describe the interactions in solids under normal conditions. The model can be used for highly compressed solids and liquids and for temperatures lower than the temperature of electron gas degeneration.

The equation of state in the Thomas–Fermi form is derived under the following assumptions (Landau and Lifshitz, 1958; Flügge and Marschall, 1952; Knopoff, 1963b; Boschi and Caputo, 1969):

1. The medium under consideration, electrically neutral, is composed of positively charged nuclei with the charge Ze and of free electron gas; Z is the atomic number of the medium, and e is the electron charge.

2. The temperature of electron gas is equal to absolute zero, then the

electrons occupy the lowest possible energy states. The electrons are governed by Fermi–Dirac statistics.

3. The charge of electrons is distributed continuously with the density ϱ_e around a nucleus.

4. The potential $\varphi = \varphi(r)$ of the field generated by the nucleus and the surrounding electrons fulfils the spherically symmetrical equation of Poisson

$$\frac{1}{r}\frac{\mathrm{d}^2}{\mathrm{d}r^2}[r\varphi(r)] = 4\pi e\varrho_e \tag{1.9.2}$$

From the second assumption it follows that the number of states $\mathrm{d}N$ of a phase cell with volume V, filled by electrons with absolute values of the momentum in the interval between P and $P+\mathrm{d}P$, is

$$\mathrm{d}N = 2V\frac{1}{h^3}4\pi P^2\mathrm{d}P \tag{1.9.3}$$

where factor 2 results from the electron spin, $h = 6.626\times10^{-34}$ J·s is the Planck constant, and h^3 is the volume of an elementary phase cell occupied by one electron. From the integration of relation (1.9.3), within the interval from $P = 0$ to the maximum value of momentum P_0, we obtain the relationship between the volume V, the number N of electrons filling this volume and their maximum momentum P_0 (or their maximum kinetic energy $\mathscr{E}_0 = P_0^2/2m$, where m is the electron mass)

$$P_0^3 = \frac{3h^3}{8\pi}\frac{N}{V}, \quad \mathscr{E}_0 = \left(\frac{3}{\pi}\right)^{2/3}\frac{h^2}{8m}\left(\frac{N}{V}\right)^{2/3} \tag{1.9.4}$$

The formula describing the maximum energy $\mathscr{E}_0$ permits the determination of the temperature of degeneration $T_{\text{deg}} = \mathscr{E}_0/k$, where k is the Boltzmann constant. The total kinetic energy of free electron gas is

$$E = \int_0^P \frac{P^2}{2m}\mathrm{d}N = \frac{4\pi}{5}\frac{P_0^5}{mh^3}V = \frac{3}{40}\left(\frac{3}{\pi}\right)^{2/3}\frac{h^2}{m}\left(\frac{N}{V}\right)^{5/3}V \tag{1.9.5}$$

When electrostatic interactions are neglected this energy is equal to the internal energy of electron gas. Since at the temperature of absolute zero the pressure p is connected with the internal energy by the relation $p = -\mathrm{d}E/\mathrm{d}V$, the pressure of electron gas in volume V filled by N electrons, after the differentiation of relation (1.9.5), can be expressed as

$$p = \frac{1}{20}\left(\frac{3}{\pi}\right)^{2/3}\frac{h^2}{m}\varrho_e^{5/3}, \quad \varrho_e = \frac{N}{V} = \frac{Z}{v} \tag{1.9.6}$$

where ϱ_e is the density of electron gas, Z is the atomic number of the given medium, and v is the volume corresponding to one atom.

Although formula (1.9.6) was derived for free homogeneous electron gas, it can also be used when the density ϱ_e varies with the distance from the nucleus. The essence of the Thomas–Fermi method consists in applying the Poisson eqn. (1.9.2) to determine of electron gas density $\varrho_{e,0} = \varrho_e(r_0)$ on the border of an area occupied by a single atom and then substituting this density in eqn. (1.9.6) defining the pressure.

The kinetic energy $\mathscr{E}(r)$ of an electron, placed at a distance r from a nucleus with the charge Ze, fulfils the condition

$$\mathscr{E}(r) \leqslant e\varphi(r) = \frac{Ze^2}{r}\,\Phi(r) \tag{1.9.7}$$

where $\varphi(r)$ is the potential of a nucleus shielded by the volume charge of electrons placed at distances smaller than r. The dimensionless function $\Phi(r)$ takes into account the lessening effect of electron gas on the interaction energy of a single electron and nucleus. Since the maximum value $\mathscr{E}(r)$ is equal to the maximum kinetic energy $\mathscr{E}_0$ of electrons contained in a volume V, with a radius r, the following equation is obtained from formulas (1.9.4) and (1.9.7):

$$\left(\frac{3}{\pi}\right)^{2/3} \frac{h^2}{8m} \left(\frac{N}{V}\right)^{2/3} = \frac{Ze^2}{r}\,\Phi(r) \tag{1.9.8}$$

and therefore the density of electron gas is

$$\varrho_e(r) = \frac{\pi}{3}\left(\frac{8me^2}{h^2}\right)^{3/2} \left(\frac{Z\Phi}{r}\right)^{3/2} \tag{1.9.9}$$

Thus, to find the density ϱ_e and, from relation (1.9.6), the pressure $p \propto \varrho_e^{5/3}$, it is necessary to determine the function $\Phi(r)$ at the border of the area occupied by a single atom, i.e., at the point $r = r_0$. Accordingly, relation (1.9.9) should be substituted in the Poisson equation (1.9.2), then, after conversion to dimensionless variables, we obtain the Thomas–Fermi equation for the function Φ

$$\frac{d^2\Phi}{dx^2} = x^{-1/2}\Phi^{3/2} \tag{1.9.10}$$

where

$$x = \frac{r}{\mu}, \quad \mu = \left(\frac{3}{32\pi^2}\right)^{2/3} \frac{h^2}{2me^2}\, Z^{-1/3} \tag{1.9.11}$$

The constant μ having the dimension of length depends only upon the nucleus charge Z.

Equation (1.9.10) should be solved under proper boundary conditions for the field potential and intensity.

In the near vicinity of the nucleus the potential $\varphi(r)$ is Coulombic, $\varphi(r) = Ze/r$, and thus for $r \to 0$

$$\Phi(0) = 1 \tag{1.9.12}$$

At the borders of areas occupied by single electrically neutral atoms, i.e., for $r = r_0$, the field potential $\varphi(r_0)$ and intensity $(\mathrm{d}\varphi/\mathrm{d}r)_{r=r_0}$ must be equal to zero. Differentiation of (1.9.7) gives

$$-e\left.\frac{\mathrm{d}\varphi}{\mathrm{d}r}\right|_{r=r_0} = \frac{Ze^2}{r_e^2}\Phi(r_0) - \frac{Ze^2}{r_0}\left.\frac{\mathrm{d}\Phi}{\mathrm{d}r}\right|_{r=r_0} = 0$$

From this and directly from eqn. (1.9.7) we obtain the boundary conditions for the dimensionless potential Φ at the border of an atom:

$$\Phi(x_0) = 0 \quad \text{and} \quad \left.\frac{\mathrm{d}\Phi}{\mathrm{d}x}\right|_{x=x_0} = 0 \tag{1.9.13}$$

From relations (1.9.6), (1.9.9) and (1.9.11) it follows that the equation of state is of the form

$$pv = \frac{2}{15}\frac{Z^2e^2}{\mu}x_0^{1/2}\Phi^{5/2}(x_0), \qquad v = \tfrac{4}{3}\pi r_0^3 \tag{1.9.14}$$

March (1955) presented the series

$$\Phi(x_0) = \frac{3^{2/3}}{x_0}\left(1 - \frac{3^{1/3}x_0}{10} + \ldots\right) \tag{1.9.15}$$

which is the solution of eqn. (1.9.10) at point $x = x_0$ and under boundary conditions (1.9.12) and (1.9.13). Substituting this series in eqn. (1.9.14), we obtain the equation of state in the form of a relation between $pZ^{-10/3}$ and Zv

$$pZ^{-10/3} = \frac{1}{20}\left(\frac{3}{\pi}\right)^{2/3}\frac{h^2}{m}(Zv)^{-5/3}\left\{1 - \frac{2me^2}{h^2}(4Zv)^{1/3} + \ldots\right\} \tag{1.9.16}$$

The first term of expansion (1.9.16) is identical with eqn. (1.9.6). Further terms correspond to deviations from an ideal electron gas, i.e., they take into account the electrostatic electron–electron or electron–nucleus interactions. The diagram of relation (1.9.16) is shown in Fig. 1.9.1.

The Thomas–Fermi equation of state can be made more precise by taking into account the interactions of electron spins, i.e., considering the exchange energy (the Thomas–Fermi–Dirac method). In this case the total energy of a nucleus–electron gas system decreases, and the pressure

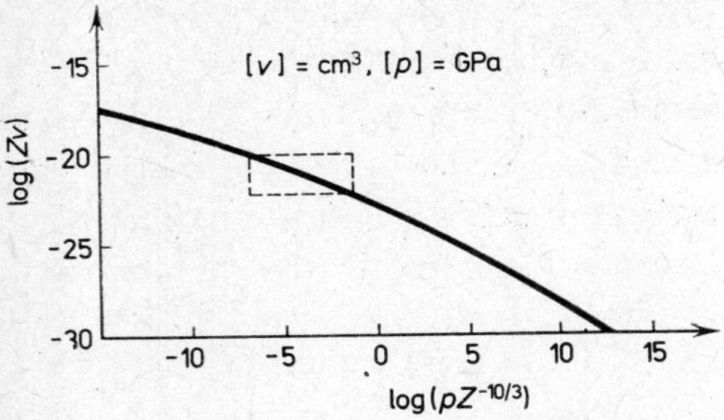

Fig. 1.9.1. Relation $\log(pZ^{-10/3}) = f[\log(Zv)]$ as given by the Thomas–Fermi equation (1.9.16). The area marked by a rectangular frame is shown in detail in Fig. 1.9.2 (Knopoff, 1963b).

and density of electron gas are given by the following corrected formulas (Knopoff, 1963b):

$$p = \frac{1}{20}\left(\frac{3}{\pi}\right)^{2/3}\frac{h^2}{m}\varrho_e^{5/3}(r_0) - \frac{1}{4}\left(\frac{3}{\pi}\right)^{1/3} e^2\varrho_e^{4/3}(r_0) \tag{1.9.17}$$

where

$$\varrho_e = \left(\frac{8me}{h^2}\right)^{3/2}\frac{\pi}{3}[\tau_0 + (\varphi-\varphi_0+\tau_0^2)^{1/2}] \tag{1.9.18}$$

$$\tau_0 = \frac{e\sqrt{2me}}{h}, \qquad \varphi_0 = \text{const}$$

The Poisson equation applied to the corrected density (1.9.18) takes the form

$$\Delta(\varphi-\varphi_0) = \frac{32\pi^2 e}{3h^2}(2me)^{3/2}[\tau_0+(\varphi-\varphi_0+\tau_0^2)^{1/2}]^3 \tag{1.9.19}$$

or the dimensionless form

$$\frac{d^2\Phi}{dx^2} = x\left[\mathscr{E} + \left(\frac{\Phi}{x}\right)^{1/2}\right]^3, \qquad \mathscr{E} = \left(\frac{3}{32\pi^2 Z^2}\right)^{1/3}$$
$$\varphi = \varphi_0 + \tau_0^2 = \frac{Ze}{r}\Phi, \qquad r = \mu x \tag{1.9.20}$$

Substituting the density ϱ_e found from eqns. (1.9.18), (1.9.19) and (1.9.20) in eqn. (1.9.17), the following equation of state is obtained:

$$pv = \frac{2}{15}\frac{Z^2e^2}{\mu}x_0^3\left\{\left[\frac{\Phi(x_0)}{x_0}\right]^{1/2} + \mathscr{E}\right\}^5\left\{1 - \frac{\frac{5}{4}}{\left[\frac{\Phi(x_0)}{x_0}\right]^{1/2} + \mathscr{E}}\right\} \tag{1.9.21}$$

where $\Phi(x_0)$ is the value of the function Φ taken at the boundary of the atom. It should be noted that for $\mathscr{E} \to 0$ (the exchange energy can be neglected only for large atomic numbers Z) the Thomas–Fermi–Dirac equation is reduced to the Thomas–Fermi equation. The relations $\log(pZ^{-10/3}) = f[\log(Zv)]$, drawn on the basis of eqn. (1.9.21), are shown by continuous lines in Fig. 1.9.2. The role of a parameter is played by the atomic number

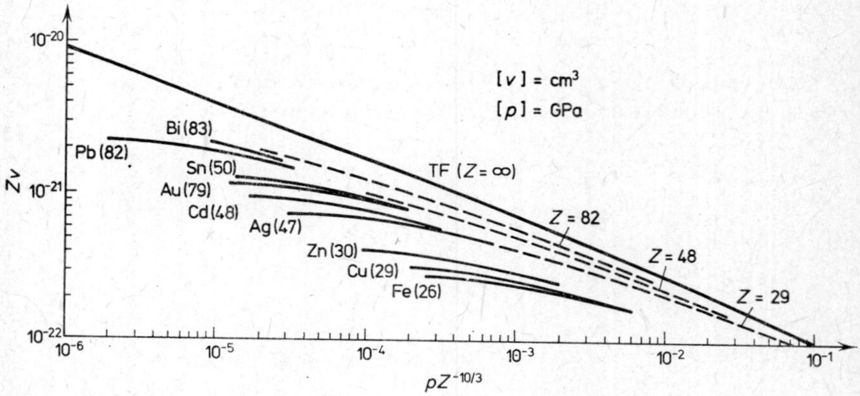

Fig. 1.9.2. A comparison of theoretical results, marked by dashed lines, found from the Thomas–Fermi–Dirac equation (1.9.21), with the experimental data for various metals shown by continuous lines (Knopoff, 1963b).

Z. The line for $Z = \infty$ corresponds to the Thomas–Fermi equation of state. Experiments with different metals carried out with the shock waves indicate good agreement between observation data (marked by dashed lines) and theoretical results.

Another possibility of correcting the Thomas–Fermi equation of state is based on the introduction of corrections dependent on temperature (Knopoff (1963b) lists several papers considering this problem), thus abandoning the initial assumption that $T = 0$ K. In this case it is possible to connect formally the Fermi–Dirac equation of state with the Grüneisen equation, as the expression $(1/\varrho c_V)(\partial p/\partial T)_V$ is analogous to the Grüneisen parameter γ (in this case the relation given by the corrected equation of Thomas–Fermi is taken for the pressure $p = p(V, T)$).

The relations $\log\varrho = f(\log p)$ drawn for silicates Mg_2SiO_4, Fe_2SiO_4 and SiO_2 and for iron, i.e., for media typical for the Earth's interior, are shown in Fig. 1.9.3. Diagrams found from the Murnaghan–Birch and Thomas–Fermi equations of state are given. The experimental points for fayalite Fe_2SiO_4 and forsterite Mg_2SiO_4 are also marked.

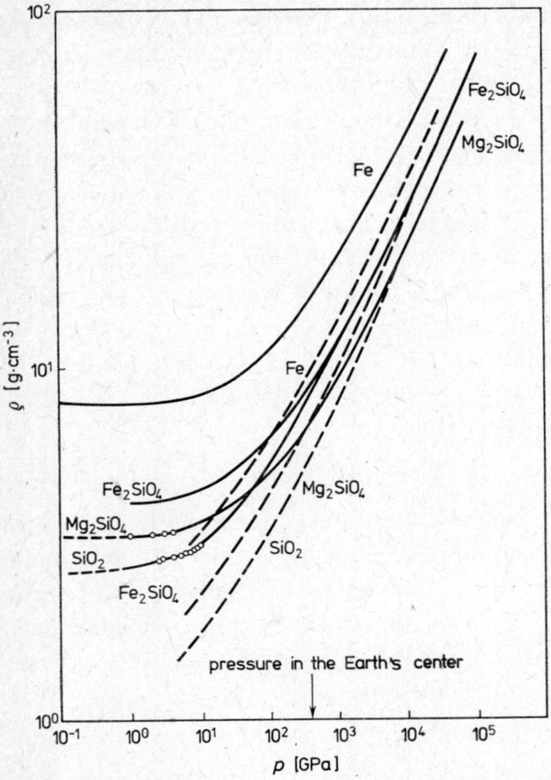

Fig. 1.9.3. Relation $\log\varrho = f(\log p)$ for the four media. The Thomas–Fermi model is marked by a dashed line, the Murnaghan–Birch model is shown by a continuous line, and experimental data are from Boschi and Caputo (1969).

1.10 GEOPHYSICAL EQUATIONS OF STATE

In studies of the Earth's interior, thermodynamic parameters such as density ϱ, pressure p and temperature T cannot be measured directly. Indirect data are used for their determination, especially the velocity distributions of longitudinal waves $v_P(r)$ and shear waves $v_S(r)$ found by seismic methods. The equations called "geophysical equations of state"* are based on empirical relations between density ϱ, elastic wave velocities v_P and v_S, hydrodynamic sound velocity $\phi^{1/2}$, bulk modulus K and mean atomic mass $\bar{A}$ in a given medium. Theoretical attempts to justify different relations of the "geophysical equation of state" type are based on various

* "Seismic equation of state" usually means eqn. (1.10.3) given by Don Anderson (1967).

models of a solid, and especially on the theory of Debye and on the equation of state derived from an electrostatic interaction potential of the Born–Mie type.

There are a number of various relations connecting the parameters ϱ, v_P, v_S, ϕ, K and $\bar{A}$ with one another, but in general it is impossible to establish which of these relations is the best. It is only possible to determine the usefulness of a particular relation for a given group of rocks or minerals, for a given range of thermodynamic parameters or for the most convenient description of a given phenomenon or group of phenomena. The review of "geophysical equations of state" given below cannot, therefore, be entirely complete or objective.

The following equations seem to be the most important:

1. Linear equation of Birch (1961a)

$$v_i = a_i + b_i \varrho \tag{1.10.1}$$

where v_i denotes v_P, v_S or $\phi^{1/2}$, and the coefficients a_i and b_i are experimental constants suitably selected. Relation (1.10.1) for the velocity v_P of longitudinal waves was established by Birch (1961a), for the velocity v_S of shear waves by Simmons (1964), and for the hydrodynamic sound velocity $\phi^{1/2}$ by McQuenn *et al.* (1964).

2. Equation of Anderson and Nafe (1965)

$$\ln K = a - b \ln \frac{2\bar{A}}{\varrho}, \qquad a, b = \text{const} \tag{1.10.2}$$

connecting the bulk modulus K with the density ϱ and mean atomic mass $\bar{A}$ of the medium.

3. "Seismic equation of state" (Anderson, 1967), usually written in the form

$$\frac{\varrho}{\bar{A}} = a\phi^n \tag{1.10.3}$$

where a and n are constants,

$$\phi = v_P^2 - \tfrac{4}{3} v_S^2 = \frac{K}{\varrho} = \left(\frac{\partial p}{\partial \varrho}\right)_S \tag{1.10.4}$$

is the seismic parameter (equal to the square of the hydrodynamic sound velocity), and the derivative of the pressure p is taken for a constant entropy S.

4. Equation of Soga (1971)

$$v_i = c_i \frac{\varrho^{1.5}}{\bar{A}^2} \tag{1.10.5}$$

where v_i denotes v_P, v_S or $\phi^{1/2}$, and c_i is a constant taking, accordingly, three different values.

5. Equation of Shankland (1972)

$$\left(\frac{\partial \ln v_i}{\partial \ln \varrho}\right)_{\bar{A}} = \gamma_i - \tfrac{1}{3} \qquad (1.10.6)$$

where v_i denotes v_P or v_S, and γ_i is the Grüneisen parameter relating either to longitudinal or to shear vibrations. Equation (1.10.6) results from the theory of Debye describing the vibrations of the crystalline lattice in a solid. A similar equation to (1.10.6), but only for the velocity v_S, was given by O. L. Anderson and Nafe (1965).

A comparison of the results obtained from equation (1.10.1) of Birch, equation (1.10.2) of Anderson and Nafe, and the seismic equation of state (1.10.3) of Don Anderson was made by Simmons and England (1969), who used experimental data relating to several rocks and minerals. They found many deviations from these equations, and this prevents the selection of one relation distinctly preferable to the others.

A detailed analysis of various forms of the geophysical equation of state was made by O. L. Anderson (1973), who concluded that power relations between the hydrodynamic sound velocity $\phi^{1/2}$ and the density are more realistic than the linear relation of Birch. He also indicated that the exponent λ_i in relations of the $v_i = \varrho^{\lambda_i}$ type should be dependent on the density, and that, moreover, this exponent might be different for different polymorphic forms of the same medium. In particular, O. L. Anderson carried out a comparison of experimental and theoretical data relating to the hydrodynamic sound velocity in α- quartz, coesite and stishovite.

Equation (1.10.1) was proposed by Birch on the basis of experimental data relating to longitudinal wave velocities in several rock samples under pressure of 1 MPa. It was found that the coefficients a and b have different values for different mineral compositions. In particular, from data obtained from 46 samples with a mean atomic mass $\bar{A} \approx 21$ g·mole^{-1} the following relation was accepted:

$$v_P = -2.55 + 3.31\varrho \qquad (1.10.1a)$$

where v_P is in km·s^{-1} and ϱ is in g·cm^{-3}. This relation was frequently used by various authors, e.g., for modelling the density distribution in the Earth's interior. Birch (1961b) has modified eqn. (1.10.1) into the following form taking into account the changes of longitudinal wave velocities caused by the changes of the mean atomic mass $\bar{A}$:

$$v_P = a + b\varrho + c\bar{A} \qquad (1.10.1b)$$

where $a+c\bar{A} = -2.55, -5.7$, and -8.6 for $\bar{A} = 21, 25$ and 30, respectively, and the constant $b \approx 3$.

In a number of papers (beginning from 1961) Birch published several experimental diagrams of the dependence of the seismic parameter ϕ on the density ϱ for various materials and under different pressures. Measurements of the seismic parameter ϕ against the density ϱ, where the density in turn depends on the pressure $\varrho = \varrho(p)$ applied to a sample, make it possible to draw a family of curves $\phi = \phi[\varrho(p, \bar{A})]$, each curve presenting the relation $\phi = \phi(\varrho)$ for a given element, mineral or rock, the mean atomic mass $\bar{A}$ being the parameter. Such a diagram is called the *diagram of Birch*. The diagram for various materials, especially for metals, is shown in Fig. 1.10.1, where the areas of ϱ and ϕ variations in the Earth's mantle and core are also marked. The numbers placed beside the symbols are the atomic numbers (or mean atomic numbers). Dunite Fo_{90} and dunite Fo_{50} denote dunites whose main component olivine contains no less than 90% and 50% of

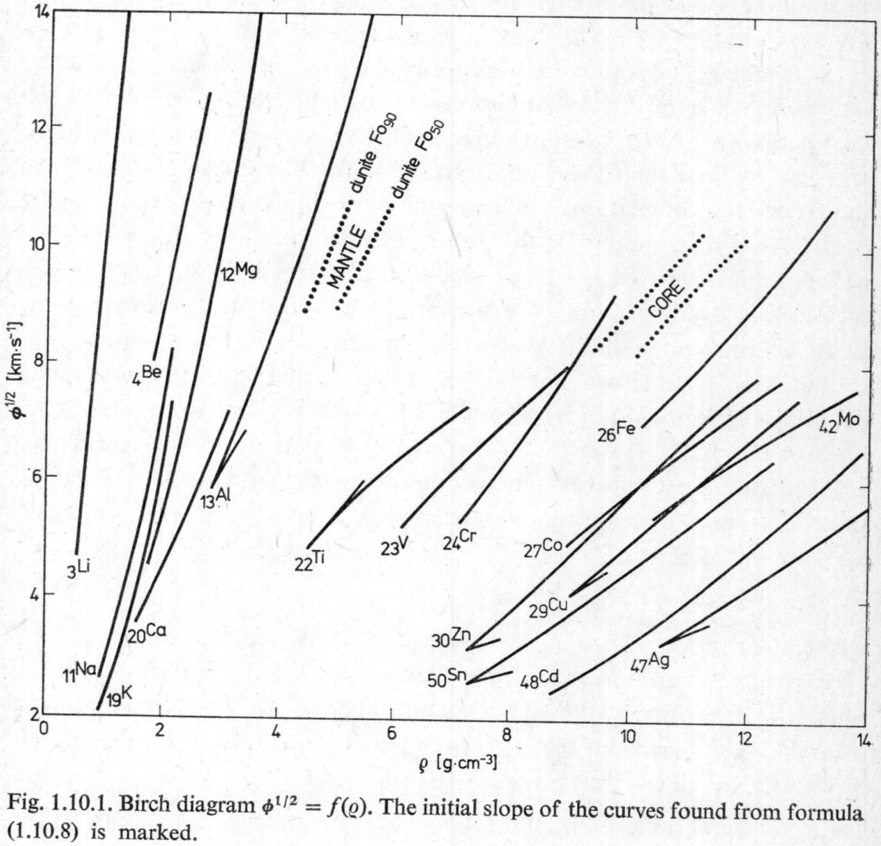

Fig. 1.10.1. Birch diagram $\phi^{1/2} = f(\varrho)$. The initial slope of the curves found from formula (1.10.8) is marked.

forsterite Mg_2SiO_4, respectively. On the diagrams relating to metals the initial slopes (i.e., for the pressure $p_0 = 0$) are marked according to Liu (1974), who finds that these slopes do not depend on an equation of state but result directly from the definitions of the modulus K and the seismic parameter ϕ. Indeed, from the seismic parameter definition (1.10.4) it follows after differentiation that

$$\frac{d(\phi^{1/2})}{d\varrho} = \frac{1}{2}\left(\frac{K}{\varrho^3}\right)^{1/2}(K'-1) \tag{1.10.7}$$

where $K' = dK/dp$, and the slope of the diagram $\phi = \phi(\varrho)$ at a point where $\varrho = \varrho_0$, or where the external pressure is equal to zero, is

$$\left.\frac{d(\phi^{1/2})}{d\varrho}\right|_{\varrho=\varrho_0} = \frac{K_0'-1}{2}\left(\frac{K_0}{\varrho_0^3}\right)^{1/2} \tag{1.10.8}$$

The dimensionless value of K_0' lies between 3.60 and 6.43 for various elements listed by Birch, and thus according to formula (1.10.7), the initial slopes range from $2.47\,\varrho_0^{-3/2}$ to $6.88\,\varrho_0^{-3/2}$.

Equation (1.10.2) of Anderson and Nafe, similarly to eqn. (1.10.1) of Birch, results directly from experimental data. Its authors concluded that the experimental data relating to several oxides and silicates, presented on the plane $\ln(2\bar{A}/\varrho)$, $\ln K$, are contained between the straight lines

$$\ln K = 17.3 - 4\ln\frac{2\bar{A}}{\varrho} \tag{1.10.2a}$$

$$\ln K = 15.1 - 3\ln\frac{2\bar{A}}{\varrho} \tag{1.10.2b}$$

where K is in MPa and ϱ is in $g \cdot cm^{-3}$. It should be noted that eqn. (1.10.2) written in a power form is, by (1.10.4),

$$\frac{\varrho}{\bar{A}} = \text{const}\ \varrho^{1/b}\phi^{1/b} \tag{1.10.2c}$$

(where $0.25 < 1/b < 0.33$), which differs by the factor $\varrho^{1/b}$ from the seismic equation of state (1.10.3) of Don Anderson, who chose the numerical coefficients in eqn. (1.10.3) from laboratory data relating to 31 selected minerals and rocks, and thus obtained the relation

$$\frac{\varrho}{\bar{A}} = 0.048\,\phi^{0.323} \pm 0.12 \tag{1.10.3a}$$

where ϱ is in $g \cdot cm^{-3}$ and $\phi^{1/2}$ is in $km \cdot s^{-1}$. The results of Don Anderson are shown in Fig. 1.10.2 together, for comparison, with the results given

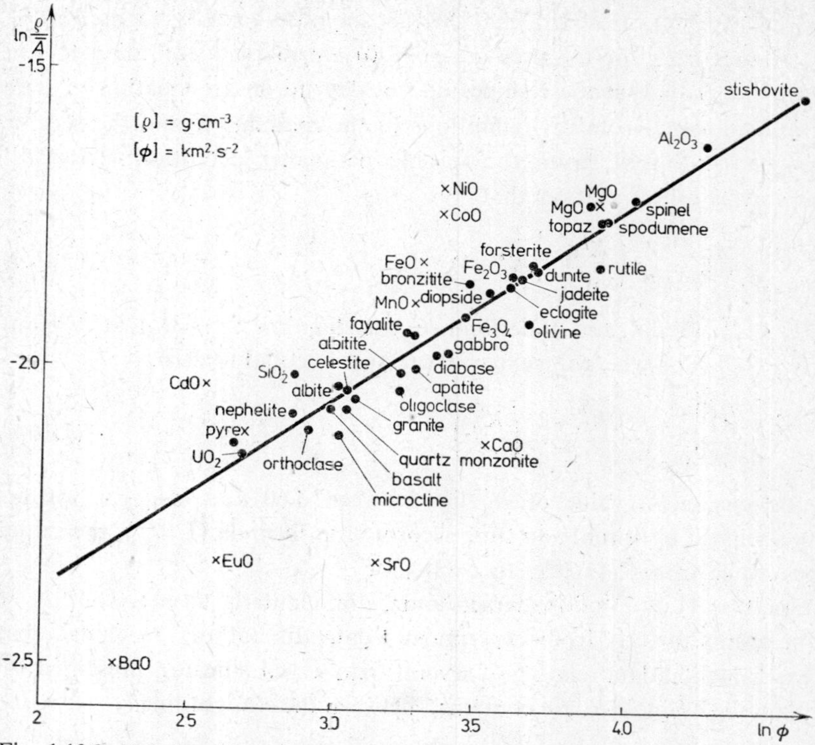

Fig. 1.10.2. Relation (1.10.3a) of Don Anderson. The empirical points (•) and the corresponding straight line are compared with the data of Simmons and England (×) for different oxides.

by Simmons and England (1969), which are distinctly different from relation (1.10.3a). It should be noted that Simmons and England are highly critical of the generality of all the geophysical equations of state which are in use at present. A further modification of the seismic equation of state was carried out (Anderson and Jordan, 1970) to make it also valid for densely packed structures of the oxides MgO, Al_2O_3 and SiO_2 (stishovite). The equation obtained

$$\frac{\varrho}{\bar{A}} = 0.0492\,\phi^{1/3} \tag{1.10.3b}$$

differs only slightly from relation (1.10.3a).

Equation (1.10.5) of Soga written for the velocities v_P, v_S and $\phi^{1/2}$, respectively, has the following form:

$$v_P = (613 \pm 36)\,\frac{\varrho^{3/2}}{\bar{A}^2} \tag{1.10.5a}$$

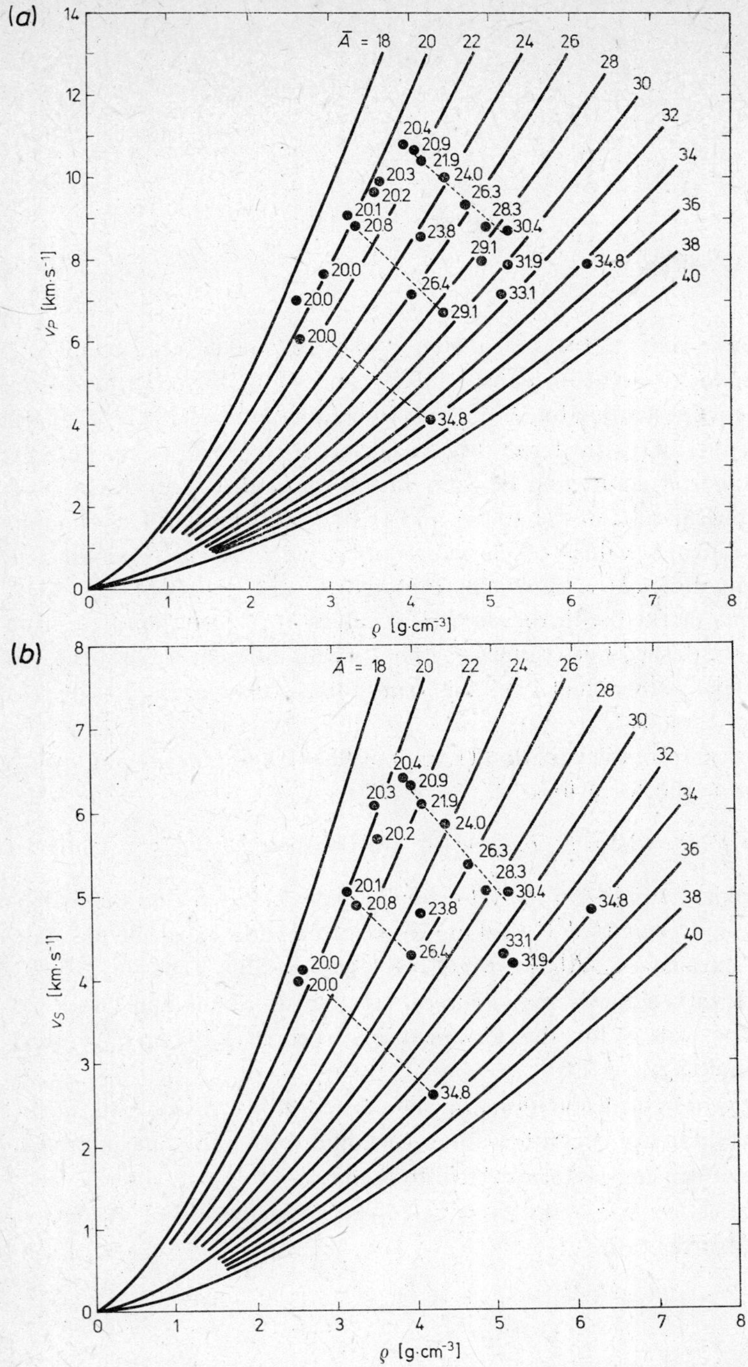

Fig. 1.10.3. (*a*) Velocity $v_P = v_P(\varrho)$ according to eqn. (1.10.5a). (*b*) Velocity $v_S = v_S(\varrho)$ according to eqn. (1.10.5b). The mean atomic mass $\bar{A}$ is the parameter. The points (•) and the numbers $\bar{A}$ placed beside them denote the values of ϱ and v_P (or v_S) found empirically for various minerals or crystals. Dashed lines connect different isomorphic crystals. According to Soga (1971).

$$v_S = (361 \pm 23)\frac{\varrho^{3/2}}{\bar{A}^2} \tag{1.10.5b}$$

$$\phi^{1/2} = (449 \pm 65)\frac{\varrho^{3/2}}{\bar{A}^2} \tag{1.10.5c}$$

where ϱ is in $g \cdot cm^{-3}$, $\bar{A}$ is in $g \cdot mole^{-1}$ and v_P, v_S and $\phi^{1/2}$ are in $km \cdot s^{-1}$. The numerical constants in eqns. (1.10.5a) and (1.10.5b) were found from measurements pertaining to 22 minerals (including α-quartz SiO_2, β-quartz SiO_2, coesite SiO_2, forsterite Mg_2SiO_4, periclase MgO, and Fe_2SiO_4 with an olivine structure and Fe_2SiO_4 with a spinel structure). A comparison of empirical relations (1.10.5a) and (1.10.5b) drawn (with continuous lines) for different values of the parameter $\bar{A}$ with the experimental data for different minerals is shown in Fig. 1.10.3*a* and 1.10.3*b* (after Soga, 1971). Points marked with the numerical values of $\bar{A}$ correspond to given minerals. The dashed lines connecting points with different values of $\bar{A}$ and approximately perpendicular to the continuous ones connect different isomorphic crystals.*

After integration the equation (1.10.6) of Shankland takes the following power form (valid for a fixed $\bar{A}$):

$$v_i = a_i \varrho^{b_i}, \quad \text{where } a_i,\ b_i = \gamma_i - \tfrac{1}{3} \text{ are constants} \tag{1.10.6a}$$

which Shankland applies to the velocities v_P and v_S, and also to the hydrodynamic sound velocity $\phi^{1/2}$. Taking into account the experimental data related to six media (α-Al_2O_3, MgO, Mg_2SiO_4, α-SiO_2, coesite, $MgO \cdot 2.6Al_2O_3$) with a mean mass number $\bar{A} \approx 20.2$, Shankland proposed the following values for the exponent b_i in eqn. (1.10.6a): $b_P = 1.39$, $b_S = 1.12$ and $b_{\phi^{1/2}} = 1.85$.

All the geophysical equations of state listed above, except the linear equation of Birch, can be written, for any medium with a fixed mean atomic mass $\bar{A}$, in a dimensionless form related to the pressure $p_0 = 0$, under which $\varrho = \varrho_0$, $\phi = \phi_0$, $v_P = v_{P_0}$ and $v_S = v_{S_0}$. Thus the equation of Anderson and Nafe takes the form

$$\frac{\varrho}{\varrho_0} = \left(\frac{\phi}{\phi_0}\right)^{1/(b-1)}, \quad \frac{1}{3} < \frac{1}{b-1} < \frac{1}{2}$$

---

* Isomorphism is the possibility of replacing of some atoms or atom groups in a crystalline structure by others with almost the same ionic radii. During the replacement, crystallographic structures and lattice constants do not change. For example, Mg_2SiO_4 and Fe_2SiO_4 crystals are isomorphic, KCl and NaCl crystals are not.

The seismic equation of state of Don Anderson is

$$\frac{\varrho}{\varrho_0} = \left(\frac{\phi}{\phi_0}\right)^n$$

where $n = 0.323$ (Anderson) or $n = 1/3$ (Anderson and Jordan); the equation of Soga is

$$\frac{\varrho}{\varrho_0} = \left(\frac{v_i}{v_{io}}\right)^{2/3}, \qquad v_i = v_P,\ v_S \text{ or } \phi^{1/2}$$

and the equation of Shankland is

$$\frac{\varrho}{\varrho_0} = \left(\frac{v_i}{v_{io}}\right)^{n_i}, \qquad n_P = 0.719,\ n_S = 0.893,\ n_{\phi^{1/2}} = 0.541$$

It can therefore be concluded that a dimensionless geophysical equation of state always takes the form of a power relation $\varrho \propto v_i^{n_i}$ independent of the atomic mass of the medium. The only unrestricted parameter of this equation is the exponent either determined empirically or resulting from a specified equation of state for a solid. It should be noted, for example, that the relation between the density and seismic parameter results directly from equation of state (1.6.3) of Murnaghan

$$p = \frac{K_0}{K_0'}\left[\left(\frac{\varrho}{\varrho_0}\right)^{K_0'} - 1\right]$$

the differentiation of which gives the following equation:

$$\phi = \frac{dp}{d\varrho} = \frac{K_0}{\varrho_0}\left(\frac{\varrho}{\varrho_0}\right)^{K_0'-1} = \phi_0\left(\frac{\varrho}{\varrho_0}\right)^{K_0'-1} \tag{1.10.9}$$

identical with the seismic equation of state of Don Anderson.

Using the Born–Mie interaction potential, which gives an isothermal equation of state $p = p(\varrho)$ in the form of (1.7.7), Don Anderson (1967) after differentiation found the following relation:*

$$\frac{\phi}{\phi_0} = \frac{1}{n-m}\left[(n+3)\left(\frac{\varrho}{\varrho_0}\right)^{n/3} - (m+3)\left(\frac{\varrho}{\varrho_0}\right)^{m/3}\right] \tag{1.10.10}$$

of a more general character than formula (1.10.9). This gives, after expansion into a series in relation to density changes $(\varrho-\varrho_0)/\varrho_0$, the equation

$$\frac{\phi}{\phi_0} = \left(\frac{\varrho}{\varrho_0}\right)^{n/3+m/3+1}\left\{1 + O\left[\left(\frac{\varrho-\varrho_0}{\varrho_0}\right)^2\right]\right\}$$

---

* Formula (1.10.10) is called the "generalized seismic equation of state" by some authors.

or approximately

$$\frac{\varrho}{\varrho_0} = \left(\frac{\phi}{\phi_0}\right)^{3/(n+m+3)} \tag{1.10.11}$$

which is a relation identical with eqn. (1.10.9). For $n = 2$ and $m = 4$ (the Murnaghan–Birch equation of state corresponds to these values of n and m) the exponent in eqn. (1.10.11) is equal to 1/3, which is the same value as that in the equation used by Anderson and Jordan (1970) to interpret experimental data.

If an assumption could be made that the Earth is homogeneous with respect to its chemical composition and that surfaces of phase discontinuities are not present in its interior, then the seismic equation of state together with data relating to changes of the seismic parameter ϕ with the depth z would be sufficient to find the density distribution $\varrho = \varrho(z)$ inside the Earth.

An interesting discussion concerning the relation between the seismic parameter ϕ and the isothermic equation of state $p = p(\varrho)$ is given by Chung *et al.* (1970). They noticed that the parameter ϕ, which results from a simple differentiation of an equation of state, is much more sensitive to the choice of the equation of state than is the density ratio ϱ/ϱ_0. The estimation was made by computing ϕ and ϱ/ϱ_0, under a fixed pressure p, from the Murnaghan equation of state and then from the Murnaghan–Birch equation. It was found that as long as the pressure is no higher than about $p = 0.5K_0$, the dimensionless densities ϱ/ϱ_0 computed from these equations differ from each other at most by 1%, while the corresponding values of the seismic parameter ϕ differ by as much as 10%. Thus, measurements of seismic wave velocities leading to the determination of the seismic parameter form a basis for the choice of the best equation of state $p = p(\varrho)$ to describe the Earth's interior.

An important conclusion which can be drawn from geophysical equations of state concerns the composition of the Earth. When the density distribution $\varrho(r)$ is known (or, more precisely, when it is postulated from a given model of the Earth's structure), it is possible to find the mean mass number $\bar{A}(r)$ of the medium forming the various layers of the Earth. Using different models of the density distribution $\varrho(r)$ (seven models, including the Birch 1, Birch 2, pyrolite and eclogite models) and different tables of the seismic parameter (three distributions—of Jeffreys, of Gutenberg and of Jonson, Anderson and Julian), Anderson and Jordan (1970) found the mean atomic mass of the lower mantle. From eqn. (1.10.3a) they obtained $\bar{A} = 23$ and from eqn. (1.10.3b) they had $\bar{A} = 21.6$. The

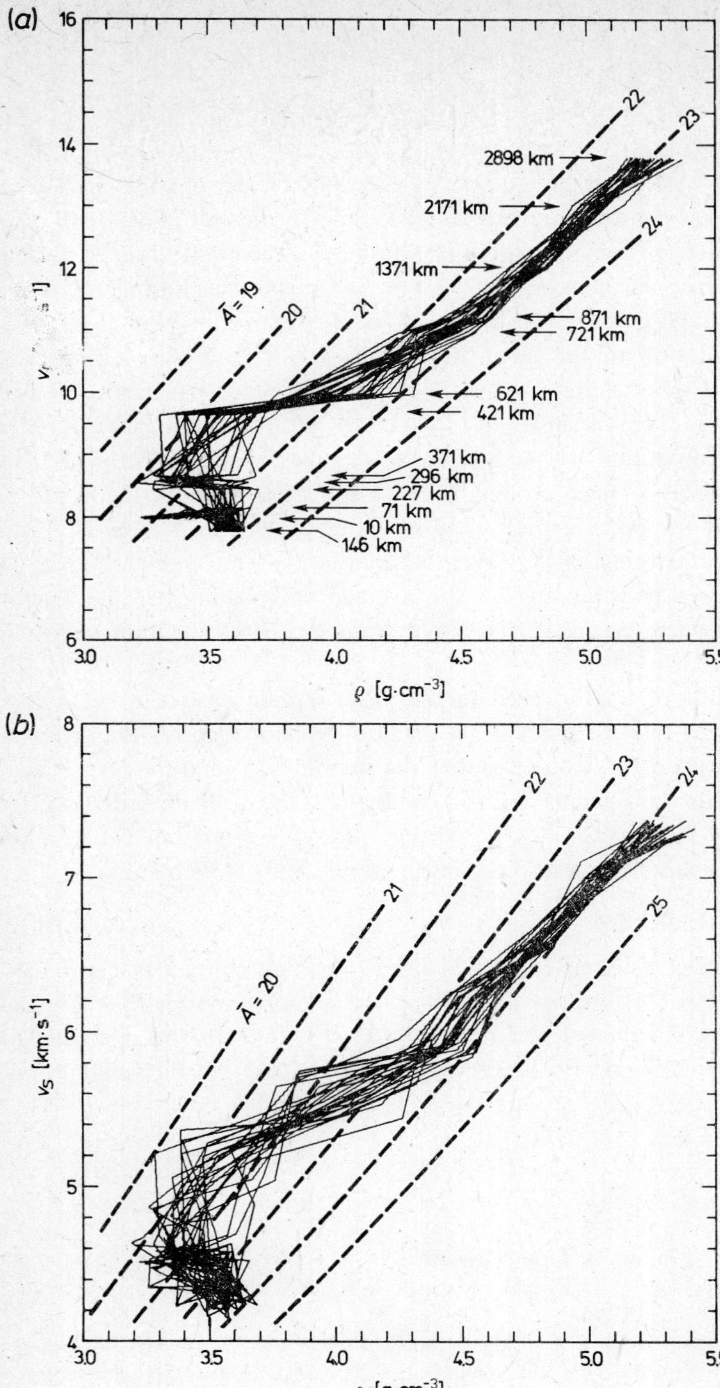

Fig. 1.10.4. (*a*) Velocity $v_P = v_P(\varrho)$ according to eqn. (1.10.5a). (*b*) Velocity $v_S = v_S(\varrho)$ according to eqn. (1.10.5b). The mean atomic mass $\bar{A}$ is the parameter. The relations $v_P(\varrho)$ and $v_S(\varrho)$ found by Press (1970) using the Monte Carlo method are shown. According to Soga (1971).

difference of 1.4 gives an opinion of the value of the possible error connected with the choice of numerical constants in the seismic equation of state. The dispersion of the values of $\bar{A}$ resulting from the application of various distributions of the density and seismic parameter, has a range of $\Delta\bar{A} \approx \pm 0.5$. Soga (1971) also used his geophysical equation of state (1.10.5) for the determination of the mean atomic mass. An interesting comparison of relations (1.10.5a) and (1.10.5b) with the models of the Earth's density given by Press (1970) is shown in Fig. 1.10.4*a* and 1.10.4*b* (after Soga, 1971). As can be seen in Fig. 1.10.4*a* relating to *P* wave velocities, the most probable value of the mean atomic mass in the area between the depth of 421 km and the boundary of the outer core (2898 km) is $\bar{A} \approx 23$. From Fig. 1.10.4*b*, relating to *S* waves, it follows that the value of $\bar{A} \approx 24$. Similar information relating to the average composition of the Earth's mantle and core can also be inferred from the Birch diagram shown in Fig. 1.10.1.

The results of Soga related to the mean atomic mass can be used to determine the relative content of magnesium and iron in the pyrolitic mantle of the Earth. Assuming that the mantle is a mixture of pyroxene and olivine in a proportion of 1:2, or that its chemical formula is of the form $(Mg_x, Fe_{1-x})SiO_3 + 2(Mg_x, Fe_{1-x})_2SiO_4$, the relation between the composition parameter x and the mean atomic mass $\bar{A}$ is found

$$x = 3.42 - 0.120\bar{A} \tag{1.10.12}$$

For values of $\bar{A}$ between 23 and 24 the magnesium content is $x \approx 0.66 \div 0.54$. This result is clearly underestimated as compared with the commonly accepted value of the ratio Mg/Fe $\approx 0.9/0.1$ in the mantle. The value $x = 0.9$ corresponds, according to eqn. (1.10.12), to the mean atomic mass $\bar{A} = 20.9$.

REFERENCES

Ahrens T. J., Anderson D. L. and Ringwood A. E. (1969), "Equations of state and crystal structures of high-pressure phases of shocked silicates and oxides", *Rev. Geophys.*, **7**, 4, 667–707.

Ahrens T. J. and Thomsen L. (1972), "Application of the fourth-order anharmonic theory to the prediction of equation of state at high compressions and temperatures", *Phys. Earth Planet. Inter.*, **5**, 4, 282–294.

Anderson D. L. (1967), "A seismic equation of state", *Geophys. J. R. Astron. Soc.*, **13**, 9–30.

Anderson D. L. and Jordan T. (1970), "The composition of the lower mantle", *Phys. Earth Planet. Inter.*, **3**, 23–35.

Anderson O. L. (1970), "Elastic constants of the central force model for three cubic structures: pressure derivatives and equations of state", *J. Geophys. Res.*, **75**, 14. 2719–2740.

Anderson O. L. (1973), "Comments on the power law representation of Birch's law", *J. Geophys. Res.*, **78**, 23, 4901–4914.

Anderson O. L. (1979), "The high-temperature acoustic Grüneisen parameter in the Earth's inerior", *Phys. Earth Planet. Inter.*, **18**, 3, 221–231.

Anderson O. L. and Nafe J. E. (1965), "Bulk modulus–volume relationship for oxide compounds and related geophysical problems", *J. Geophys. Res.*, **70**, 3951–3953.

Baumgardner J. R. and Anderson O. L. (1980), "Using the thermal pressure to compute the physical properties of terrestrial planets", presented on the XXIII COSPAR meeting, Budapest.

Bernardes N. and Swenson C. A. (1963), "The equation of state of solids at low temperature", in: *Solids under Pressure*, W. Paul and M. D. Warschauer (eds.), McGraw-Hill, New York.

Birch F. (1952), "Elasticity and construction of the Earth's interior", *J. Geophys. Res.*, **57**, 227–286.

Birch F. (1961a), "The velocity of compressional waves in rocks to 10 kbars", Part 2, *J. Geophys. Res.*, **66**, 7, 2199–2224.

Birch F. (1961b), "Composition of the Earth's mantle", *Geophys. J. R. Astron. Soc.*, **4**, 295–311.

Born M. and Huang K. (1954), *Dynamical Theory of Crystal Lattices*, Clarendon Press, Oxford.

Boschi E. and Caputo M. (1969), "Equations of state at high pressure and the Earth interior", *Nuovo Cimento*, Ser. I, **1**, 441–513.

Brush S. G. (1966), "Theories of the equation of state of matter at high pressure and temperatures", in: *Progress in High Temperature Physics and Chemistry*, Vol. 1, C. A. Russe (ed.), Pergamon Press, Oxford.

Bullen K. E. (1975), *The Earth's Density*, Chapman and Hall, London.

Burke J. J. and Weiss V. (eds.) (1971), *Shock Waves and the Mechanical Properties of Solids*, Syracuse University Press.

Chung D. H., Wang H. and Simmons G. (1970), "On the calculation of the seismic parameter at high pressure and high temperatures", *J. Geophys. Res.*, **75**, 26, 5113–5120.

Davies G. F. (1973), "Quasi-harmonic finite strain equations of state of solids", *J. Phys. Chem. Solids*, **34**, 1417–1429.

Demiański M. (1978), *Relativistic Astrophysics*, PWN, Warszawa (in Polish); English transl. PWN Warszawa, Elsevier, Amsterdam 1983.

Dugdale J. S. and MacDonald D. K. C. (1953), "The thermal expansion of solids", *Phys. Rev.*, **89**, 832–834.

Dziewonski A. M., Hales A. L. and Lapwood E. R. (1975), "Parametrically simple Earth model consistent with geophysical data", *Phys. Earth Planet. Inter.*, **10**, 1, 12–48.

Flügge S. and Marschall H. (1952), *Rechenmethoden der Quantentheorie dargestellt in Aufgaben und Lösungen*, 2nd ed., Springer-Verlag, Berlin.

Fung Y. C. (1965), *Foundation of Solid Mechanics*, Prentice-Hall, Englewood Cliffs.

Gilvarry J. J. (1969), "Equation of state at high pressure from Thomas–Fermi atom model", in: *The Application of Modern Physics to the Earth and Planetary Interiors*, S. K. Runkorn (ed.), Wiley-Interscience, London.

Harrison B. K., Thorne K. S., Wakano M. and Wheeler J. A. (1965), *Gravitation Theory and Gravitational Collapse*, University of Chicago Press, Chicago–London.

Knopoff L. (1963a), "Solids: equations of state of solids at moderately high pressures", in: *High Pressure Physics and Chemistry*, Vol. 1, R. S. Bradley (ed.), Academic Press, London–New York.

Knopoff L. (1963b), "Equations of state of matter at ultrahigh pressures", in: *High Pressure Physics and Chemistry*, Vol. 1, R. S. Bradley (ed.), Academic Press, London–New York.

Knopoff L. and Shapiro J. N. (1969), "Comments on the interrelationships between Grüneisen's parameter and shock and isothermal equations of state", *J. Geophys, Res.*, **74**, 6, 1439–1450.

Landau L. D. and Lifshitz E. M. (1958), *Statistical Physics*, Pergamon Press, London (transl. from the Russian).

Liu Lin-Gun (1974), "Birch's diagram: some new observations", *Phys. Earth Planet. Inter.*, **8**, 56–62.

Mao H. K. and Bell P. M. (1978), "High-pressure physics: sustained static generation of 1.36 to 1.72 Megabars", *Science*, **200**, 1145–1147.

March N. H. (1955), "Equation of state of elements from the Thomas–Fermi theory", *Proc. Phys. Soc. Lond.* **A68**, 726–734.

McQueen R. G., Fritz J. N. and Marsh S. P. (1964), "On the composition of the Earth's interior", *J. Geophys. Res.*, **69**, 2947–2965.

Murnaghan F. D. (1944), "The compressibility of media under extreme pressure", *Proc. Nat. Acad. Sci.*, **30**, 244.

Murnaghan F. D. (1951), *Finite Deformation of an Elastic Solid*, John Wiley and Sons, New York.

Press F. (1970), "Earth models consistent with the geophysical data", *Phys. Earth Planet. Inter.*, **3**, 3–22.

Prieto F. E. and Renero C. (1974), "L'équation d'état des solides sous pressions", *C. R. Acad. Sc. Paris*, **279**.

Prieto F. E. and Renero C. (1976), "The equation of state of solids", *J. Phys. Chem. Solids*, **37**, 151–160.

Prieto F. E. and Renero C. (1980), private communication.

Ragan C. A. (1981), "Shock wave experiments at pressures from 10 to 70 Mbar", in: *Topical Conference of Materials under Extreme Conditions*, Los Alamos National Laboratory.

Rees M., Ruffini R. and Wheeler J. A. (1974), *Black Holes, Gravitational Waves and Cosmology: An Introduction to Current Research*, Gordon and Breach, New York.

Shankland T. J. (1972), "Velocity–density systematics: derivation from Debye theory and the effect of ionic size", *J. Geophys. Res.*, **77**, 20, 3750–3758.

Simmons G. (1964), "Velocity of shear waves in rocks to 10 kilobars", *J. Geophys. Res.*, **69**, 1123–1130.

Simmons G. and England A. W. (1969), "Universal equations of state for oxides and silicates", *Phys. Earth Planet. Inter.*, **2**, 2, 69–76.

Slater J. C. (1939), *Introduction to Chemical Physics*, McGraw-Hill, New York.

Soga N. (1971), "Sound velocity of some germanate compounds and its relation to the law of corresponding states", *J. Geophys. Res.*, **76**, 17, 3983–3989.

Stacey F. D. (1977), "A thermal model of the Earth", *Phys. Earth Planet. Inter.*, **15**, 4, 341–348.

Stevenson D. J. (1978), "The outer planets and their satellites" in: *The Origin of the Solar System*, S. F. Dermott (ed.), J. Wiley and Sons, Chichester.

Stiller H. and Frank S. (1980), "A generalization of the Vashchenko–Zubarev formula for the Grüneisen parameter", *Phys. Earth Planet. Inter.*, **22**, 3–4, 184–188.

Thomsen L. (1970), "On the fourth-order anharmonic equation of state of solids", *J. Phys. Chem. Solids*, **31**, 2003–2016.

Tossi M. P. (1964), "Cohesion of ionic solids in the Born model", in: *Solid State Physics*, **16**, F. Seitz and D. Turnbull (eds.), Academic Press, New York–London.

Ullmann W. and Pankov V. L. (1976), "A new structure of the equation of state and its application in high-pressure physics and geophysics", *Veröffentlichungen des Zentralinstituts für Physik der Erde*, Nr 41, Potsdam.

Ullmann W., Pankov V. L. and Maaz R. (1977), "Tables and diagrams for prediction of compressional properties of different substances under high pressure", *Veröffentlichungen des Zentralinstituts für Physik der Erde*, Nr 42, Potsdam.

Vashchenko V. Ya. and Zubarev V. N. (1963), "Concerning the Grüneisen constant", *Sov. Phys. Solid State*, **5**, 653.

Zeldovitch J. B. and Rayzer J. P. (1966), *Physics of Shock Waves and High Temperature Hydrodynamic Phenomena*, 2nd ed., Nauka, Moskva (in Russian).

Zharkov V. N. and Kalinin V. A. (1971), *Equations of State for Solids at High Pressures and Temperatures*, Academic Press, New York–London (transl. from the Russian).

2. Distribution of Selected Physical Parameters in the Earth and Planetary Interiors

2.1 OBSERVATIONAL BASES

In this chapter models of the Earth's interior structure, constructed under the assumption that the Earth is spherically symmetric, are considered. The Earth's rotation is neglected. The only variable on which all the paramaters are dependent is the distance r from the Earth's centre. Because of strong regional heterogeneities in the crust, spherically symmetric models of the Earth are usually built for areas placed below the Mohorovičić boundary, thus for depths $z \geqslant 33$ km ($z = R - r$, where R is the average radius of the Earth). The value $z_{\text{Moho}} = 33$ km corresponds to the average thickness of the crust, or to the average depth of the Moho discontinuity. Assuming after Birch (1964) that the average density of the crust is $2.79\ \text{g} \cdot \text{cm}^{-3}$, its mass and moment of inertia are found to be 4.7×10^{22} kg and $1.26 \times 10^{36}\ \text{kg} \cdot \text{m}^2$, respectively.

The determination of a model of the Earth means the determination of such paramaters as the density $\varrho(r)$, pressure $p(r)$, gravity acceleration $g(r)$, temperature $T(r)$, bulk modulus $K(r)$ and shear modulus $\mu(r)$, and also the estimation of the chemical composition and phase state as a function of r. Since the first three parameters (ϱ, p, g) are connected with one another by simple relations, the values of g and p are determined by integration when the distribution $\varrho(r)$ is known (see Section 2.2). Different modern models relating to the Earth's interior below the upper mantle ($r \lesssim 6000$ km) generally provide very similar distributions of ϱ, p and g. The determination of the temperature and chemical and mineral composition is much more difficult and the results based on different assumptions and given by different authors show greater dispersion.

The following observational data are the basic initial data used for the construction of models of the Earth's interior:

1. The parameters characterizing the Earth as a planet, namely the Earth's mean radius R, total mass M and moment of inertia I. According to the recommendation of the International Union of Geodesy and Geo-

physics the following values are accepted at present (listed by Bullen, 1975):

$$R = 6.371023 \times 10^6 \text{ m}$$
$$GM = 3.98603 \times 10^{14} \text{ m}^3 \cdot \text{s}^{-2} \qquad (2.1.1)$$
$$J_2 = 0.0010807$$

The gravity constant is

$$G = (6.670 \pm 0.004) \times 10^{-11} \text{ m}^3 \cdot \text{kg}^{-1} \cdot \text{s}^{-2} \qquad (2.1.2)$$

The constant J_2 is connected with the dimensionless moment of inertia reduced to a spherically symmetric shape of the Earth (I/MR^2) and with the precession constant $H = 0.003273$

$$\frac{I}{MR^2} = \frac{J_2}{H} = 0.3308 \qquad (2.1.3)$$

This relation indicates that the Earth's density is not constant, but increases with depth. For a homogeneous sphere, $I = 0.4\ MR^2$, while a smaller value of the moment of inertia indicates concentration of the mass as the centre is approached. The total mass of the Earth, its moment of inertia and the mean density are

$$M = 5.976 \times 10^{24} \text{ kg}$$
$$I = 0.3308\ MR^2 = 8.024 \times 10^{37} \text{ kg} \cdot \text{m}^2 \qquad (2.1.4)$$
$$\varrho = 5.517 \text{ g} \cdot \text{cm}^{-3}$$

Analogous values for the Earth's crust are (Birch, 1964)

$$M_{\text{crust}} = 4.7 \times 10^{22} \text{ kg}$$
$$I_{\text{crust}} = 1.26 \times 10^{36} \text{ kg} \cdot \text{m}^2 \qquad (2.1.5)$$
$$\varrho_{\text{crust}} = 2.79 \text{ g} \cdot \text{cm}^{-3}$$

2. Seismic data relating to:

(a) travel times of various body waves, especially of longitudinal P waves and shear S waves coming to the observer directly from the source as well as those passing through the core;

(b) velocities of surface waves;

(c) free oscillations and tides of the Earth.

3. Laboratory experiments carried out on samples of rocks collected from accessible layers of the Earth, especially relating to:

(a) chemical and mineral composition;

(b) physical properties, especially mechanical properties (measurements of bulk modulus K changes as functions of pressure, measurements of elastic wave velocities v_P and v_S as functions of density), leading to various geophysical equations of state.

4. Laboratory experiments related to physical properties of various materials, especially to those of metals and silicates of iron and magnesium, under high pressures and temperatures (up to pressures in the range of hundreds of gigapascals obtainable by shock wave methods); observations relating to melting processes and phase transfomations.

5. Analyses of meteorite composition (under the assumption that different types of meteorites have compositions analogous to those of various layers of the Earth).

6. Observations of the Earth's magnetic field, whose properties are undoubtedly related to the structure of the Earth's core.

Not all the observational data listed here are used to the same extent by the authors of different models of the Earth. The planetary paramaters of the Earth and seismic observations relating to velocity distributions of *P* and *S* waves together with certain results of laboratory experiments should be considered as the most basic. During the last decade observations of free oscillations of the Earth have become more and more important. Assumptions about the spherical symmetry of the Earth and neglecting its rotation are of special importance for the analysis of these data. As a result, Earth models of SNREI (Spherical Non Rotating Elastic Isotropic) type are constructed. In SNREI models the eigen modes are degenerated; any deviations from SNREI (e.g., nonsphericity, rotation) lead to the splitting of levels, that is to the appearance of new eigenfrequencies. This phenomenon is very similar to the effect of splitting of energy levels of an atom placed under the influence of an external electric or magnetic field.

2.2 BASIC EQUATIONS

Let us assume that the density distribution $\varrho(r)$ is known. Then the mass contained in a sphere with the radius r is

$$m(r) = 4\pi \int_0^r r'^2 \varrho(r')\mathrm{d}r', \quad m(R) = M \tag{2.2.1}$$

and the gravity acceleration is given by the formula

$$g(r) = \frac{Gm(r)}{r^2} = \frac{4\pi G}{r^2} \int_0^r r'^2 \varrho(r')\mathrm{d}r' \tag{2.2.2}$$

Assuming that the pressure $p(r)$ is given by the hydrostatic formula

$$\mathrm{d}p(r) = -\varrho(r)g(r)\mathrm{d}r \tag{2.2.3}$$

and that at the Earth's surface $p(R) = 0$, a formula describing the distribution of the pressure is found

$$p(r) = \int_{R-r}^{R} \varrho g \,\mathrm{d}r \equiv \int_{0}^{z} \varrho g \,\mathrm{d}z \tag{2.2.4}$$

This formula can be rewritten taking into account the average pressure p_{Moho} at the depth z_{Moho}

$$p(r) = p_{\mathrm{Moho}} + \int_{r_{\mathrm{Moho}}-r}^{r_{\mathrm{Moho}}} \varrho g \,\mathrm{d}r = p_{\mathrm{Moho}} + \int_{z_{\mathrm{Moho}}}^{z} \varrho g \,\mathrm{d}z \tag{2.2.5}$$

Thus the distributions of the gravity acceleration and pressure, when the density distribution is known, can be determined from formulas (2.2.2) and (2.2.5).

According to Bullen (1975), the equation describing density changes with depth $\varrho = \varrho(z)$ is (in those ranges of z in which density is a continuous function) of the form

$$\frac{\mathrm{d}\varrho}{\mathrm{d}z} = \frac{\partial \varrho}{\partial p}\frac{\mathrm{d}p}{\mathrm{d}z} + \frac{\partial \varrho}{\partial s}\frac{\mathrm{d}s}{\mathrm{d}z} + \sum_i \frac{\partial \varrho}{\partial n_i}\frac{\mathrm{d}n_i}{\mathrm{d}z} + \sum_j \frac{\partial \varrho}{\partial q_j}\frac{\mathrm{d}q_j}{\mathrm{d}z} \tag{2.2.6}$$

where s is the specific entropy, and n_i and q_j are variables describing the chemical composition and phase state, respectively. In those of the Earth's layers which, as regards their chemical composition and phase state, can be considered as homogeneous both sums can be neglected and eqn. (2.2.6) takes the simpler form

$$\frac{\mathrm{d}\varrho}{\mathrm{d}z} = \left(\frac{\partial \varrho}{\partial p}\right)_s \frac{\mathrm{d}p}{\mathrm{d}z} + \left(\frac{\partial \varrho}{\partial s}\right)_p \frac{\mathrm{d}s}{\mathrm{d}z} \tag{2.2.7}$$

In the layers in which distinct changes* of chemical composition and phase state occur, certainly within the range 400 km $< z <$ 1000 km, the sums in eqn. (2.2.6) are not small enough to be neglected, but at the same time the equation becomes practically impossible to solve.

For the determination of density in these heterogeneous layers, therefore, different types of interpolations are used, connecting the solutions of eqn. (2.2.7) found in the neighbouring homogeneous areas, in which this equation is applicable.

* These changes can be deduced from observations of seismic wave velocities: discontinuities in v_P and v_S or their sharp changes within small ranges of depth are connected with discontinuities or sharp changes in the density ϱ of the medium and its elastic constants K and μ.

If the density ϱ and temperature T are expressed as functions of the pressure p and entropy s

$$\varrho = \varrho(p, s), \qquad T = T(p, s) \tag{2.2.8}$$

where $p = p(z)$ and $s = s(z)$, then by differentiating relations (2.2.8) eqn. (2.2.7) for the density gradient is obtained, and a similar equation for the temperature gradient is

$$\frac{dT}{dz} = \left(\frac{\partial T}{\partial p}\right)_s \frac{dp}{dz} + \left(\frac{\partial T}{\partial s}\right)_p \frac{ds}{dz} \tag{2.2.9}$$

This equation indicates that changes of temperature are the sum of adiabatic effects (s = const) and nonadiabatic ones ($s = s(z)$). The difference between the total temperature gradient and its adiabatic part

$$\frac{dT}{dz} - \left(\frac{\partial T}{\partial p}\right)_s \frac{dp}{dz} = \left(\frac{\partial T}{\partial s}\right)_p \frac{ds}{dz} \equiv \tau \tag{2.2.10}$$

is called the superadiabatic temperature gradient. The right-hand side of eqn. (2.2.7) can be transformed as follows:

1. Using the seismic parameter ϕ

$$\left(\frac{\partial \varrho}{\partial p}\right)_s = \frac{1}{\phi} \qquad \text{where} \qquad \phi = v_P^2 - \tfrac{4}{3} v_S^2 = \frac{K}{\varrho}$$

and taking into account the hydrostatic formula

$$\frac{dp}{dz} = \varrho g$$

it is easy to see that the first term of eqn. (2.2.7) is $\varrho g/\phi$.

2. The second term can be expressed by the superadiabatic gradient τ (2.2.10) and by the coefficient of thermal expansion

$$\alpha = -\frac{1}{\varrho}\left(\frac{\partial \varrho}{\partial T}\right)_p$$

Then the second term of the right-hand side of eqn. (2.2.7) is

$$\left(\frac{\partial \varrho}{\partial s}\right)_p \frac{ds}{dz} = \left(\frac{\partial \varrho}{\partial T}\right)_p \left(\frac{\partial T}{\partial s}\right)_p \frac{ds}{dz} = -\varrho\alpha\tau$$

Equation (2.2.7) thus takes the final form

$$\frac{d\varrho}{dz} = \frac{\varrho g}{\phi} - \varrho\alpha\tau \tag{2.2.11}$$

called the *Birch–Williamson equation* (Birch, 1939). From this equation and from eqn. (2.2.2) for the gravity acceleration, together with proper assumptions relating to the term $\varrho\alpha\tau$ and boundary conditions, numerical determination of the density distribution is possible.

Instead of eqn. (2.2.11) its simplified form is often used (the Williamson–Adams equation, 1923), namely

$$\frac{d\varrho}{dz} = \frac{\varrho g}{\phi} \tag{2.2.12}$$

The omission of the term $\varrho\alpha\tau$, connected with the temperature gradient, introduces into the density distribution in the mantle an error not greater than 0.07 $g \cdot cm^{-3}$, i.e., much smaller than the error caused by the assumption that the particular layers of the Earth are homogeneous as regards their composition (Bullen, 1963). From eqns. (2.2.2) and (2.2.11) the density distribution can be determined for any range of the depth z for which we know the seismic parameter ϕ_0 and the values $m_0 = m(r_0)$ and $\varrho_0 = \varrho(r_0)$ of the mass and the density corresponding to a boundary value of the radius r_0. Equations (2.2.11) and (2.2.12) cannot be used in areas where the density ϱ or its derivative $d\varrho/dz$ are not continuous, e.g., in a layer in which chemical composition heterogeneities and phase transformations might occur. It can be seen from eqn. (2.2.11) that discontinuities in ϱ or in $d\varrho/dz$ would appear at points at which discontinuities in the seismic parameter ϕ are observed. Equation (2.2.11) therefore can be integrated by segments within those intervals in which the seismic parameter is continuous. The main difficulty here is the estimation of the boundary values m_0 and ϱ_0 for the individual intervals. In particular, the application of eqn. (2.2.12) with boundary conditions at the Moho discontinuity

$$\begin{gathered} m_0 = m(r_{\text{Moho}}) = M - M_{\text{crust}} = 5.93 \times 10^{24} \text{ kg} \\ 3.31 \text{ g} \cdot \text{cm}^{-3} \leqslant \varrho_0 = \varrho(r_{\text{Moho}}) \leqslant 3.55 \text{ g} \cdot \text{cm}^{-3\star} \end{gathered} \tag{2.2.13}$$

leads to a density distribution in the whole Earth which becomes steeper, with the density increasing towards the centre, the smaller the value of ϱ_0. The distribution obtained must also fulfil the condition for the moment of inertia

$$I_0 = I(r_{\text{Moho}}) = I - I_{\text{crust}} = 7.878 \times 10^{37} \text{ kg} \cdot \text{m}^2 \tag{2.2.14}$$

where $I = \frac{8}{3}\pi \int_0^R r^4 \varrho(r) dr$ denotes the moment of inertia of the spherical symmetric Earth.

Two specific approaches to the Williamson–Adams equation (2.2.12), discussed by Bullen, are given below.

★ See Table 2.3.1, models of Clark and Ringwood.

TABLE 2.3.1

Models of the Earth's internal structure proposed by various authors. The depth corre-
km · s^{-1}. The subscript at ϱ denotes the corresponding depth. The column "Layer name
necessarily related to the same intervals of depth

Layer name and symbol: Bullen (1963)	Layer name and symbol: others see Fig. 5.3.4	Clark and Ringwood (1964) "pyrolite"	Clark and Ringwood (1964) "eclogite"	Birch 1 (1964)
Crust A				ϱ = const = 2.79
		33	33	33
Upper mantle B	Upper mantle	$\varrho_{33} = 3.31$ same as for pyroxene and olivine in proportion 1:2	$\varrho_{33} = 3.55$ same as for natural eclogite	$\varrho_{33} = 3.425$ $\varrho = a+bv_P = 0.768+0.328v_P$
		Distribution $\varrho(z)$ is found from the theory of finite deformations taking into consideration the thermal expansion		Constant $b = 0.328$ found from experiments on rocks and crystals with a low value of the mean atomic mass
		400		
Transition layer C		Linear increase of density with depth is assumed		
		1000		1047
Lower mantle D		$\frac{d\varrho}{dz} = \frac{g\varrho}{\phi} - \alpha\varrho\tau$, $\tau \lesssim 1\ \text{K} \cdot \text{km}^{-1}$		$\frac{d\varrho}{dz} = \frac{g\varrho}{\phi}$
		$\varrho_{2920} = 5.71$	2920 $\varrho_{2920} = 5.49$	$\varrho_{2898} = 5.51$ 2898
		$\varrho_{2920} = 9.85$	$\varrho_{2920} = 10.11$	$\varrho_{2898} = 10.05$
Outer core E	Outer core	$\frac{d\varrho}{dz} = \frac{g\varrho}{\phi}$		$\frac{d\varrho}{dz} = \frac{g\varrho}{\phi}$
Transition layer F				Calculations of Bullard (1957), based on this equation, are used
		5080	5080	
Inner core G		ϱ = const = 12.861	ϱ = const = 13.258	$\varrho_{6371} = 12.62$

* The model of Wang is made to fulfil different conditions relating to mass, moment of inertia,
undergoes various disturbances.

sponding to layer boundaries is in km, density in $g \cdot cm^{-3}$ and velocities v_P and v_S in and symbol " indicates that the descriptions do not have an univocal sense and are not

Birch 2 (1964)	Models of Bullen of type A: Model A″ Bullen and Haddon (1967a)	Models of Bullen of type B: Model B$_2$ Bullen and Haddon (1967b)	Wang (1972)*
$\varrho = \text{const} = 2.79$		$\varrho = \text{const} = 2.84$	
33	33	33	16
$\varrho_{33} = 3.320$ $\varrho = a + bv_P = 0.252 + 0.378v_P$ Constant $a = 0.252$ found from the condition that $\varrho_{33} = 3.320$	$\varrho_{33} = 3.32$ $\frac{d\varrho}{dz} = \frac{g\varrho}{\phi}$	$\varrho_{33} = 3.32$ $\frac{d\varrho}{dz} = \frac{g\varrho}{\phi}$	$\varrho = 0.613 + 0.328v_P$ under room conditions; this relation is corrected (according to the theory of finite deformations) for the pressure and temperature
	413	245	
	$\varrho_{413} = 3.64$ Parabolic increase of density with depth is assumed	$\varrho_{245} = 3.51$ $\frac{d\varrho}{dz} = (1 \div 2)\frac{g\varrho}{\phi}$	
1047	984	984	≈ 1000
$\frac{d\varrho}{dz} = \frac{g\varrho}{\phi}$	$\varrho_{984} = 4.55$ $\frac{d\varrho}{dz} = \frac{g\varrho}{\phi}$	$\varrho_{984} = 4.49$ D′ $\frac{d\varrho}{dz} = \frac{g\varrho}{\phi}$ 2700 $\varrho_{2700} = 5.40$ D″ $\frac{d\varrho}{dz} = 3\frac{g\varrho}{\phi}$	$\varrho = 0.742 + 0.424\phi^{1/2}$
$\varrho_{2898} = 5.58$ 2896 $\varrho_{2898} = 9.96$	$\varrho_{2898} = 5.56$ 2898 $\varrho_{2898} = 9.98$	$\varrho_{2886} = 5.69$ 2886 $\varrho_{2886} = 9.95$	2890
$\frac{d\varrho}{dz} = \frac{g\varrho}{\phi}$ Calculations of Bullard (1957), based on this equation, are used	$\frac{d\varrho}{dz} = \frac{g\varrho}{\phi}$	E′ $\frac{d\varrho}{dz} = \frac{g\varrho}{\phi}$ 4000 $\varrho_{4000} = 11.87$ E″ $\frac{d\varrho}{dz} = (3 \div 4)\frac{g\varrho}{\phi}$ 4710 $\varrho_{4710} = 12.30$ $\frac{d\varrho}{dz} = (1 \div 2)\frac{g\varrho}{\phi}$	$\frac{d\varrho}{dz} = \frac{g\varrho}{\phi}$
	5120	5160	5160
$\varrho_{6371} = 12.48$	$\varrho_{5120} = 12.25$ $\frac{d\varrho}{dz} = \frac{g\varrho}{\phi}$ $\varrho_{6371} = 12.51$	$\varrho_{5120} = 12.74$ $\frac{d\varrho}{dz} = \frac{g\varrho}{\phi}$ $\varrho_{6371} = 13.03$	$\frac{d\varrho}{dz} = \frac{g\varrho}{\phi}$ $v_S = \text{const} = 3.3$

eigenperiods, velocites of surface waves measured along the great circles. The given distribution

Bullen (1975) considers the relationship between the seismic equation of state of Anderson and the construction of a mathematical model of the Earth's interior structure based on the Williamson–Adams equation. He notices that assuming a linear relationship between the bulk modulus K and density p

$$\frac{\mathrm{d}K}{\mathrm{d}p} = c_1, \quad c_1 = \text{const} = 3.3^{\star} \tag{2.2.15}$$

as an average for the whole Earth, and using the definition of the modulus K

$$K = \varrho \frac{\mathrm{d}p}{\mathrm{d}\varrho}$$

the following relation is obtained:

$$\frac{\dfrac{\mathrm{d}\varrho}{\mathrm{d}p}}{\dfrac{\mathrm{d}K}{\mathrm{d}p}} = \frac{\varrho}{Kc_1}$$

i.e., after integration,

$$K = K_0 \left(\frac{\varrho}{\varrho_0}\right)^{c_1}, \quad K_0 = \text{const}, \quad \varrho_0 = \text{const} \tag{2.2.16}$$

After substituting $K = \varrho\phi$, relation (2.2.16) is transformed into the seismic equation of state of Anderson

$$\varrho = \varrho_0 \left(\frac{\phi}{\phi_0}\right)^n, \quad n = \frac{1}{c_1 - 1}, \quad \phi_0 = \text{const} \tag{2.2.17}$$

This is a simplified form of the seismic equation of state (1.10.3), valid in a medium with a fixed mean atomic mass. These relations may be used to make further transformation of the Williamson–Adams equation (2.2.12). Let us write (2.2.12) using r as an independent variable

$$\frac{\mathrm{d}\varrho}{\mathrm{d}r} = -\frac{G\varrho m(r)}{\phi r^2} \tag{2.2.18}$$

Multiplying this relation by r^2/ϱ, differentiating with respect to r ($\mathrm{d}m = 4\pi r^2 \varrho \mathrm{d}r$) and substituting the seismic equation of state in the form

---

⋆ According to Bullen (1975):
$K = 229 + 3.16p$ for the lower mantle, and
$K = 184 + 3.44p$ for the core
where K and p are in gigapascals.

(2.2.17), the following equation is found:

$$\frac{\mathrm{d}}{\mathrm{d}r}\left(r^2\varrho^{1/n-1}\frac{\mathrm{d}\varrho}{\mathrm{d}r}\right) = -A^2r^2\varrho, \quad A^2 = \frac{4\pi G\varrho_0^{1/n}}{\phi_0} \tag{2.2.19}$$

After substituting new variables ϑ and ξ for ϱ and r

$$\varrho^{1/n} = \frac{1}{n}\vartheta \quad \text{and} \quad r = \frac{1}{A}n^{n/2}\xi \tag{2.2.20}$$

eqn. (2.2.19) takes the final form of the Emden equation

$$\frac{\mathrm{d}^2\vartheta}{\mathrm{d}\xi^2}+\frac{2}{\xi}\frac{\mathrm{d}\vartheta}{\mathrm{d}\xi}+\vartheta^n = 0 \tag{2.2.21}$$

The only empirical parameter here is the coefficient $n = 1/(\mathrm{d}k/\mathrm{d}p-1) \approx 0.33 \div 0.43$. The value $n = 0.33$ is adopted in the seismic equation of state, while the value $n = 0.43$ results from Bullen's assumption (eqn. (2.2.15)) concerning the bulk modulus K. The Emden equation (2.2.21) is of no practical value for the determination of density distributions in the Earth's interior. It is used, however, in astrophysics for the study of stellar structure, for which a polytropic equation of state is assumed

$$p = \text{const}\cdot\varrho^{1+1/n}, \quad n = \text{const} \tag{2.2.22}$$

For $1+1/n = c_1$ the equation of polytropy (2.2.22) is equivalent to the seismic equation of state (2.2.17) for a medium with a fixed mean atomic mass. The Emden equation is presented here just to show that even under highly simplifying assumptions, such as the constancy of the mean atomic mass and the applicability of the seismic equation of state in the whole Earth, we obtain a nonlinear equation which can be integrated in a finite form only for some specific values of the parameter n (but not for n from the $0.33 \div 0.43$ range).

An other approach to the Williamson–Adams equation is connected with Earth models of the type B (Bullen, 1963, 1975). Calculating the derivative with respect to p of the relation $K = \varrho\phi$ we obtain

$$\frac{\mathrm{d}K}{\mathrm{d}p} = \phi\frac{\mathrm{d}\varrho}{\mathrm{d}z}\frac{\mathrm{d}z}{\mathrm{d}p}+\varrho\frac{\mathrm{d}\phi}{\mathrm{d}z}\frac{\mathrm{d}z}{\mathrm{d}p} \tag{2.2.23}$$

Since $\mathrm{d}p = \varrho g\mathrm{d}z$, for density changes with depth the following equation is obtained:

$$\frac{\mathrm{d}\varrho}{\mathrm{d}z} = -\eta\frac{\varrho g}{\phi} \tag{2.2.24}$$

where the parameter η is defined by the formula

$$\eta = \frac{dK}{dp} - \frac{1}{g}\frac{d\phi}{dz} \tag{2.2.25}$$

This parameter, or more precisely the difference $\eta - 1$, can be considered as a measure of nonadiabaticity and heterogeneity of the medium. Indeed, rewriting the modified Williamson–Adams equation (2.2.24) in the form

$$\eta = \phi \frac{d\varrho}{dz}\frac{1}{\varrho g} = \phi \frac{d\varrho}{dz}\frac{dz}{dp} = \frac{\frac{d\varrho}{dp}}{\left(\frac{\partial \varrho}{\partial p}\right)_s}$$

the following expression is obtained:

$$\eta - 1 = \frac{\frac{d\varrho}{dp} - \left(\frac{\partial \varrho}{\partial p}\right)_s}{\left(\frac{\partial \varrho}{\partial p}\right)_s} \tag{2.2.26}$$

in which the numerator describes the density change related to the deviation of the medium from an adiabatic and chemically homogeneous state. The numerical value of the parameter η was estimated by Bullen, who used the empirical relation between the bulk modulus and pressure

$$K = 234 + 3.00\, p + 0.0010\, p^2 \tag{2.2.27}$$

where p and K are in gigapascals. According to Bullen this relation is valid inside a pressure interval from 40 GPa to 400 GPa, i.e., for depths $z > 1000$ km.

2.3 REVIEW OF MODELS OF THE STRUCTURE OF THE EARTH'S INTERIOR

All the existing models of the Earth's interior, and notably the models of the density distribution $\varrho = \varrho(r)$, can be divided into three groups.

1. Early models based on data relating to the planet Earth and taking into account seismic observations only on a very limited scale. The models, discussed by Bullen (1975) in detail, were developed before the introduction of the Williamson–Adams equation and are now only of historical value.

2. Models based on the Williamson–Adams equation and on various observations excluding those relating to free oscillations of the Earth. A review of the most important models from this group, especially of those presented in the sixties and in principle still valid, is given in Table 2.3.1.

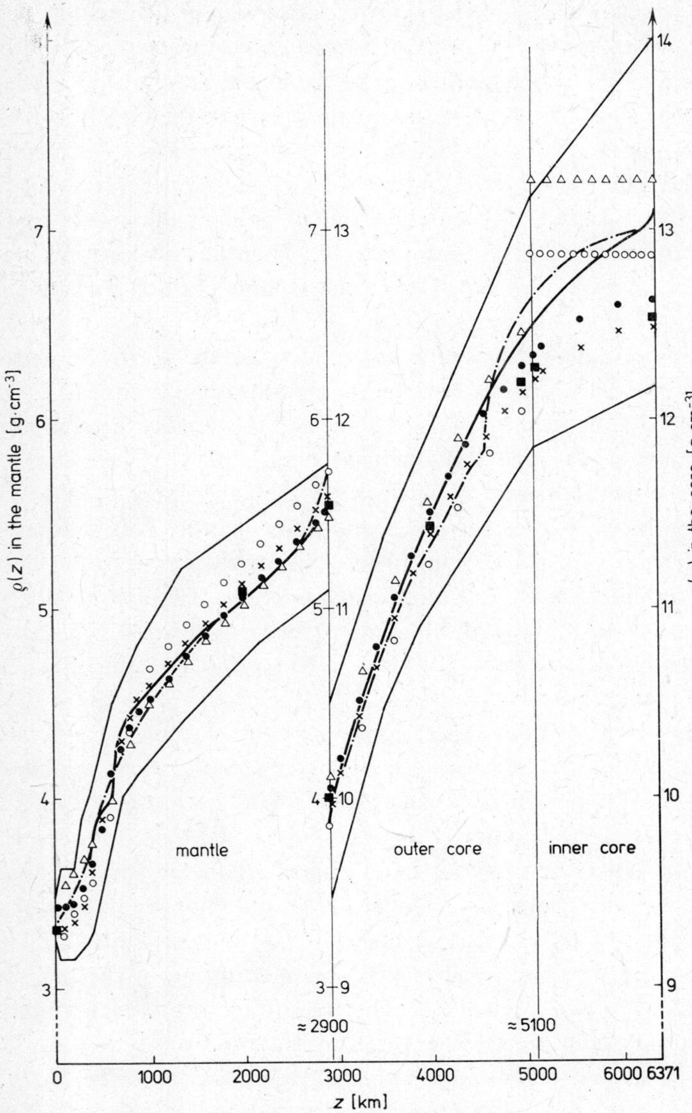

Fig. 2.3.1. Density distributions according to various models: Bullen A (solid square), Bullen B_2 (dashed-dot line), Birch 1 (solid circle), Birch 2 (cross), Clark and Ringwood—pyrolite (open circle), Clark and Ringwood—eclogite (open triangle), Wang (thick continuous line). The thin continuous lines denote the range of density within which the distributions $\varrho(z)$ were selected by Monte Carlo method in calculations of Press.

3. Models which adopt as a first approximation one of the models from the second group and then correct it by various numerical methods and by including data relating to free oscillations of the Earth. Several such models have been constructed in recent years, but in principle they do not differ significantly from models taken as a first approximation (for depths $z > 500$ km the differences between these two kinds of models are as a rule smaller than 1%). In this chapter only a limited number of models based on free oscillations of the Earth is considered. Their detailed review and their mathematical bases will be given in the volume entitled *Gravity and Low-Frequency Geodynamics.*

The methods of constructing different models of the Earth's interior are listed in Table 2.3.1. As can be seen, many authors use eqn. (2.2.11), and more often its simplified form (2.2.12), to find the density distributions in the lower mantle and core. The individual models differ from one another first of all by the methods used to determine $\varrho(z)$ in the upper mantle and in the transition layer between the upper and lower mantle. This is an area about 1000 km thick in which various phase transformations occur, leading to crystalline lattices of more and more dense packing. The density distributions calculated with the use of different models are shown in Fig. 2.3.1, which is an illustration to Table 2.3.1. As can be seen, the results obtained from all these models are very similar.

The models listed in Table 2.3.1, constructed by several authors, are based on more and more precise values of the seismic wave velocities v_P and v_S and on slightly differing values of the constants M and I of the Earth. The changes of these initial values from one model to another model are insignificant, however, and the most important factor for the final distribution $\varrho(z)$ is the choice of a model of density changes in the area to a depth of about 1000 km. Below this depth, as a rule eqns. (2.2.11) or (2.2.12) are used. Further, we give a concise description of the models listed in Table 2.3.1 and of a few other, more recent ones, for which a schematic representation such as in Table 2.3.1 is not useful.

Clark and Ringwood (1964) considered different models of the Earth, based on two models of the upper mantle composition.

A pyrolite model (the name pyrolite derives from the names of the minerals pyr-oxene + oli-vine) in which it is assumed that the upper mantle, at a first approximation, is composed only of pure oxides MgO and SiO_2 occuring in such quantites as to produce a mixture of olivine Mg_2SiO_3 and pyroxene $MgSiO_3$ in a molecule ratio of 2: 1. The density of this material is $\varrho_0 = 3.321\ g \cdot cm^{-3}$ under the pressure $p = 0$. A mixture of olivine Mg_2SiO_4 (forsterite) and pyroxene $MgSiO_3$ in a molecule ratio of 2 : 1

requires that the molecule ratio of the oxides MgO and SiO_2 be 5: 3, or 62.5% of MgO and 37.5% of SiO_2. Ringwood (1966) assumes the following composition of pyrolite: 45% of SiO_2, 3.5% of Al_2O_3, 8% of FeO, 37.5% of MgO, 3% of CaO and others 3%.

An eclogite model with the density $\varrho_0 = 3.55\ \mathrm{g \cdot cm^{-3}}$ under the pressure $p = 0$. The composition of this model, found from analyses of real rocks, is: 47% of SiO_2, 12% of Al_2O_3, 10% of Fe, 17% of MgO, 11% of CaO and $\approx$ 3% of other substances.

In both the models of Clark and Ringwood it was assumed that the density of the inner core is constant. The density in the upper mantle was found from the Birch equation (1.6.8) of the theory of finite deformations

$$p = \frac{3}{2} K_0 \left[\left(\frac{\varrho}{\varrho_0} \right)^{7/3} - \left(\frac{\varrho}{\varrho_0} \right)^{5/3} \right] \tag{2.3.1}$$

where K_0 is the bulk modulus under the pressure $p = 0$. The density calculated from this equation* was corrected for thermal expansion, and this permitted the determination of density along a geotherm. The influence of pressure on changes of the coefficient of thermal expansion was also taken into consideration. In the pyrolite model the correction corresponding to a density increase by $0.06\ \mathrm{g \cdot cm^{-3}}$ in the area of $z = 150 \div 200$ km, in which a phase transformation from pyroxenic pyrolite to garnetiferous pyrolite takes place, was also introduced. The density distributions in the upper mantle for both models of Clark and Ringwood are shown in Fig. 2.3.2. They are distributions averaged from several curves corresponding to different geotherms of the mantle under the oceanic and continental crust.

The Earth's models of Clark and Ringwood were frequently updated by their authors. The alternations were based first of all on new laboratory data relating to mineral properties under thermodynamic conditions close to those existing in the Earth's mantle. The pyrolite model (Ringwood, 1975, 1976) was extensively used to explain mantle structure to a depth of about 1000 km. In particular, it was possible to explain the nature of seismic discontinuities by considerably sharp changes (in layers several tens of kilometres thick) of mantle density at certain depths, caused by phase transitions. These problems are considered in detail in Chapter 7.

* Clark and Ringwood do not describe the method by which they obtained the distribution $\varrho(z)$ from $\varrho(p)$. For this purpose they had to assume as a zero approximation a certain distribution $p(z)$ resulting from other models of the Earth's structure.

In Figs. 2.3.1 and 2.3.2, besides various density distributions, their envelope is also marked. The envelope corresponds to a density range in which Press (1970) selected density distributions by the Monte Carlo method. Randomly selected distributions were then tested in such a way as to fulfil conditions relating to: (1) the mass and moment of inertia of the whole Earth, (2) the distribution of *P* wave velocity in the mantle, (3) the distribution of *S* wave velocity below 800 km, (4) eigenperiods of free

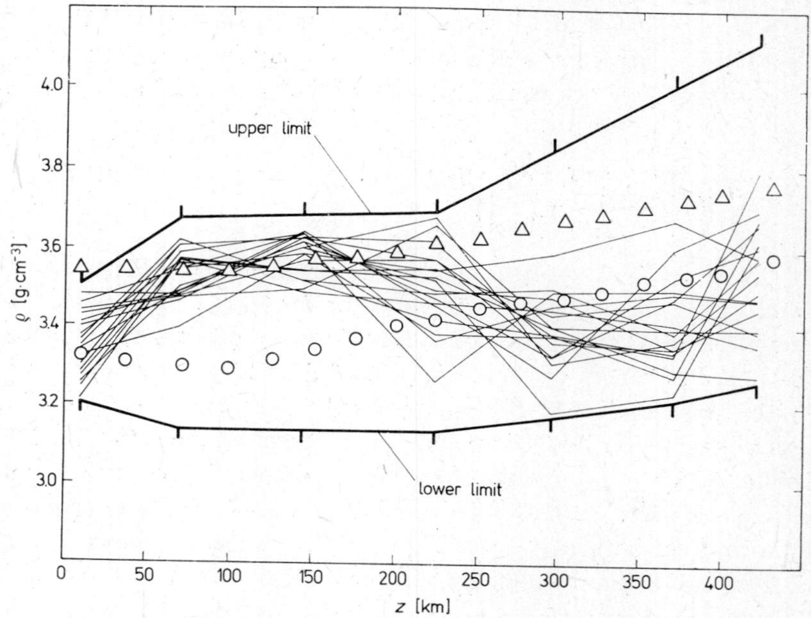

Fig. 2.3.2. Density distributions in the upper mantle: "pyrolite" model (open circle), "eclogite" model (open triangle), various correct models of Press (thin continuous line), density range within which the distributions were selected by Monte Carlo method in the calculations of Press (thick continuous line). According to Press (1970).

Earth oscillations, (5) phase velocities of Rayleigh surface waves with periods from 125 to 325 s and Love surface waves with periods from 80 to 340 s. Eighteen density distributions fulfilling these conditions are shown in Fig. 2.3.2. Their characteristic feature is density decrease, as depth increases, from $\varrho = 3.5 \div 3.6 \text{ g} \cdot \text{cm}^{-3}$ at a depth of 100 km to $\varrho = 3.3 \div 3.5 \text{ g} \cdot \text{cm}^{-3}$ at a depth of 300 km. A comparison of the results of Press with the models of Clark and Ringwood shows good agreement with the eclogite model in the area between 80 and 150 km and slightly worse agreement with the pyrolite model at depths between about 250 and 400 km.

Birch (1964) used his own relation (the Birch equation (1.10.1))

$$\varrho = a + b v_P \tag{2.3.2}$$

where a and b are constants, for the determination of the density in the upper mantle and transition layer in his models 1 and 2. In the "Birch 1" model it was assumed that $b = 0.328$ $(\text{g}\cdot\text{cm}^{-3})/(\text{km}\cdot\text{s}^{-1})$, the value resulting from laboratory measurements relating to a number of rocks and crystals with a small value of the mean atomic number. The constant a and the mass of the core were found from the condition for the whole mass M and the whole moment of inertia I of the Earth. In the "Birch 2" model it was assumed that the density of the upper mantle close to the Moho boundary is 3.32 $\text{g}\cdot\text{cm}^{-3}$. The parameters a and b in eqn. (2.3.2)

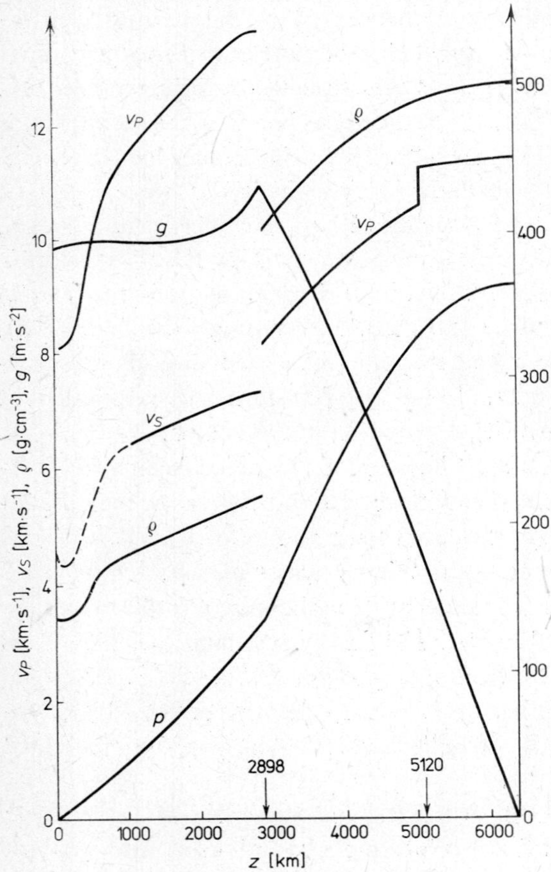

Fig. 2.3.3. Model "Birch 1": distributions of the density $\varrho(z)$, gravity acceleration $g(z)$, pressure $p(z)$, and velocity $v_P(z)$ and $v_S(z)$ of P and S waves. The segment of the curve $v_S(z)$ drawn by the dashed line was not used in the calculations of Birch.

and the mass of the core were determined from this assumption and from the known values of M and I. The distribution of selected parameters of the Earth (ϱ, p, g, v_P, v_S) in the "Birch 1" model are shown in Fig. 2.3.3.

The models of Bullen (1963, 1975) can be divided into type A models, with a continuous density distribution in the Earth's mantle, and type B models, in which the occurrence of discontinuities in the mantle is assumed. In the oldest versions of these models (A, 1942 and B, 1950) it was assumed that in the inner core $v_S = 0$, i.e., it was assumed that the inner core is liquid.

The Bullen models of type A are based on the Williamson equation (2.2.12), on the parabolic interpretation of density in the transition layer between the upper and lower mantle (in this layer eqn. (2.2.12) was not used), and on the assumption that the density in the Earth's centre is $\varrho_{6371} > 12.3 \text{ g} \cdot \text{cm}^{-3}$. In his first model A of 1942 Bullen took the range $12.3 \text{ g} \cdot \text{cm}^{-3} < \varrho_{6371} < 22.3 \text{ g} \cdot \text{cm}^{-3}$ as acceptable for the central density. According to Bullard (1957), however, the upper boundary of the density is considerably lower, being $13.6 \text{ g} \cdot \text{cm}^{-3}$. This value is in good agreement with the experimental data obtained for iron by shock wave methods. Bullen considered various type A models based on different initial data, especially on different travel times. In type A models the Earth's concentric layers are denoted by $A, \ldots, G$. Model A of Bullen and Haddon (1967a) is presented in Fig. 2.3.1 and in Table 2.3.1. According to Bullen (1975), model A was constructed in 1965 as a preliminary step towards incorporating free Earth oscillation data. Models of type A form a basis for different models with a continuous density distribution in the mantle.

The Bullen models of type B are based on the modified Adams–Williamson equation (2.2.24), i.e., on an empirical relation between the bulk modulus K and pressure p. The most characteristic feature of type B models is the existence of density discontinuities in the upper mantle, i.e., in layers B and C. These discontinuities are interpreted as boundary surfaces between solid phases with different densities. The first type B model of 1950 gives the central density of the fluid core as $\varrho_{6371} = 18.1 \text{ g} \cdot \text{cm}^{-3}$. Like type A models, models of type B have also been frequently corrected by Bullen and other authors. Model B_2 of Bullen and Haddon (1967b) is presented in Fig. 2.3.1 and in Table 2.3.1.

The model of Wang (1972) is the last one presented in Table 2.3.1; this model is a synthesis of the previous models, and takes into account several corrections not considered so far. The density distribution in the upper mantle and transition layer was found from the petrological data of Clark and Ringwood and from the Birch equation in the form

$\varrho = 0.613 + 0.328\, v_P$*. In the lower mantle, Wang's linear relation between density and hydrodynamic sound velocity was used

$$\varrho = 0.742 + 0.424\, \phi^{1/2} \tag{2.3.3}$$

In the core the Williamson equation (2.2.12) was used, and it was assumed that the inner core is solid and the shear wave velocity there is $v_S = 3.3\, \mathrm{km \cdot s^{-1}}$. An initial model determined in this way was then numerically perturbed so as to obtain the final model, fulfilling conditions relating to the mass, the moment of inertia, the eigenperiods of free Earth oscillations, and the velocities of surface waves measured along the great circles.

Jordan and Anderson (1974) in model B1 calculated the density distribution and the seismic wave velocity v_P and v_S distributions from 219 data, of which 178 are eigenperiods of free Earth oscillations, 39 are the differences of arrival times of different waves and the remaining two are the mass M and the moment of inertia I of the Earth. As in the model of Wang (1972), Jordan and Anderson started with model "B", which was transformed into their final model "B1" by the application of a special numerical algorithm taking into account the above 219 data. The initial model "B" is very close to the "Birch 1" model. The differences between the functions $v_P(z)$, $v_S(z)$ and $\varrho(z)$ calculated by Jordan and Anderson and those found from the initial model B are shown in Fig. 2.3.4.

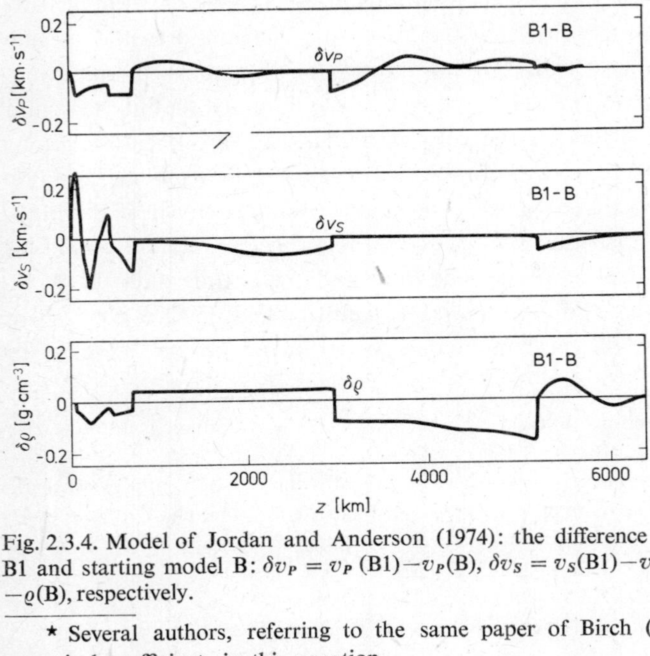

Fig. 2.3.4. Model of Jordan and Anderson (1974): the difference between final model B1 and starting model B: $\delta v_P = v_P\,(\mathrm{B1}) - v_P(\mathrm{B})$, $\delta v_S = v_S(\mathrm{B1}) - v_S(\mathrm{B})$ and $\delta\varrho = \varrho(\mathrm{B1}) - \varrho(\mathrm{B})$, respectively.

* Several authors, referring to the same paper of Birch (1961), give different numerical coefficients in this equation.

A number of models based on eigenperiods of free Earth oscillations recorded during large earthquakes were developed by Dziewonski, Gilbert and Hales. Similarly to Bullen in his constructions of models A and B, these authors also consider continuous and discontinuous distributions of parameters in the mantle. Thus, for example, model 508 and model 1066A are models with a continuous density distribution, while model 1066B is a model with discontinuities in the upper mantle.

Model 508 is based on a set of 506 normal modes plus the Earth's mass and moment of inertia. This set of data was included in model C196 (Alaska II) as a starting model to obtain model 508. For the construction of models 1066A and 1066B a set of 1463 gross Earth data (eigenperiods) of which 1066 are distinct was used (Gilbert and Dziewonski, 1975). Two inversions were performed with the set of 1463 gross Earth data. The two initial models are model 508 and model B1 (Jordan and Anderson, 1974). The two final models are model 1066A derived from model 508 and model 1066B derived from model B1. In model 1066B discontinuities of parameters occur at depths of 421 and 671 km.

Dziewonski *et al.* (1975) developed models of the Earth based on the assumption that the distributions of velocities and density in the lower mantle and core can be properly described in segments by polynomials of the second and third degree as functions of the radius. The coefficients of the polynomials were fitted by the least-squares method to the specific parameters of model 1066B. Discontinuities in the upper mantle were assumed to be at depth of 670 and 420 km. It was assumed that in the region between these discontinuities the velocity and density are linear functions of the radius. For the area shallower than 420 km two models, oceanic and continental, were constructed. To establish the final form of the polynomials describing distributions v_P, v_S and ϱ an inverse method was used, in which a set of 1064 normal-mode data and travel-time data for *P*, *S*, *SKS*, *PKIKP* waves and differential travel times of *SKKS–SKS* were applied.* Regional data relating to the dispersion of surface waves were also included.

---

* The symbols denoting different types of body waves whose source is in the mantle are as follows (see e.g., Bullen, 1975):

SKS — a wave generated as transversal, partially transformed into longitudinal at the mantle-outer core boundary, travelling along an curve–line chord of the outer core, and transformed again into a transversal wave after crossing the core–mantle boundary.

SKKS — the same as *SKS* except that within the outer core the wave travels along two different curve–line chords between which the point of reflection of the wave from the lower surface of the mantle is situated.

Finally, Dziewonski *et al.* (1975) presented three different models of the Earth, called the *parametric Earth models* (PEM). They are the PEM-O (oceanic model), the PEM-C (continental model), and the PEM-A (average Earth model). Beneath a depth of 420 km the three models are identical. The results relating to the average PEM-A model were obtained using weighted averages with weights 2/3 or 1/3 corresponding approximately to real proportions of the Earth's surface covered by oceans and continents. The distributions of p, ϱ, g, v_P and v_S corresponding to the PEM model are shown in Fig. 2.3.5, where for depths $z < 420$ km the curves correspond to

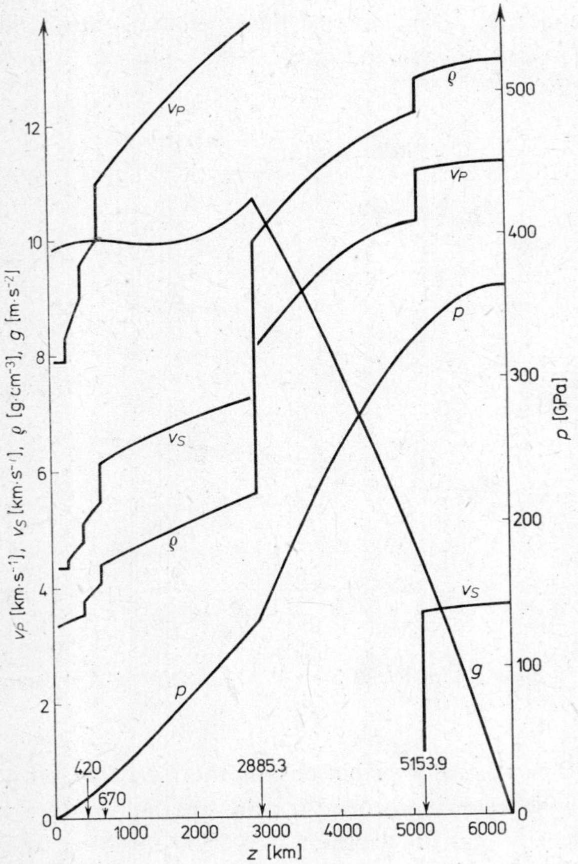

Fig. 2.3.5. Distributions of p, ϱ, g, v_P and v_S according to the PEM model of Dziewonski *et al.* (1975).

PKIKP — a longitudinal wave passing in succession through the mantle, the outer core, the inner core, the outer core and the mantle.

the average PEM-A model. Dziewonski *et al.* (1975) remark that in the area $z > 670$ km the density distribution in model PEM differs by less than 0.2% from that resulting from the Adams–Williamson equation. Indeed, in comparing Fig. 2.3.5 with Fig. 2.3.3, which present the distributions of the same parameters obtained in the Birch 1 model, only small differences can be noted. It should be remembered that the Birch 1 model is based on calculations made in 1957, that is at a time when the number of data was much smaller and especially when the Earth's free oscillations were not taken into account. The PEM model is an idealized model, especially the discontinuities at 420 and 670 km corresponding to phase transitions are treated in a very simplified manner, and the assumed linear relations are highly approximate. The distributions of ϱ, v_P and v_S for depths down to $z = 600$ km are shown in Fig. 2.3.6.

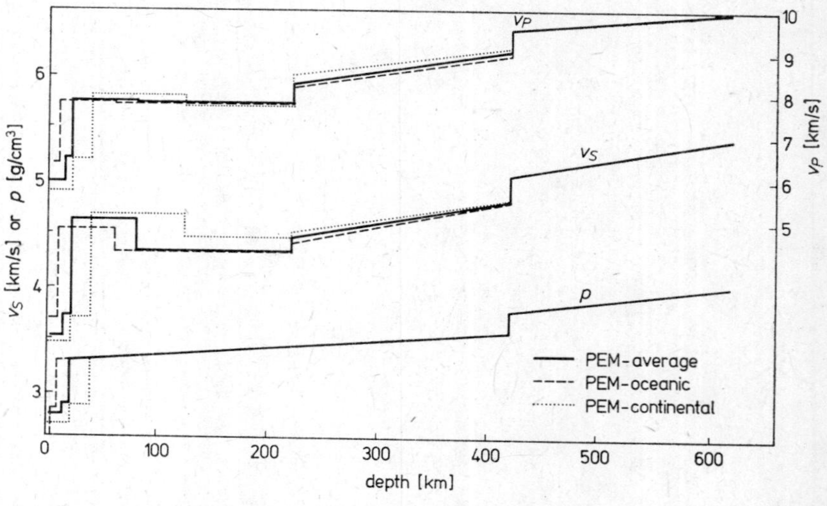

Fig. 2.3.6. Distribution of ϱ, v_P and v_S in models PEM-O, PEM-C and PEM-A at depths down to 600 km, after Dziewonski *et al.* (1975).

From the models of parameter distributions in the Earth's interior presented in this chapter the following general conclusions can be drawn:

1. Distributions of density $\varrho(z)$ for depths $z \gtrsim 700$ km resulting from different models are very similar. The scattering of the density values at various depths does not exceed a few per cent.

2. Distributions of density in the upper mantle and transition layer is still an incompletely solved problem. Models of the Earth in these layers can be divided into models with discontinuous and models with continuous

density distributions. Discontinuous distributions (the most characteristic example being the PEM model) can be classified as idealized models of the mathematical type, while models with continuous distributions (example the pyrolitic model of Ringwood) are models of the geological-mineralogical type. Since the problem of mineral composition of the Earth's mantle and the phase transformations occurring there still has many uncertain points (see Chapter 7), it can be expected that density models will undergo further, perhaps even substantial, modifications.

3. Distributions of pressure $p(z)$ and gravity acceleration $g(z)$ are known quite accurately in the entire Earth. These distributions, resulting from the integration of relations containing the density $\varrho(z)$—formulas (2.2.2) and (2.2.4)— have a smoothing effects on possible discontinuities in the density distribution. The depths at which density is not continuous are depths at which the first derivatives of pressure and gravity acceleration are also discontinuous. The pressure is a function monotonically increasing with depth; it can be assumed that in the upper mantle we have approximately $dp/dz \approx 1/30$ GPa · km^{-1}. The gravity acceleration in the entire mantle is close to 10 m · s^{-2} and in the core it decreases to zero according to a relation close to linear.

4. The elastic constants K, μ, λ and E are discontinuous at the same depths as the density.

5. Since each of the models is based, at least to some extent, on assumptions relating to the composition of the Earth layers and these assumptions are impossible to check, none of the models can uniquely explain the physical sense of all the density discontinuities appearing in the Earth's interior. The question whether a given discontinuity is a discontinuity of phase state or of composition, or of both these factors is often unanswerable. A classical example is the upper and lower boundary surface of the outer core. It is known for certain that they are phase boundaries of the liquid phase–solid phase type, but the problem of chemical composition on both sides of the boundary is still under discussion (see Chapter 8).

2.4 MODELS OF PLANETARY INTERIORS

2.4.1 Classification of the planets

Up to a dozen or so years ago, astronomical observations carried out from the Earth's surface were the only source of information about the planets.*

* The term "planet" in this chapter is used for planets as well as satellites.

From these observations we can determine the size and shape of the planets (equatorial radius R_e, polar radius R_p, mean radius R and oblateness) and their mass M, and thus also the average density $\bar{\varrho}$. Data related to a planet's gravity field, obtained from observations of the motion of its natural or artificial satellites, lead to the determination of its moment of inertia I and permit to study of deviations from spherically symmetric mass distribution inside the planet.

These data, coming from the field of celestial mechanics, are supplemented by astrophysical observations relating first of all to the composition of the atmosphere of the planets and to the optical properties of their surface. In particular, energy balances of the planets are measured. The energy balance is the excess of the energy radiated by the planet in the whole spectral range (practically only infrared radiation) over the energy received from the Sun:

$$\begin{aligned}&\text{excess luminosity } L\\&\quad = (\text{reflected energy} + \text{radiated energy})\\&\quad - \text{the incident energy of solar radiation}\end{aligned} \tag{2.4.1}$$

The energy radiated by a planet is composed of the reemitted solar radiation and the energy originating from sources inside the planet, e.g., gravity or nuclear energy (radioactivity, thermonuclear fusion). If $L > 0$, then such sources of energy must exist.

For the Earth the average heat flow from radioactive sources, directed from the centre towards the surface, is known ($\approx 0.063\ \mathrm{J \cdot m^{-2} \cdot s^{-1}}$). Thus the total energy radiated outside the globe is $0.063 \times 4\pi R^2 = 3.2 \times 10^{13}\ \mathrm{J \cdot s^{-1}}$, where $R = 6.371 \times 10^6$ m is the Earth's radius. The energy coming from the Sun and completely reemitted is $Q_{\mathrm{Earth}} = R^2 q = 1.7 \times 10^{17}\ \mathrm{J \cdot s^{-1}}$, where $q = 1.33 \times 10^3\ \mathrm{J \cdot m^{-2} \cdot s^{-1}}$ is the solar constant. Thus, for the Earth, $L_{\mathrm{Earth}} = 3.2 \times 10^{13}\ \mathrm{J \cdot s^{-1}} = 1.9 \times 10^{-4}\ Q_{\mathrm{Earth}}$. It can be expected that also for other planets of similar structure excess luminosity is insignificant in comparison to the incident solar energy.

Entirely different results relating to excess luminosity were found for Jupiter and Saturn. From the data compiled by Stevenson (1978), based partly on the experiments carried out by the probe Pioneer 11, the excess luminosity for Jupiter is $L_{\mathrm{Jupiter}} = 5 \times 10^{17}\ \mathrm{J \cdot s^{-1}}$ and for Saturn $L_{\mathrm{Saturn}} = 1.7 \times 10^{17}\ \mathrm{J \cdot s^{-1}}$. These values are twice as high as the value of the solar energy absorbed by these planets. From calculations of excess luminosity per mass unit for the Earth, Jupiter and Saturn it follows that the efficiency of the internal energy sources of Jupiter and Saturn is 49 and 56 times that of the Earth. One cannot see any reason why the concentration of radioactive elements on Jupiter and Saturn should be so much higher than it is on the Earth; the possibility of nuclear reactions inside the giant planets is not indicated by their thermodynamical parameters. Thus the only significant source of internal energy in Jupiter and Saturn can be energy of gravitational origin, e.g., the energy of phase transformations induced by hydrostatic compression.

One of the most important characteristics of a planet is its average density $\bar{\varrho}$. It can be determined with relatively high precision (as a rule better than 5%) from astronomical measurements. The average density $\bar{\varrho}$, the composition of the surface of the planet or of its atmosphere and a general knowledge of equations of state form the basis for the following classification of planets (see e.g., Stevenson, 1978; Stiller and Frank, 1980):

1. Rocky formed planets with high average density $\bar{\varrho}$, composed mainly of iron, magnesium, silicon and their oxygen compounds. Terrestrial planets—Mercury with the average density $\bar{\varrho} = 5.4$ (in $g \cdot cm^{-3}$), Venus with $\bar{\varrho} = 5.253$, the Earth with $\bar{\varrho} = 5.517$, Mars with $\bar{\varrho} = 3.93$, and Moon-like bodies—the Moon with $\bar{\varrho} = 3.34$, two of Jupiter's satellites, Io with $\bar{\varrho} = 3.4$ and Europe with $\bar{\varrho} \approx 3$, and also other satellites and asteroids belong to this group.

2. Icy formed planets composed of volatiles such as oxygen, nitrogen, carbon and hydrogen appearing in chemical compounds H_2O, CH_4, NH_3. Uranus with $\bar{\varrho} = 1.3$, Neptune with $\bar{\varrho} = 1.65$, and two of Jupiter's satellites, Ganymede with $\bar{\varrho} = 1.9$ and Callisto with $\bar{\varrho} = 1.7$, and also Saturn's satellite Titan with $\bar{\varrho} = 1.6$ belong to this group.

3. The hydrogen-helium giant planets: Jupiter with $\bar{\varrho} = 1.333$ and Saturn with $\bar{\varrho} = 0.705$, composed basically of hydrogen with an admixture of helium.

In spite of the fact that the various planets differ greatly in composition, mass and size, models of their structure indicate that all the planets probably have a central core composed of rocky or metallic material in a solid or liquid state, and being in temperatures of the order of several thousand kelvins. The larger the mass of a planet, the higher is the core temperature. Further on, models of the planets are described according to their classification.

2.4.2 Rocky formed planets

Rocky planets and meteorites are the most dense bodies of the solar system. They are bodies of nontypical composition with respect to the hydrogen–

helium composition of the Sun and the giant planets. This fact and the complexity of composition are an essential obstacle to the determination of the internal structure of terrestrial planets. On the other hand, terrestrial planets are the only planets known not only from astronomical observations but also from observations carried out directly on their surfaces. Measurements relating to surface composition and to seismicity have been carried out not only on the Earth, but also on the Moon and on Mars.

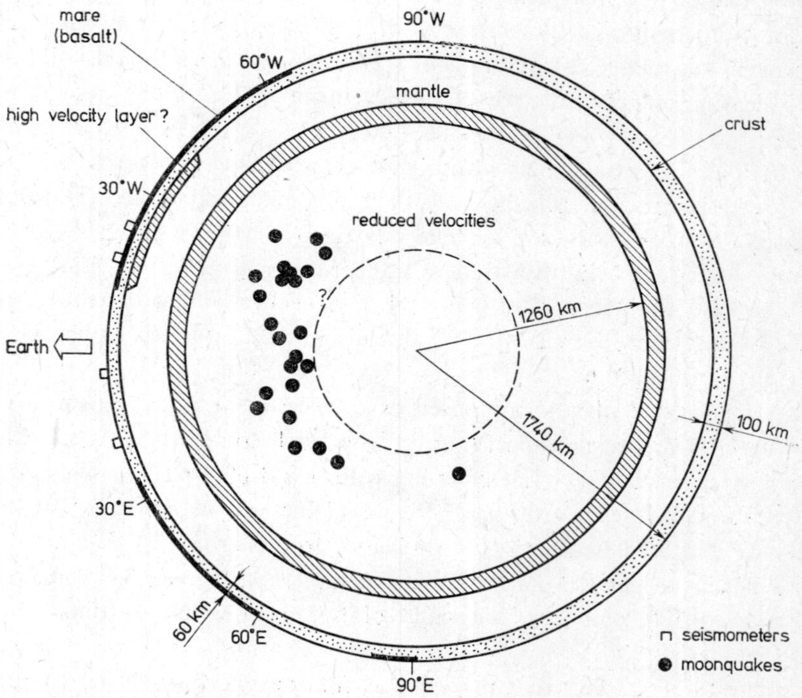

Fig. 2.4.1. Schematic diagram of the lunar interior along an equatorial cross-section, after Toksöz (1979). The crust is dotted. The transition zone between the upper and lower mantle is shaded. There is increased attenuation inside the dashed circle. Only deep moonquakes are shown (black circles).

From 21 July 1969 to 12 December 1972 the Apollo 11, 12, 14, 15, 16 and 17 missions installed six seismic stations at different sites on the Moon's surface (see Toksöz, 1979). All stations except the first were in operation until 30 September 1977. Passive and active seismic experiments, gravity measurements and lunar surface profiling experiments were carried out. Active seismic experiments at the Apollo 14, 16 and 17 stations were made to determine the shallow structure of the lunar crust by using explosives and refraction techniques.

Studies of the seismicity of Mars were made between 4 September 1976 and 2 April 1978 by means of the seismometer on the Viking II lander. No events of distinct seismic origin were found.

Toksöz (1979) distinguishes three types of moonquake events:

1. Thermal events occurring near the surface, connected with daily changes of temperature and with thermal stresses induced by these changes.

2. High frequency teleseisms caused by the release of frozen-in tectonic stresses at depths down to 100 km.

3. Deep-focus moonquakes. During eight years of seismic recording about 80 moonquake foci were located. Their magnitudes were never greater than 3.

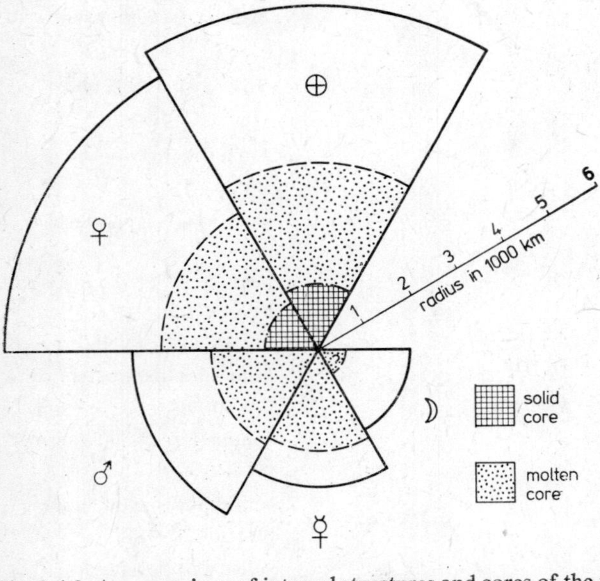

Fig. 2.4.2. A comparison of internal structures and cores of the terrestrial planets—Earth, Venus, Mars, Mercury and the Moon, after Toksöz (1979). Core sizes and conditions of planets other than the Earth are based on theoretical modelling.

Seismic measurements carried out by the Apollo network permitted to establish velocity distributions of P and S waves, and consequently the layered model of the Moon structure, shown in Fig. 2.4.1. For the side directed towards the Earth, which is the side where seismic observations were made, a more detailed structure was proposed.

The existence and structure of the Moon's core is an unsolved problem. The central pressure in the Moon equal to about 0.5×10^{10} Pa, corresponds to the pressure at a depth of about 150 km in the Earth (Lyttleton, 1977).

TABLE 2.4.1
Differential equations for computing the model of a spherical planet, after Baumgardner and Anderson (1980). Variables: ϱ—density, p—pressure, T—temperature, M—mass inside radius r, I—moment of inertia inside radius r, Q—heat flow, α—thermal expansivity, γ—Grüneisen parameter, g—gravity acceleration, H—heat generation parameter, R—spherical planet radius, D—thickness of a surface layer of the order of 10 km, k—thermal conductivity, K_0, K_0'—isothermal bulk modulus and its pressure derivative at $p = 0$, K_s—adiabatic bulk modulus, θ—Debye temperature

1	$p - aT + b = \dfrac{3K_0}{3K_0'-8}[(\varrho/\varrho_0)^{K_0'-4/3} - (\varrho/\varrho)^{4/3}]$	EOS, $T > \theta$
2	$\dfrac{dp(r)}{dr} = -g(r)\varrho(r)$	hydrostatic equilibrium
3	$\dfrac{dM}{dr} = 4\pi r^2\varrho(r)$	M in a spherical body
4	$\dfrac{dI}{dr} = \dfrac{8\pi}{3} r^4\varrho(r)$	I in a spherical body
5	$g(r) = \dfrac{GM(r)}{r^2}$	g in a spherical body
6	$\dfrac{dQ}{dr} = H - \dfrac{2Q(r)}{r}$	heat flow in a spherical body
7	$H_s = H_0 \exp[-(R-r)/D]$	distribution of the heat generation parameter near the surface
8	$\dfrac{dT}{dr} = -\dfrac{1}{k(r)}Q$	conducting shells: thermal gradient
9	$\dfrac{dT}{T} = \gamma\dfrac{d\varrho}{\varrho}$	convecting shells: temperature distribution
10	$\gamma/\gamma_0 = (\varrho/\varrho_0)^q$	γ distribution with density; γ_0, q constants
11	$\alpha(\Delta T)_{sa} = \int_{\varrho_1}^{\varrho_2}\left(\dfrac{d\varrho}{\varrho} - \dfrac{dp}{K_s}\right)dp$	superadiabatic ΔT across virtually adiabatic shell
12	$T_m = 1500 + 10p$ (p in GPa, T_m in kelvin)	solidus curve for mantle

Similarly, the central pressure in Mars, equal to about 3×10^{10} Pa, corresponds to the pressure at a depth of about 750 km in the Earth. The central pressure in Mercury corresponds also to that in the upper mantle of the Earth. Thus it is certain that the Moon, Mars and Mercury do not have a liquid metallic core. If a liquid core exists in these planets, it is composed of matter similar in composition to that of the Earth's mantle, and the mantle–core boundary is determined by the temperature and pressure of the solidus. The schematic diagram of the structure of terrestrial planets and the Moon, compiled by Toksöz (1979), is shown in Fig. 2.4.2.

A very interesting paper on the structure of terrestrial planets has recently been published by Baumgardner and Anderson (1980). The equations used for the construction of models of the Moon, Mars and Mercury are listed in Table 2.4.1. The equations were tested by their application in determining the structure of the Earth's before they were accepted for other planets. It was found that such a test model agrees very well with the PEM model of Dziewonski *et al.* (1975). Thus the application of the equations listed in Table 2.4.1 to the planets is well justified. An essential feature

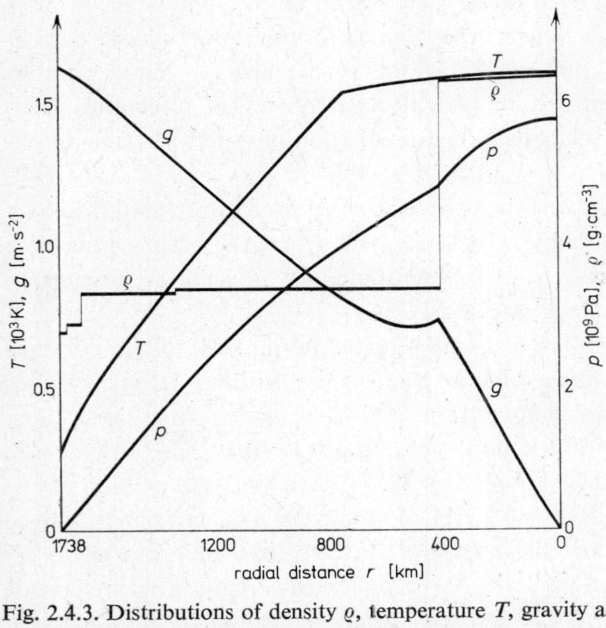

Fig. 2.4.3. Distributions of density ϱ, temperature T, gravity acceleration g and pressure p in the lunar interior, according to Baumgardner and Anderson (1980).

of the equation of state used (Table 2.4.1, eqn. 1) is the fact that this equation expresses in a simple way the total pressure by the thermal pressure and elastic pressure resulting from the application of the Born–Mie potential.* The coefficients a and b in the thermal part of the equation of state are expressed by the coefficient of thermal expansion α and by the bulk modulus K. The use of the equations from Table 2.4.1 in numerical calculations relating to a given planet requires the assumption of several boundary conditions relating to various physical quantities, not always known with sufficient accuracy. Since some of the numerical boundary values were selected very arbitrarily, the Baumgardner and Anderson models may undergo substantial changes in the future. The distributions of density, temperature, gravity acceleration and pressure in the Moon's interior, drawn from the calculations of Baumgardner and Anderson (1980), are shown in Fig. 2.4.3.

2.4.3 Hydrogen–helium and icy formed giant planets

The models of Jupiter and Saturn presented here are based primarily on papers by Slattery (1977) and Stevenson (1978). The models of Uranus and Neptune are considered rather superficially since the number of observations relating to these planets is too limited for their more precise description. Moreover, the composition of icy type planets is more complex than that of hydrogen–helium planets and theoretical difficulties begin already at the stage of selecting a proper equation of state describing matter of the icy type.

The basic initial data for the construction of models of Jupiter, Saturn, Uranus and Neptune are given in Table 2.4.2, where the basic assumptions and the average values (from different models) of selected parameters characterizing these planets are also listed.

The principal assumption usually made in models of Jupiter and Saturn is an assumption about their composition. It is assumed that these planets are composed of hydrogen and helium in a ratio close to that observed on the Sun, where the mass fractions of hydrogen and helium are $X = 0.78$ and $Y = 0.22$, respectively. The values of the total mass and momentum of inertia of these planets, however, and the equation of state for metallic hydrogen indicate that the cores of Jupiter and Saturn (as well as of the icy type planets) must contain heavier elements than hydrogen and helium. They are, therefore, cores of the rocky type, constituting at least a few per cent of the mass of the entire planet. A core with such a large mass contradicts the assumption about a hydrogen–

* If the values $m = 1$ and $n = 3K_0'-7$ are substituted in eqn. (1.7.7), then eqn. 1 from Table 2.4.1 is obtained.

TABLE 2.4.2

Basic observational data, typical assumptions and the most important results relating to the structure of the giant planets

	Planet's parameter	Jupiter	Saturn	Uranus	Neptune
Initial data*	Mean radius R [km]	69800 ± 50	57800 ± 100	25200 ± 300	24600 ± 200
	Mass $M(M_{\text{Earth}} = 1)$	317.9	95.2	14.6	17.2
	Average density [$g \cdot cm^{-3}$]	1.333 ± 0.003	0.705 ± 0.005	1.3 ± 0.05	1.65 ± 0.03
	Main atmospheric constituents	H_2, He also NH_3, CH_4, H_2O	H_2, He? also NH_3, CH_4	H_2, CH_4, He?	H_2, CH_4, He?
	Excess luminosity L [$J \cdot s^{-1}$]	5×10^{17}	1.7×10^{17}		
	Effective temperature [K]	125 ± 5	97 ± 4	57 ± 4	45 ± 3
Typical assumptions	Temperature distribution in the core	isothermal	isothermal	isothermal	isothermal
	Temperature distribution in the envelope	adiabatic	adiabatic	adiabatic	adiabatic
	Envelope composition	≈ 0.78 H, ≈ 0.22 He by weight	≈ 0.78 H, ≈ 0.22He by weight	H_2, He, CH_4, H_2O, NH_3	H_2, He, CH_4, H_2O, NH_3
	Temperature in the atmosphere (in K, at a level of $p = 10^5$ Pa)	160	160		
The most important results	Core mass M_c ($M_{\text{Earth}} = 1$)	$14 \div 16$	$15 \div 17$	≈ 4	models close to Uranus
	Central pressure [Pa]	$\approx 10 \times 10^{12}$	7×10^{12}	$\approx 1.6 \times 10^{12}$	
	Central temperature [K]	≈ 19000	≈ 10000	≈ 3000	
	Central density [$g \cdot cm^{-3}$]	$13 \div 26$	$10 \div 22$	≈ 12	
	Rock core radius ($R = 1$)	0.13*	0.21*	0.33*	
	Radius of metallic–molecular boundary ($R = 1$)	0.76*	0.40*		
	Temperature of metallic–molecular boundary [K]	10000	8500		

* According to Stevenson (1978).

helium composition of the planet in its entire volume. Consequently, the models of Jupiter and Saturn assume the existence of a hydrogen–helium envelope and a rocky core.

According to the estimations presented in Chapter 1, the relations between the radius R and the mass M are of the form

$$R \propto M^{1/3}$$

for ordinary matter, and

$$R \propto M^{-1/3}\left(\frac{Z}{A}\right)^{5/3}$$

for degenerated matter described by the Thomas–Fermi–Dirac equation, where Z is the mass number and A is the atomic number of the medium. The relation $R = R(M)$ in the areas where both the above relations are valid and in the intermediate area are shown in Fig. 2.4.4, with the medium

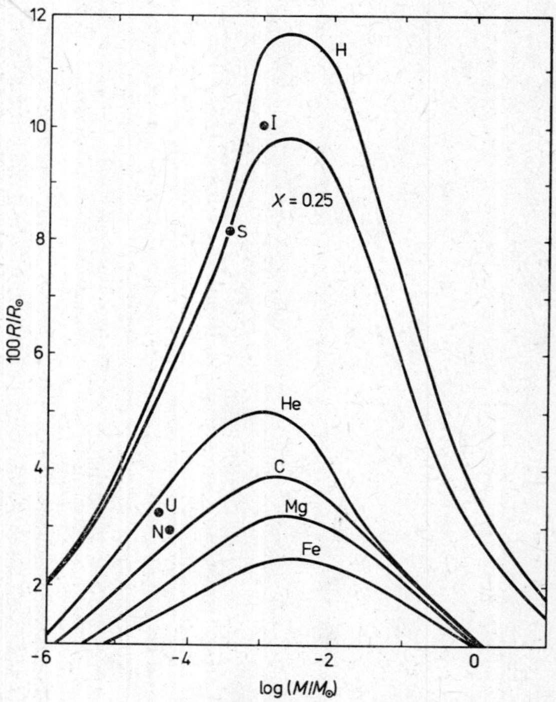

Fig. 2.4.4. Mass–radius relationship for zero-temperature spheres of various chemical compositions, based on the Thomas–Fermi–Dirac theory, from Stevenson (1978). $X = 0.25$ refers to a 75 per cent H and 25 per cent He mixture by mass. The symbols J, S, U, N represent the giant planets. $R_\odot$ and $M_\odot$ are the solar radius and mass.

composition taken as a parameter. Under the assumption that a given body is composed of only one element, the respective curves in Fig. 2.4.4 represent the mass-radius relations for such homogeneous* one-component spheres. Thus from R and M of a given planet (or a star) an attempt can be made to determine its composition. Such a procedure is not unique since the parameter describing the medium composition has the sense of an average value and can, therefore, represent a mixture of different elements in different ratios. The most certain information which can be found from Fig. 2.4.4 relates to the hydrogen–helium composition of Jupiter and Saturn; Jupiter is more "hydrogenous" while Saturn is more "helium-like" character. It can also be said more generally that Saturn has a larger admixture of heavy elements.

According to Slattery (1977), for a hydrogen–helium admixture within an interval of density from 0.005 to 0.3 $g \cdot cm^{-3}$, the equation of state based on the intermolecular potential for the interaction pairs H_2–H_2, H_2–He and He–He can be used. For higher densities a three-body interaction should be taken into account. From the density $\varrho = 1.3\ g \cdot cm^{-3}$, corresponding to a hydrogen phase transition from the molecular to the metallic state, it follows the appropriate equation is the equation of state for hydrogen metallic plasma with dissolved helium. Slattery considered two different types of models of giant planets: with a metallic layer composed only of hydrogen and helium and with a layer composed also of heavier elements.

The distributions of pressure, density and temperature inside Jupiter given by Slattery (1977) and Stevenson (1979) are shown in Fig. 2.4.5. The corresponding diagrams for Saturn are very similar.

The basic problem in giant planets is the generation of energy. As has already been stated, the energy is most probably of gravitational origin, and it is released during phase transformations of the molecular phase → metallic phase type, as a result either of the contraction of the planet or of gravitational differentiation of a heavier element (helium) dissolved in a lighter element (hydrogen) and sinking towards the centre of the planet. The solution of the problem of the existence of a sharp molecular hydrogen → metallic hydrogen phase transformation inside Jupiter and Saturn depends on the knowledge of the hydrogen phase diagram and temperature distribution inside these planets (see Fig. 1.3.3). Frank (1980) noted that the critical temperature T_c of the molecular hydrogen → metallic hydrogen

* Homogeneous in the sense that in the entire volume one specified equation of state is applied. This is, to some extent, a contradiction since near the surface, where the pressure is $p = 0$, matter is certainly in an ordinary state in spite of the fact that inside it is in the state of Fermi matter.

transformation, above which the sharp phase boundary could not exist, is $T_c = (6 \div 9) \times 10^3$ K. At the same time, a temperature of 10^4 K in Jupiter and 8.5×10^3 K in Saturn corresponds to the pressure necessary for this

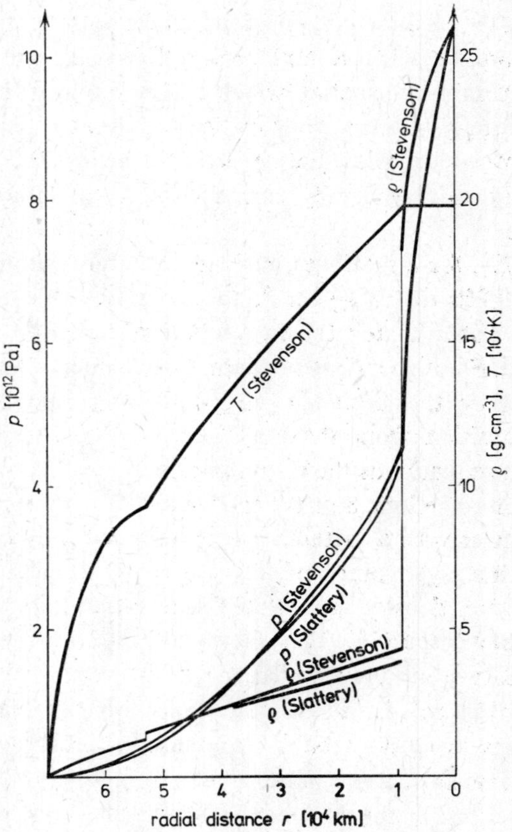

Fig. 2.4.5. Distributions of density ϱ, temperature T and pressure p in Jupiter's interior according to Slattery (1977) and Stevenson (1978).

transformation. Frank concludes that inside Jupiter a phase boundary of the molecular phase → metallic phase type could not exist. This boundary might exist in Saturn, if it did, the heat generation would result from phase transitions. The above conclusions are in good agreement with the observations made by Stevenson (1978), who concluded that "the excess luminosity of Jupiter can be explained as primordial heat slowly leaking out, but Saturn may require additional energy generation".

REFERENCES

Baumgardner J. R. and Anderson O. L. (1980), "Using the thermal pressure to compute the physical properties of terrestrial planets", presented on the XXIII COSPAR meeting, Budapest.

Birch F. (1939), "The variation of seismic velocities within a simplified Earth model in accordance with the theory of finite strain", *Bull. Seismol. Soc. Am.*, **29**, 463–479.

Birch F. (1961), "The velocity of compressional waves in rocks to 10 kilobars", *J. Geophys, Res.*, **66**, 7, 2199–2224 (Table 15, Solution 2).

Birch F. (1964), "Density and composition of mantle and core", *J. Geophys. Res.*, **20**, 69, 4377–4388.

Bullard E. C. (1957), "The density within the Earth", *Verhandel. Ned. Geol. Mijnbouwk. Genoot.*, **18**, 23–42.

Bullen K. E. (1963), *An Introduction to the Theory of Seismology*, University Press, Cambridge.

Bullen K. E. (1975), *The Earth Density*, Chapman and Hall, London.

Bullen K. E. and Haddon R. A. (1967a), "Derivation of an Earth model from free oscillation data", *Proc. Nat. Acad. Sci. Wash.*, **58**, 846–852.

Bullen K. E. and Haddon R. A. (1967b), "Earth models based on compressibility theory", *Phys. Earth Planet. Inter.*, **1**, 1–13.

Clark S. P. and Ringwood A. E. (1964), "Density distribution and constitution of the mantle", *Rev. Geophys.*, **2**, 1, 35–88.

Dziewonski A. M., Hales A. L. and Lapwood E. R. (1975), "Parametrically simple Earth models consistent with geophysical data", *Phys. Earth Planet. Inter.*, **10**, 1, 12–48.

Frank S. (1980), "Critical temperature for dielectric–metal transition in hydrogen and implications for Jupiter and Saturn", presented on the XXIII COSPAR meeting, Budapest.

Gilbert F. and Dziewonski A. M. (1975), "An application of normal mode theory to the retrieval of structural parameters and of source mechanisms from seismic spectra", *Philos. Trans. R. Soc. Lond.*, A**278**, 187–269.

Jordan T. H. and Anderson D. L. (1974), "Earth structure from free oscillations and travel times", *Geophys. J. R. Astron. Soc.*, **36**, 2, 411–459.

Lyttleton R. A. (1977), "Structure of the terrestrial planets", *Astrophys. Space Sci.*, **49**, 7, L1–L6.

Press F. (1970), "Earth models consistent with the geophysical data", *Phys. Earth. Planet. Inter.*, **3**, 3–22.

Ringwood A. E. (1966), "Chemical evolution of the terrestrial planets", *Geochim. Cosmochim. Acta*, **30**, 41–105.

Ringwood A. E. (1975), *Composition and Petrology of the Earth's Mantle*, McGraw-Hill, New York.

Ringwood A. E. (1976), "Phase transformations in descending plates and implications for mantle dynamics", *Tectonophysics*, **32**, 1/2, 129–143.

Stevenson D. J. (1978), "The outer planets and their satellites", in: *The Origin of the Solar System*, S. F. Dermott (ed.), J. Wiley and Sons, Chichester.

Slattery W. L. (1977), "The structure of the planets Jupiter and Saturn", *Icarus*, **32**, 58–72.

Stiller H. and Frank S. (1980), "State of matter and internal structure of planets", presented on the XXIII COSPAR meeting, Budapest.

Toksöz M. N. (1979), "Planetary seismology and interiors", *Rev. Geophys. Space Phys.*, **17**, 7, 1641–1655.

Wang Chi-yuen (1972), "A simple Earth model", *J. Geophys. Res.*, **77**, 23, 4318–4329.

Williamson E. D. and Adams L. R. (1923), "Density distribution in the Earth", *J. Wash. Acad. Sci.*, **13**, 413–428.

3. Creep Processes, Viscosity Models and the Quality Factor Q in the Earth's Interior

3.1 ROCK PROPERTIES UNDER HIGH PRESSURES AND TEMPERATURES

Laboratory experiments carried out with rocks under high pressures and temperatures show a wide range of changes in the rock properties and imply various physical mechanisms corresponding to these changes. Phenomena occurring under very high pressures, interesting for the problems of the Earth's interior, and rheological creep processes characterized by very long duration are inaccessible to static laboratory experiments. Thus in studies of the properties of rocks forming the Earth's interior, and especially of rocks forming the lithosphere and the Earth's mantle, it is necessary to use geophysical observations supplemented by extrapolations of laboratory data. Besides the data relating to progressive processes of deformations and mechanical fracturing, laboratory studies of minerals also provide information about phase transformations which these minerals could undergo under high pressures.

Further on the processes of creep and viscous flow are considered in detail, as they are at the root of the dynamic processes occurring in the Earth's interior, such as global convection, deformations accumulating before earthquakes, attenuation of elastic waves and others.

Problems of the mechanical properties of rocks and constitutive relations describing these properties are considered in another volume of this series. Formally the properties of rocks considered as a continuum medium can be described by constitutive relations connecting the value and rate of stresses with the value and rate of strains. The constants appearing in these relations are the material constants of the medium characterizing the type of substance by its reaction to the forces applied, including variation with time. In the simplest case the relations are linear and the constitutive relations correspond to various types of bodies, such as perfectly elastic, thermoelastic, viscous, plastic, or visco-plastic in different variants. They can also take into account the conjugation of mechanical fields with electric and magnetic fields. The time variations of material constants, for example, under long-lasting loads, are determined by the rheological properties of the medium.

Studies of mechanical properties of rocks together with data related to the Earth's composition and with equations of state lead to theories of the structure of the Earth's interior.

Under the elastic limit solids undergo deformations according to the law of Hooke, i.e., the strain is proportional to the stress. At a certain stress load close to the strength limit the material undergoes permanent deformation as a result of plastic flow or breaking.

Rocks can be considered as perfectly elastic media only under relatively low pressures ($p \lesssim 10^8$ Pa) and at temperatures considerably lower than their melting temperature. Moreover, since rocks have a grain structure, a given sample can be considered as a homogeneous and isotropic one only when its size is several times as large as the characteristic size of grains.

For several problems the material of the Earth's interior can be described reasonably well by linear relations between the stresses and strains and their time derivatives, and various models of visco-elastic bodies can be constructed from two basic elements: an elastic and a viscous one.

Firstly, the part corresponding to volumetric changes and the deviatoric part should be separated from stress and strain tensors. Secondly, an assumption is accepted that for rocks of the Earth's interior a common relation describing volumetric changes is valid, namely

$$\sigma = K\varepsilon \tag{3.1.1}$$

where $\sigma = \sigma_{ii} = \mathrm{tr}\boldsymbol{\sigma}$, $\varepsilon = \varepsilon_{ii} = \mathrm{tr}\boldsymbol{\varepsilon}$, and K is the bulk modulus. Different relations, however, describe the deviatoric components

$$\begin{aligned} \hat{\sigma}_{is} &= \sigma_{is} - \tfrac{1}{3}\delta_{is}\sigma_{nn}, \quad & \hat{\boldsymbol{\sigma}} &= \boldsymbol{\sigma} - \tfrac{1}{3}\boldsymbol{I}\,\mathrm{tr}\boldsymbol{\sigma} \\ \hat{\varepsilon}_{is} &= \varepsilon_{is} - \tfrac{1}{3}\delta_{is}\varepsilon_{nn}, \quad & \hat{\boldsymbol{\varepsilon}} &= \boldsymbol{\varepsilon} - \tfrac{1}{3}\boldsymbol{I}\,\mathrm{tr}\boldsymbol{\varepsilon} \end{aligned} \tag{3.1.2}$$

Perfectly elastic bodies are described by the following relation (spring element):

$$\hat{\boldsymbol{\sigma}} = 2\mu\hat{\boldsymbol{\varepsilon}} \tag{3.1.3}$$

where μ is the rigidity modulus, and a perfectly viscous fluid (dash-point) is described by the relation

$$\hat{\boldsymbol{\sigma}} = 2\nu\dot{\hat{\boldsymbol{\varepsilon}}} \tag{3.1.4}$$

where ν is the viscosity coefficient.

From these two relations, complex linear models of visco-elastic bodies can be constructed. A visco-elastic Maxwell body is given by

$$\frac{1}{2\mu}\dot{\hat{\sigma}}+\frac{1}{2\nu}\hat{\sigma} = \dot{\hat{\varepsilon}} \tag{3.1.5}$$

while a Voigt-Kelvin body is described by the relation

$$\hat{\sigma} = 2\mu\hat{\varepsilon}+2\nu\dot{\hat{\varepsilon}} \tag{3.1.6}$$

A Maxwell body is formed by elements of the spring and dash-point type connected in series, while a Voigt–Kelvin body is formed by parallel connections of elements (analogies to connections of elements in electric circuits are used here). A more general relation for visco-elastic bodies is given by a system of four elements

$$p\dot{\hat{\sigma}}+q\hat{\sigma} = m\dot{\hat{\varepsilon}}+n\hat{\varepsilon} \tag{3.1.7}$$

where p, q, m and n are material constants. O'Connell and Budiansky (1978) proposed to include the history of deformation into the relation between the scalar measures of stress and strain

$$\varepsilon(t) = \int_{-\infty}^{t} J(t-\xi)\mathrm{d}\sigma(\xi), \qquad \sigma(t) = \int_{-\infty}^{t} M(t-\xi)\mathrm{d}\varepsilon(\xi) \tag{3.1.8}$$

In further considerations of deformation processes and wave propagation in the Earth's interior, a significant role is played by the values of the stored elastic energy w and energy dissipation Φ. For perfectly elastic incompressible bodies the stored energy is

$$w = \frac{1}{2}\hat{\sigma}\hat{\varepsilon} = \frac{1}{4\mu}\hat{\sigma}\hat{\sigma} \tag{3.1.9}$$

while $\Phi = 0$. Conversely, for viscous fluid $w = 0$, while

$$\Phi = \frac{1}{2\nu}\hat{\sigma}\hat{\sigma} \tag{3.1.10}$$

For a Maxwell visco-elastic body the energy and dissipation are

$$w = \frac{1}{4\mu}\hat{\sigma}\hat{\sigma}, \qquad \Phi = \frac{1}{2\nu}\hat{\sigma}\hat{\sigma} \tag{3.1.11}$$

and for a Voigt–Kelvin body they are

$$w = \mu\hat{\varepsilon}\hat{\varepsilon}, \qquad \Phi = 2\nu\dot{\hat{\varepsilon}}\dot{\hat{\varepsilon}} \tag{3.1.12}$$

For constitutive relations of the (3.1.7) and (3.1.8) type more complex expressions are obtained, but the general relation of energy conservation connects these quantities (Bland, 1960)

$$\dot{w}+\Phi = \hat{\sigma}\dot{\hat{\varepsilon}} \tag{3.1.13}$$

Of course, mechanical properties of rocks change under the influence of temperature. In models, rocks are usually considered as a homogeneous medium. More realistic theories, however, must take into account rock granularity and porosity. Domany *et al.* (1975), for example, proposed a theory of self-consistent approximation, and used it for the description of metals and rocks, which they considered as aggregates composed of chaotically orientated grains. Domany compared the theoretical results obtained with the earlier theory of Kröner and with the experimental results listed by Thomsen (1972).

Plastic effects in rocks are well known. The theory of creep in brittle rocks was developed by Cruden (1970), who verified the theoretical results by laboratory studies of sandstone and marble. Nonelastic volumetric deformations in dry rocks were considered by Holcomb (1977). The modulus K in eqn. (3.1.1) can depend here on an increase in the number of open cracks in the medium. Holcomb carried out laboratory experiments with rock samples and also analysed theoretically the problem of open crack formation. He took into account the fact that the rock material in seismic regions was, in its geological history, frequently submitted to variable stresses. Accordingly, he applied an initial load to fatigue the rock samples so as to bring their mechanical properties as close as possible to those of rocks in seismic regions. Under natural conditions, rocks in seismic regions are characterized by a certain network of potential microcracks, which open up under a uniaxial load and after the removal of the load partially close again. The cracks open up in the directions of minimum compressive stresses. By loading the rock samples (Westerly granite) up to 75–80% of their strength limit, Holcomb found that after a dozen or so loading–unloading cycles the stress–strain hysteresis curve remains unchanged. Thus a dilatation mechanism is also recoverable. The recovery of cycles can be explained by the fact that cracks are formed along the boundaries of the grains of which the rock is composed (Brace *et al.*, 1966). This model is similar to the model of crack formation when dislocation planes are present (Teisseyre, 1970).

A linear relation between the acoustic emission (seismoacoustic cracking) and dilatation, found from experiments, implies that the opening of the cracks is a step process, while their closing is a continuous process.

In general, solids can be classified either as brittle or as plastic, depending on the course of their deformation process above their elastic limit. Brittle substances suffer destruction without significant deformation, while plastic substances are considerably deformed during creep processes There are several criteria of failure, mainly phenomenological, which describe the destruction of various material subject to various stresses. The same crtiteria are used to describe the destruction of rocks and minerals under laboratory conditions. At the same time, there is quite a suitable theory of material failure based on the presence of material defects in the form of microcracks (see e.g., *Fracture*—an excellent monograph on the failure theory edited by Liebovitz, 1968), according to which in each material microcracks are always present and the fracture process is initiated when a certain critical value of the coefficient of stress intensity is exceeded at the edge of a crack. At high confining pressures rocks are deformed by yielding. According to the theory of Griffith, during the yielding process microdefects propagate slowly and the whole process can be stopped by a decrease in the acting stresses. Orowan (1960) suggested that a transition from brittle to plastic deformation takes place when the friction on the surface of the defect is equal to or greater than the shear strength of the rock. Other studies (Byerlee, 1968) seem to confirm the Orowan hypothesis.

The transition from brittle to plastic deformation is, for a given material, a function of pressure and temperature. Under high pressures brittle fracturings are usually replaced by plastic deformations.

At temperatures and pressures dominant in the uppermost parts of the crust, almost all rocks are brittle. At room temperature and under atmospheric pressure the compressive strength of granite is 140 MPa and its tensile strength is 4 MPa. The shear strength usually has an intermediate value. The strength of rocks increases with the increase of the confining pressure, and the point of plastic yielding is more distinct. The strength limit decreases with an increase in temperature. Laboratory studies of rocks and minerals under high pressures and temperatures suggest that the strength of rocks in the crust most probably increases with depth, to a depth of a few kilometres, but at depths greater than $5 \div 10$ km the influence of temperature becomes dominant and the strength decreases, and brittle fracturings are replaced by plastic deformations. In plastic deformations diffusion plays a substantial role; microfracturing phenomena, however, also occur at higher temperatures.

Two substantial changes in the mechanical properties of rocks most probably take place at depths of $10 \div 25$ km. Firstly, there is a transition from brittle fracturing to plastic deformation. Griggs and Handin (1960) demonstrated that above 500 MPa and 800 K brittle fracturing does not occur in any rock except quartzite. Such conditions are dominant at depths of approximately 20 km. Secondly, the compressive strength probably decreases with the depth because of the dominant effect of temperature. The compressive strength of dunite, pyroxene and granite, for example, is 2 GPa under a confining pressure of 500 MPa and a temperature of 300 K, and it is 1 GPa under a pressure of 700 MPa and a temperature of 1100 K.

Under sufficiently high confining pressures (e.g., about 68.5 MPa for marbles of Carrara) complete plasticity takes place, i.e., instead of brittle fracturing a very high increase in strain with a simultaneous small stress drop is observed. Under still higher confining pressures (about 165÷326 MPa for the same marbles) the process of brittle hardening is observed. The weakening effect of water present in the rock pores has a substantial influence on the rock strength. The pressure of water in pores reduces the consolidating influence of the confining pressure, and, for example, the Navier–Stokes criterion in this situation takes the form

$$\hat{\sigma} = \hat{\sigma}_0 + \varphi(p_n - p) \tag{3.1.14}$$

where $\hat{\sigma}$ is the shear stress, $\hat{\sigma}_0$ is a constant, φ is the coefficient of friction, p is the normal stress acting on the slip surface and p is the water pressure in the pores of a given material.

The consolidation of grains in polycrystals is also weakened by the action of water penetrating into the free spaces between the grains. This mechanism favours brittle failure. Thus, besides the reducing influence of water pressure in pores, the mechanical properties of a porous medium are also changed, facilitating internal slippage and microfracturing. The recrystallization of grains is also facilitated, as well as the hydrolysis of the Si–O–Si and Si–O–Mg bonds in silicates, which is revealed in creep processes (Griggs, 1967; Murrell, 1965; Jaeger and Cook, 1971).

The dehydration of serpentinites with increasing temperature causes a transition from plastic to brittle behaviour as a result of the filling of pores by water vapour. This phenomenon is contrary to the usually observed behaviour of other minerals previously described (Raleigh and Paterson, 1965). Considering that the strength of minerals under shear stresses is approximately equal to half of the confining pressure, that is about 10÷20 GPa for conditions corresponding to sources of deep earthquakes, it becomes clear that theories of tectonic earthquakes should take into consideration the phenomena diminishing mineral strength, such as for example the dehydration of serpentinites (Raleigh, 1967). It should also be noted that fracturing along a clearly marked slip surface is possible only under sufficiently high confining pressures (Griggs and Handin, 1960). But even there the influence of penetrating fluids on the decrease of the friction coefficient along slip surfaces should also be taken into account.

The plastic behaviour of the material between the point of creep and the strength limit is caused by the movement of dislocations. It was shown by Pratt (1967) that dislocations in ionic crystals are equally mobile at high and at low temperatures. In crystals with covalent bonds (SiO_2, TiO_2, Al_2O_3) the dislocations become mobile only at a temperature above $0.5\,T_m$ (T_m is the temperature of melting).

The temperature at which a material reaches a state called total plasticity is identical with that at which all possible crystallographic slip surfaces are unlocked. In monocrystals any plastic deformations preserving volume are possible only when at least five independent slip planes are unlocked. Most crystals, however, do not have five independent slip planes. In this sense, monocrystals usually show tendencies to certain specific plastic deformations, even at temperatures lower than the temperature of transition from brittle fracturing to total plasticity. Polycrystals display greater brittleness, and the missing slip systems are probably replaced by the formation of cracks (Pratt, 1967). Thus plasticity is observed only when all possible slip systems are active. The picture, however, is not complete since such factors as the deformation homogeneity within one slip system and the migration and slippage of grain boundaries are not taken into consideration. The primary reasons for the transition of material into a brittle state is the locking of dislocations on barriers, caused by an increasing strain, and such an association of dislocations that the slip plane between them becomes a crack (Stroh, 1957). The transition from dislocation systems to cracks is of particular importance. These processes are activated thermally as a result of the presence of certain barriers, and they should be sensitive to pressure. They are important in creep phenomena as well as in phenomena of internal stress concentrations and in mechanisms leading to earthquakes.

The influence of the hydrostatic component of stresses on plastic creep can be explained by a decrease of the ratio of the number of vacancies to the number of dislocations, while at the same time the dislocation mobility connected with nonconservative movements (dislocation climb) is limited; the conservative movements along dislocation slip planes remain. Moreover, hydrostatic pressure causes a certain redistribution of dislocations, revealed, for example, by a slight change of grain orientations in the crystal, found from measurements carried out on aluminium polycrystals. The observed change of orientation results from a change of position of dislocation planes and probably also from a change of dislocation density on these planes. Taking into account these phenomena, Ohnami (1972) proposed a general criterion of creeping in the form

$$f(\hat{\sigma}) = g\left(\sigma, \int \frac{\mathrm{d}}{\mathrm{d}t}\,\varepsilon[t]\mathrm{d}t\right) \tag{3.1.15}$$

where σ is the first invariant of the stress tensor, i.e., the hydrostatic component of stresses $\sigma = \sigma_{ii}$, and $\hat{\sigma}$ is the second invariant of the stress deviator $\hat{\sigma} = \hat{\sigma}_{is}\hat{\sigma}_{is}$, commonly used as a measure of shear stresses under isotropic conditions; $\int \frac{\mathrm{d}}{\mathrm{d}t}\,\varepsilon[t]\,\mathrm{d}t$ describes the strain history (hardening). In particular, Ohnami proposed the following form of the criterion:

$$f(\hat{\sigma}) = \hat{\sigma}, \quad g\left(\sigma, \int \mathrm{d}\varepsilon\right) = k(k^2 + Ck\sigma + D\sigma^2)^{1/2} \tag{3.1.16}$$

where the parameter k changes monotonically with the hardening. Laboratory experiments carried out on samples undergoing normal extension at a raised temperature and under confining pressures revealed a slight increase of creep stress with the increase of hydrostatic pressure for pure iron at a temperature of about 370 K and a lack of a clear effect for polycrystalline Al at a temperature of about 470 K (Ohnami, 1972). With the development of a deformation the absolute values of the negative constants C and D increase from zero (Ohnami, 1972).

The knowledge of rheological properties of rocks and minerals under high pressures and temperatures is of considerable importance for theories of the modelling of tectonic earthquakes (Stacey, 1963, 1967; Griggs and Handin, 1960) and of seismic wave attenuation (e.g., Anderson, 1967a, b; Scheidegger, 1970, 1971), and also for theories of convection based on the hypothesis of a low viscosity layer (Weertman, 1957). It should also be added that our knowledge relating to the temperature dependence of respective coefficients is relatively sound owing to the activating character of the processes determining the rheological properties of a material, while our knowledge of their dependence on pressure is much more limited.

3.2 CREEP PROCESSES IN THE MANTLE

Elastic and anelastic deformation, viscous flow and plastic yielding of solids may occur simultaneously. When stresses are removed, anelastic deformation decreases to zero with a given relaxation time, while plastic deformation remains constant. All kind of time dependent phenomena under a constant load are called *creep*. Experiments show that creep processes are thermally activated. Fluctuations of thermal energy help to overcome barriers to creep deformation

$$\dot{\varepsilon} = f(\hat{\sigma}, p)e^{-\Delta E/KT} \tag{3.2.1}$$

whre $\dot{\varepsilon}$ is the creep rate, $\hat{\sigma}$ are the deviatoric stresses, p and T are pressure and temperature, ΔE is the activation energy and k is the Boltzmann constant. A typical creep–time curve is shown in Fig. 3.2.1; immediately after a stress

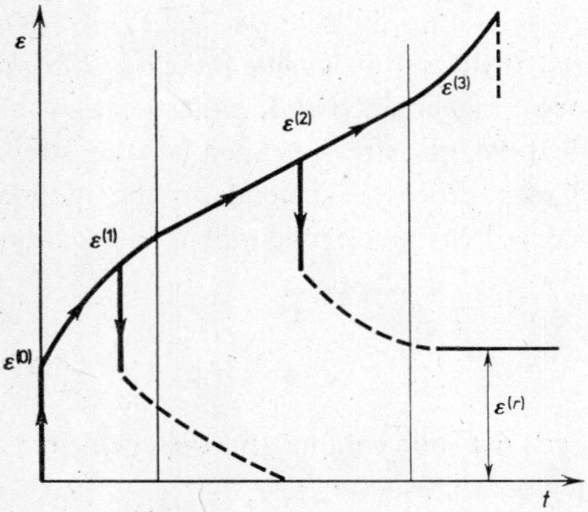

Fig. 3.2.1. Creep–time relation: $\varepsilon^{(0)}$—instantaneous elastic strain, $\varepsilon^{(1)}$—transient creep, $\varepsilon^{(2)}$—stationary creep, $\varepsilon^{(3)}$—tertiary creep, $\varepsilon^{(r)}$—remanent plastic deformation.

load is applied the elastic deformation $\varepsilon^{(0)}$ appears, later followed by a transient (primary) creep $\varepsilon^{(1)}$, stationary creep $\varepsilon^{(2)} = g(\hat{\sigma}, p, T)t$ and tertiary creep $\varepsilon^{(3)}$. Anelastic processes belong to transient creep. These processes are recoverable and will be discussed in detail in the last section of this chapter. In the case of tertiary creep a material can be brought to failure, and the corresponding time is called the creep rupture time. In general, it can be written for deformations

$$\varepsilon = \varepsilon^{(0)} + \varepsilon^{(1)}(t) + gt + \varepsilon^{(3)}(t) \tag{3.2.2}$$

Several phenomenological laws describe transient creep, the logarithmic and power laws being of particular importance.

Transient creep can be described as follows

$$\varepsilon^{(1)}(t) = \alpha(\sigma, T)\ln(1+\gamma t) + \beta(\sigma, T)t^n \tag{3.2.3}$$

where σ is the stress, and n and γ are constants. It was first noticed by Murrell (1967) that α creep, or logarithmic creep (the Lomnitz law), is dominant at temperatures $T \leqslant 0.2\ T_m$ (where T_m is the absolute temperature of melting). The second term (called the β creep) dominates at temperatures

$0.2T_m < T < 0.5T_m$ and under high stresses. Above $0.5T_m$ the creep rate under a fixed load becomes constant and independent of time, and the strain $\varepsilon(t) \propto t$. This amounts to stationary creep.

The rheological behaviour for transient creep can also be well described by the exponential law (Garofalo, 1966):

$$\varepsilon^{(1)}(t) \propto 1 - e^{-t/\tau} \tag{3.2.4}$$

where τ is the relaxation time.

Experiments show that the total dislocation density increases during the transient creep processes. Large movements and multiplication of dislocations occur above the critical Peierls stress defined as the stress required to move a dislocation. Creep processes depend on the mobile dislocation density. Li (1963) considered changes of the mobile dislocation density assuming the relation

$$\frac{dN}{dt} = k_1 N - k_2 N^2 \tag{3.2.5}$$

where N is the total dislocation length per unit volume and the coefficients k_1 and k_2 depend on stresses and temperatures.

Weertman J. and Weertman J. R. (1975) assumed that the creep rate depends on the mobile dislocation density N

$$\dot{\varepsilon} = Nbv \tag{3.2.6}$$

where b is a Burgers vector of dislocation and the density $N = (\sigma/\mu b)^2$ is computed for a steady case under the assumption that dislocation stresses are comparable to the stresses applied. This leads to a unified relation for primary $\varepsilon^{(1)}$ and stationary creep $\varepsilon^{(2)}$ (Akulov, 1964):

$$\dot{\varepsilon} = \dot{\varepsilon}^{(2)} \left[1 - \frac{\dot{\varepsilon}^{(1)} - \dot{\varepsilon}^{(2)}}{\dot{\varepsilon}^{(1)}} e^{-k_1 t} \right]^{-1} \tag{3.2.7}$$

which can be reduced to logarithmic, exponential or parabolic creeps, and for $t \to \infty$ this also leads to steady state creep. At low temperatures the dislocation movements related to creep occur along slip planes, while at higher temperatures a diffusional mass transport requires movements perpendicular to slip planes (dislocation climb).

Going back to the types of transient creep, it should be added that under atmospheric pressure the melting temperature T_m for various rock types is contained within a range of $1600 \text{ K} < T_m < 1800 \text{ K}$ (Murrell, 1967), and the values of the parameter n in formula (3.2.3) are $0.06 < n < 0.36$, reaching a value close to 1/3 (the Andrade law for metals) at temperatures T close to $0.5\,T_m$, while $\alpha(\sigma, T) \propto \sigma T$. The term β strongly depends on the stress σ and temperature T.

The logarithmic creep α and the power creep β are properly described by the theory of Mott (1953). According to this theory, α creep is caused by thermal activation of dislocations overcoming such barriers as other dislocations, grain boundaries, and others. The motion of dislocations takes place in separate segments. The height of energy barriers increases with the increase of deformations. At the moment of activation of a diffusion process the locked dislocations are allowed to abandon their slip planes by climbing or by transverse slippage, and this results in creep of the β type. A continuous decrease of creep rate is probably the result of material hardening (Orowan, 1967). This process is related to local stress concentration caused by the piling up of dislocations into arrays. Neither α nor β creep produces large strains. At temperatures sufficiently high for the number of dislocations to increase fast enough to stop the hardening process, i.e., above approximately $0.5\,T_m$, hot stationary creep takes place. This type of creeping produces, after exceeding a certain strength limit, non-Newtonian viscous flow.

The attenuation of tides and seismic waves can be explained reasonably well by the transient creep β (Jeffreys, 1967), while isostasy processes could be explained by a gradual transition to hot creeping (Murrell, 1967). Stationary creep will now be discussed, as it seems to be a substantial element of upper mantle dynamics.

The concept of creep is used here in a rather broad sense: it covers many different mechanisms of slow deformations of solid media. The creep phenomenon is closely related to viscous flow of solids and some of its mechanisms also resemble plastic yielding. The creep mechanism in some cases is referred to microfracturing and in other cases to diffusion phenomena in solids. Several aspects and mechanisms of creep will briefly be discussed, with special reference to possible creep processes in the mantle. The rheological behaviour of olivine (Fo_{85}–Fo_{95}) and its material constants is thus of special interest, since olivine is a dominant phase of the upper mantle. The symbol Fo_x denotes olivine containing x per cent of foresterite Mg_2SiO_4 and $1-x$ per cent of fayalite Fe_2SiO_4. Although our consideration of creep is limited only to the Earth's mantle, it must be remembered that in the Earth's crust and at its surface creep phenomena are also of extreme importance, for example in orogenic processes, in the formation of salt domes, in slow crustal movements and in glacier flows. The practical aspects can be found in the study of pillar behaviour in mines. The creep rate in the mantle convection is of an order of 10^{-16} s^{-1} to 3×10^{-14} s^{-1}

and for the crust uplift it is about 10^{-15} s^{-1}. Such values cannot be obtained under laboratory conditions.

The process of hot creeping (the stress power-law) could be of special importance for the Earth's mantle, since the temperature in most areas of the mantle probably exceeds $0.5\,T_m$, especially in the upper mantle, where even melting points could locally be reached. The increase of temperature above $0.5\,T_m$ probably causes a sharp increase in the creep rate, but a certain specified creep strength probably remains, even just below the melting temperature (Orowan, 1965).

An increase in confining pressure probably reduces the creep rate. Misra and Murrell (1965) estimated that the creep rate of peridotite at a temperature of 1470 K will be diminished by a factor of 100 after the application of a confining pressure of 1.2×10^9 Pa. Extrapolating their observations to the creep in peridotite for other mantle rocks, Misra and Murrell suggested that the creep rate at a temperature of 1470 K and under a confining pressure of 1.2×10^9 Pa could be still sufficient to justify convection creep in the upper mantle. It should be expected, therefore, that hot creeping takes place in those areas of the mantle where shear stresses exceed the creep strength. This phenomenon could be expected rather in the upper mantle and not deeper, since it is in the upper mantle that the temperature is close to that of melting and the confining pressures are not yet too high.

The diffusion mechanism of creeping is of great importance for the mantle. Diffusion creep in metals is observed at temperatures higher than about $0.85\,T_m$. Theory suggests that it can also occur at lower temperatures, but is either masked by other mechanisms of hot creeping or its rate is too small to be observed in laboratory experiments. It is generally believed that diffusion creeping takes place in all crystalline materials and the only question is whether its rate is high enough for the process to be of substantial geophysical importance. It should be noted that the deformation rate in many natural processes in the Earth's interior is smaller than that in laboratory experiments on metals by a factor as high as one million or even more. The strain rate in diffusion creep depends linearly on the shear stress, while in hot creep it depends by a power function on the shear stresses. Thus a decrease in strain rate sharply increases the relative value of diffusion creeping. For this reason diffusion creep could be a dominant mechanism in the stationary creep of mantle material under the influence of small stresses with long duration. Stationary creep can be called a pseudo-viscous flow of solids. The creep depends here on the temperature, pressure and deviatoric stresses. According to the theory of Eyring and according to the studies of polycrystalline metals and NaCl (Rogan and Alexander, 1966; Pratt, 1967), the stationary types of creep can be described as follows:

$$\dot{\varepsilon} = B \exp\left(-\frac{\Delta E}{kT}\right) \sinh \frac{a\sigma}{kT}, \qquad a = \text{const} \tag{3.2.8}$$

The activation energy ΔE usually depends on pressure. This relation $\dot{\varepsilon} = \varepsilon(\sigma)$ is known as the law of Nadai (1938).

In general, however, the constitutive law for creep can be expressed by the following relation between the invariant of the shear strain rate $\dot{\varepsilon}$ and the stress load $\hat{\sigma}$:

$$\dot{\varepsilon} = f(\hat{\sigma}, T, M_1(p), M_2(p)) \tag{3.2.9}$$

where $\hat{\sigma}^2 = \hat{\boldsymbol{\sigma}}\hat{\boldsymbol{\sigma}}$ ($\hat{\boldsymbol{\sigma}}$ are the stress deviators), $\dot{\varepsilon}^2 = \dot{\boldsymbol{\varepsilon}}\dot{\boldsymbol{\varepsilon}}$ ($\dot{\varepsilon}$ are the strain rates), and M_1, M_2, ... are the material constants. The relation is influenced by confining pressure only through the material parameters. A generalization of relation (3.2.9) by Prandl and Reuss can be expressed in the form

$$\dot{\varepsilon}_{ik} = f(\hat{\sigma}, T, M_1(p), M_2(p)) \frac{\hat{\sigma}_{ik}}{\hat{\sigma}} \tag{3.2.10}$$

In further considerations the creep mechanism will be related to a certain activation volume (e.g., ionic impurity volume) and to the Burgers vector of dislocation, equal to the lattice parameter. These quantities depend on the confining pressure

$$\Omega = \Omega_0 \exp\left[\frac{-(p-p_0)}{K_0}\right], \qquad b = b_0 \exp\left[\frac{-(p-p_0)}{3K_0}\right] \tag{3.2.11}$$

where K_0 is the bulk modulus at 300 K and under atmospheric pressure. The bulk and rigidity moduli depend on the pressure and temperature

$$\begin{aligned} \mu &= \mu_0\left[1 - \frac{T_m}{\mu_0}\frac{\partial \mu}{\partial T}\left(\frac{T-300}{T_m}\right)\right] + \frac{\partial \mu}{\partial p}(p-p_0) \\ K &= K_0\left[1 - \frac{T_m}{K_0}\frac{\partial K}{\partial T}\left(\frac{T-300}{T_m}\right)\right] + \frac{\partial K}{\partial p}(p-p_0) \end{aligned} \tag{3.2.12}$$

where T_m is the melting temperature under atmospheric pressure.

An excellent presentation of creep processes in the Earth's mantle was given by Ashby and Verall (1977) in a paper published in a special issue of the *Philosophical Transactions* devoted to the geophysical aspects of creep processes. Several mechanisms of microfracturing and flow in solids, depending on the stresses, confining pressures, temperatures and material constants are responsible for creeps:

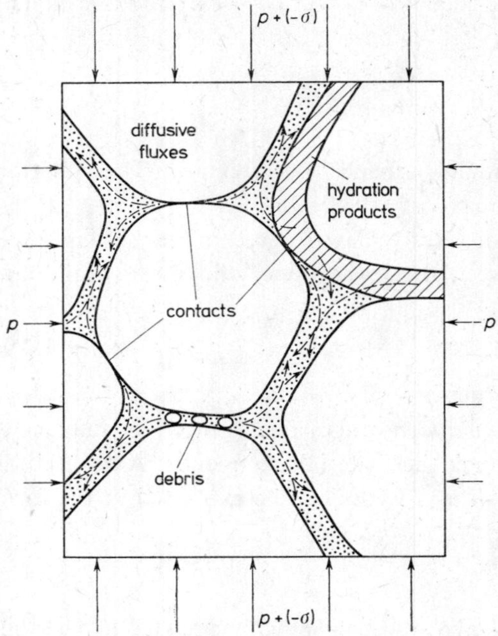

Fig. 3.2.2. Creep processes related to fluid phase transport; liquid film formed by water or oil in the crust and by basaltic melt in the upper mantle (after Ashby and Verall, 1977).

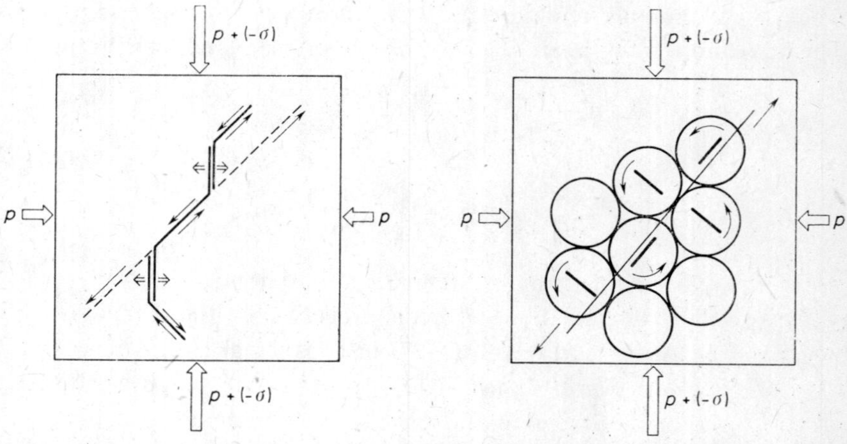

Fig. 3.2.3. Microfracturing with tensile cracking and rolling of fractured grains and sliding processes.

1. Cataclastic flow is related to repeated cleavage fracturing, it permits large shear deformations but, being sensitive to a confining pressure, can play an important role only at shallow depths.

2. Low temperature plasticity involves plastic deformations caused by dislocation sliding.

3. Diffusion flow mechanism is based on the diffusion of ions, vacancies or other impurities. It corresponds approximately to a viscous Newtonian flow.

4. Power–law creep is a nonlinear flow and diffusion related to the motion of dislocations in the case of a complex dense dislocation network.

5. Fluid phase transport can also contribute to creep in porous or partly melted media, while creep with dynamic recrystallization refers to phase transformation processes. Fig. 3.2.2 explains the mechanism of fluid phase transport, where fluid film may be formed, e.g., water in the crust and melted basalt in the mantle.

The main types of creep will now be discussed in more detail. Fig. 3.2.3 explains cataclastic flow which includes microfracturing with tensile cracking and rolling of fractured grains in sliding processes. The microfracturing process and the formation of open cracks is related to the tensile strength of a crystal. For the tensile strength of an ideal crystal we have

$$\sigma_{\text{ideal}} = \left[\frac{2\mu(3\lambda+2\mu)}{\pi(\lambda+\mu)}\frac{\gamma}{b}\right]^{1/2} \tag{3.2.13}$$

where γ is the surface energy of a crack, b is the length of the Burgers vector, and, approximately, $\sigma_{\text{ideal}} \approx 0.25\,\mu$. In reality, in each crystal a number of cracks exist and a criterion for the strength σ relates to the crack propagation. Griffith (1971) connects the strength with the stress concentration at the tip of a crack

$$\sigma_f = \left[\frac{2\mu(3\lambda+2\mu)}{\pi(\lambda+\mu)}\frac{\gamma}{l_0}\right]^{1/2} \tag{3.2.14}$$

where l_0 is the length of the crack. Repeated fracturing (cataclasis) occurs when the stresses exceed the strength limit

$$\hat{\sigma}-p \geqslant \sigma_f \tag{3.2.15}$$

where $\hat{\sigma} = \frac{1}{2}(\sigma_1-\sigma_2)$, $p = -\frac{1}{3}(\sigma_1+\sigma_2+\sigma_3)$ and $\hat{\sigma}-p$, $-p$, $-p-\hat{\sigma}$ are the principal stresses. The strength σ_f for olivine is close to $5\times 10^{-3}\mu$. The friction coefficient in the rolling and sliding mechanism enters into the respective relation (Ashby and Verall, 1977). The constitutive law for cataclasis is

$$\dot{\varepsilon} = \infty \tag{3.2.16}$$

which corresponds to the fracture character of the process.

For low temperature plasticity the constitutive relation takes the form

$$\dot{\varepsilon} = \dot{\varepsilon}_0 \exp\left[-\frac{\Delta E}{kT}\left(1-\frac{\sigma}{\sigma_0}\right)\right] \tag{3.2.17}$$

where σ_0 is the flow stress at a temperature of 0 K and k is the Boltzmann constant. Here the sliding of dislocations is tempered by the lattice resistance, the latter being controlled by the activation energy ΔE. This energy is required to overpass obstacles (e.g., impurities or other dislocations). According to Weertman (1977), the strain rate $\dot{\varepsilon}$ is connected with mobile dislocation density (3.2.6), where the motion velocity v of dislocations depends on the stresses applied.

Diffusion flow, known as the *Nabarro–Herring creep*, includes diffusion through the lattice and grain boundaries. The rate of shear creep is

$$\dot{\varepsilon} = A\left(\frac{D}{d^2}\right)\frac{\hat{\sigma}\Omega}{kT} \tag{3.2.18}$$

where D is the coefficient of effective diffusion ($D = D_0 \exp(-\Delta E/kT)$, where ΔE is the activation energy for diffusion), d is the diameter of grain, and Ω is the ionic volume.

At higher temperatures nonlinear creep contributes to deformation processes. The dislocations activated in a sliding process aggregate to form cells, while the mechanism of diffusion creep is related to dislocation climb. The size of a cell depends on the stresses (hence a nonlinearity results)

$$\frac{d'}{b} \approx \frac{\mu}{\hat{\sigma}} \tag{3.2.19}$$

Replacing d in formula (3.2.18) by d' from this expression, the following relation is obtained ($n = 3$):

$$\dot{\varepsilon} = B\left(\frac{D\mu\Omega}{b^2 kT}\right)\left(\frac{\hat{\sigma}}{\mu}\right)^n \tag{3.2.20}$$

The constants B and n are to be found from experimental data.

The influence of hydrostatic pressure on the creep of metals at room temperature, and also at higher temperatures, is considerable (Ohnami, 1972). The influence of pressure on the activation energy ΔE in mechanisms of diffusion creep can be written in the form $\Delta E = pV_c$, where V_c is the volume of activation for a given mechanism (for pure polycrystalline Al it is approximately 6.3 cm · mole^{-1}). Pressure also reduces the diffusion of vacancies in a significant degree, and this in turn causes decrease in the mobility of dislocations in climb and also in other nonconservative movements. Both these effects explain the observed form of the influence of pressure on creep

$$\dot{\varepsilon} \propto D(p) \exp\left(\frac{-pV_c}{kT}\right) \tag{3.2.21}$$

where $D(p)$ is the density of vacancies under the pressure p (Ohnami, 1972).

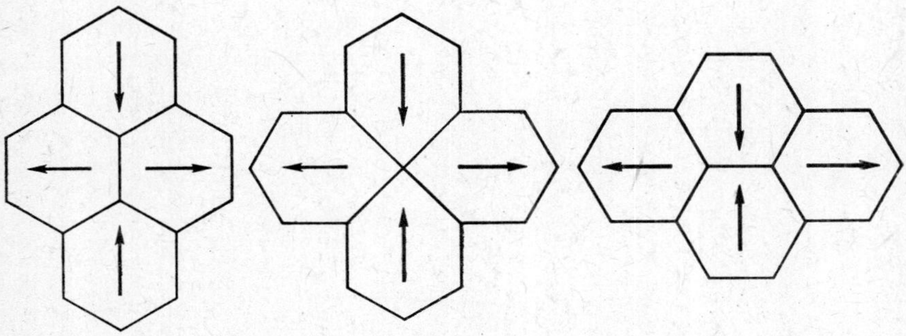

Fig. 3.2.4. Diffusion and changes of grain shapes.

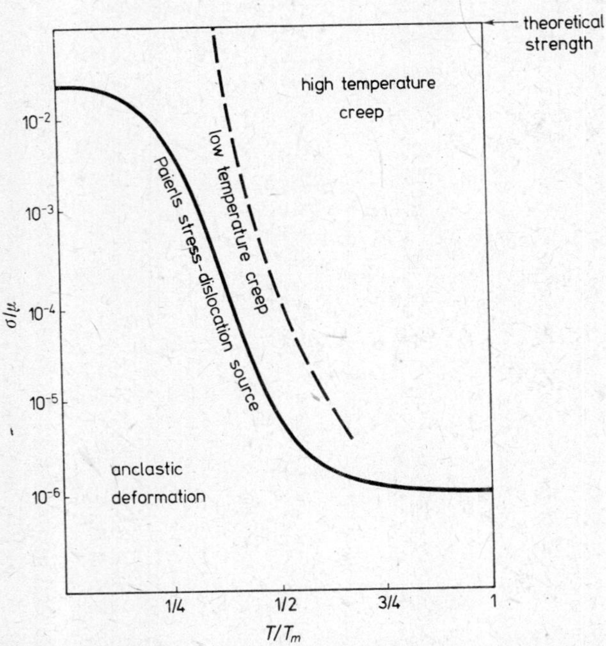

Fig. 3.2.5. Ranges of creep processes (simplified, after Weertman, 1977).

It has already been mentioned that diffusion creep is related to the mechanism of dislocation. Firstly, it relates to slide controlled creep (here the dislocation drags along an impurity atmosphere—the Cottrell cloud), while at higher temperatures it relates to the creep controlled by the dislocation climb (higher power exponents are here expected).

The constant n in formula (3.2.20) depends on the mechanism of dislocation climb. This law properly describes the creep of some alloys and metals, MgO, Al_2O_3, ice, and some rocks (Robertson, 1964). For stresses $\sigma > 10^{-3}\mu$ the power law cannot be used. Garofalo (1966) has shown, however, that a relation of the form of (3.2.8) can be applied here, covering a whole range of stresses. The diffusion mechanism becomes more effective at higher temperatures, as the change of grain shapes can facilitate creep deformation and diffusion (Fig. 3.2.4). A more complex mechanism includes the diffusion of grain boundaries (Cable, 1962).

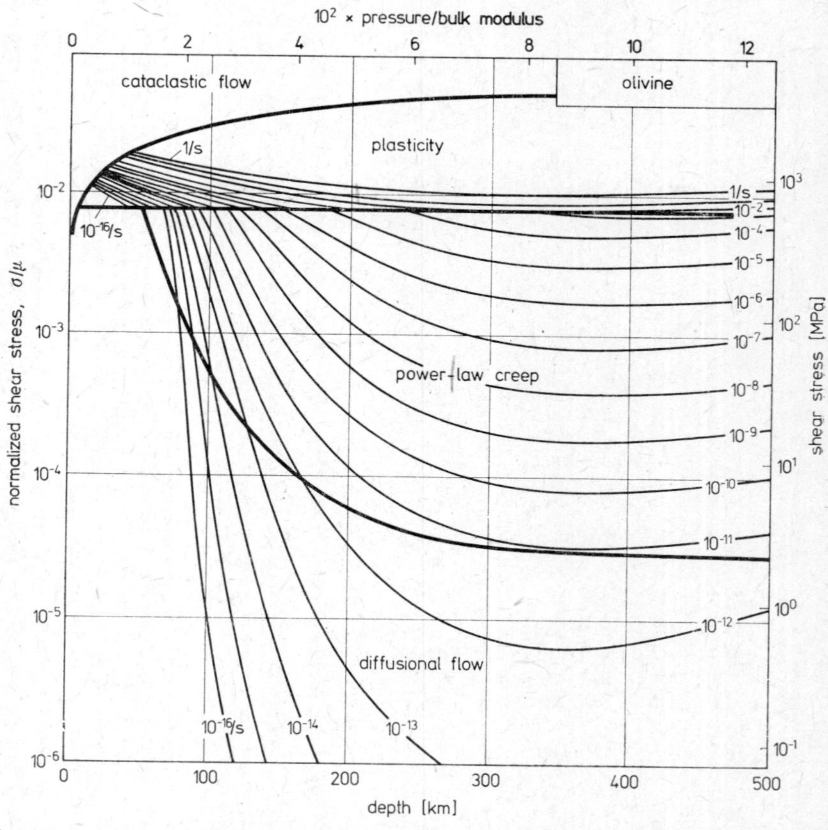

Fig. 3.2.6. Curves of constant creep rate (grain size 0.1 mm and the activation volume equal zero), after Ashby and Verall (1977).

Weertman (1977) presented the diagram of creep processes in relation to temperatures and stresses, shown in Fig. 3.2.5. Similar but more accurate diagrams were elaborated by Ashby and Verall (1977). They show the lines of the constant rate $\dot{\varepsilon}$ of shear strain as functions of the homologous temperature T/T_m and normalized stresses σ/μ, with the confining pressure and grain size as the curve parameters. For upper mantle conditions the relation between temperature and pressure (or depth) can be used. Thus the diagrams of constant creep related to pressure (or depth) and stress can be obtained. The diagram of creep with zero activation volume is shown in Fig. 3.2.6, while the diagram with the upper boundary of the activation volume equal to a single oxygen ion, which diffuses very slowly in most oxides including olivine, is presented in Fig. 3.2.7. These figures show the upper and lower boundaries of creep in the mantle. The grain

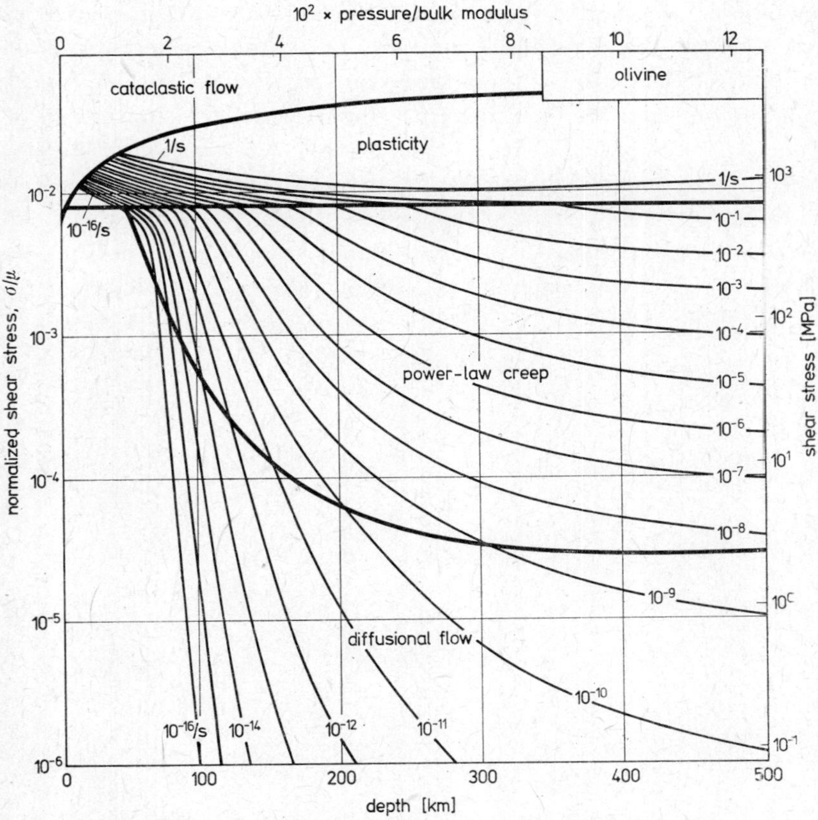

Fig. 3.2.7. Curves of constant creep rate (grain size 0.1 mm and the upper limit of activation volume), after Ashby and Verall (1977).

size was taken equal to 0.1 mm, and the lines of the creep rate are related to the stresses and depths. The contours of the constant creep rate in Fig. 3.2.6, at depths greater than 200 km, are almost flat, implying that the effective viscosity ν in the formula $\hat{\sigma} = 2\nu\dot{\varepsilon}$ at those depths could be regarded as constant.

It has already been mentioned that one more type of creep could be of significance in studies of the Earth's mantle. This is the creep connected with fluid transport in rocks. Unfortunately, a precise form of the stress–strain relation for this type of creep is not yet known.

Of course, the dominant type of creep depends on temperature, pressure, grain size and the presence of fluids in rock pores in a given area.

The knowledge of creep processes in rocks is necessary for the estimation of the energy dissipated in convection processes in the Earth's mantle. The discussion of creep processes in the mantle presented above displays the complex mechanism of deformations. Unfortunately, the estimates obtained may be highly approximate. Particularly in a porous material with some interstitial fluid the creep mechanism can be affected by fluid transport and pore pressure. This can change the depth range of fracturing processes of cataclasis, enlarging its range of application. J. Weertman and J. R. Weertman (1975) computed earlier the distribution of the effective viscosity in the Earth's interior by a method similar to that described above. They extended the calculations to different planets using the olivine and spinel models as regards the material composition. The result should be consider-

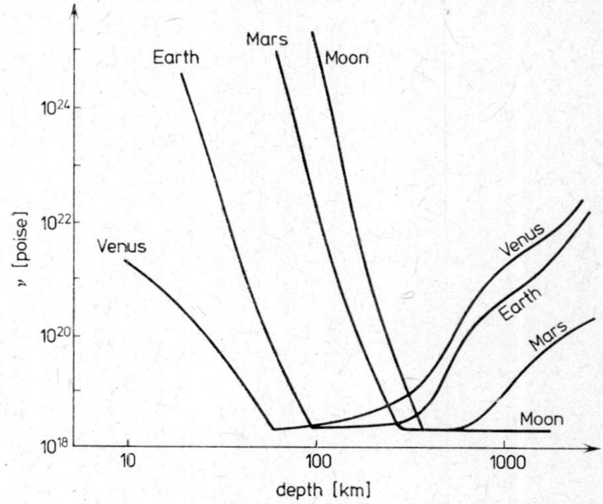

Fig. 3.2.8. Viscosity for constant creep rate ($\dot{\varepsilon} = 10^{-14}\ s^{-1}$), after Weertman and Weertman (1975). (Reproduced, with permission, from *Annual Review of Earth and Planetary Sciences*, Vol. 3, © 1975 by Annual Reviews Inc.).

ed only as tentative because of the many still doubtful parameters entering into viscosity determination, for example, the temperature distribution inside the planet. The effective viscosity computed for a constant creep rate in a depth range of 1000 km is shown in Fig. 3.2.8, and the viscosity computed for constant deviatoric stresses in the same depth range is presented in Fig. 3.2.9; the changes of the latter viscosity with a depth are larger by several orders of magnitude. The minimum in viscosity is observed in both cases at the same depth of 200 km.

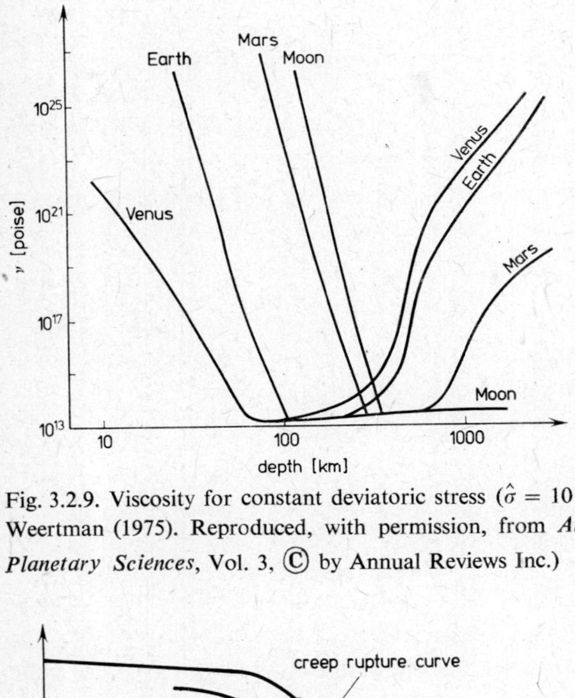

Fig. 3.2.9. Viscosity for constant deviatoric stress ($\hat{\sigma}$ = 10 bars). (After Weertman and Weertman (1975). Reproduced, with permission, from *Annual Review of Earth and Planetary Sciences*, Vol. 3, © by Annual Reviews Inc.)

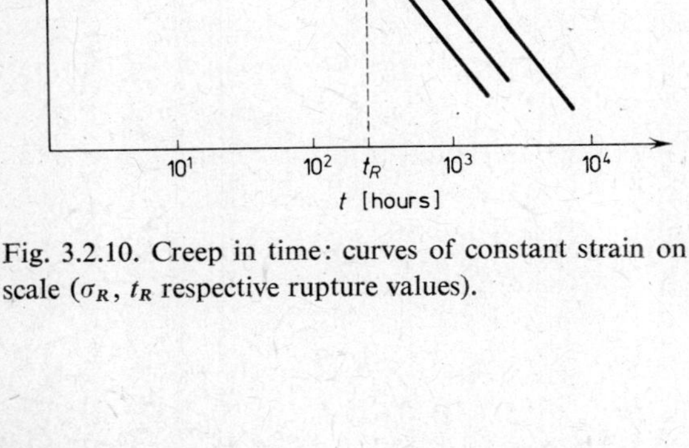

Fig. 3.2.10. Creep in time: curves of constant strain on stress–time diagram in log–log scale (σ_R, t_R respective rupture values).

Continuum creep can force the material into failure, preceded by tertiary creep (Fig. 3.2.10). This creep fracturing process is probably also very important in the Earth's crust and mantle, since stresses with very long duration are present in it. According to experimental data, processes of creep failure can be described by the curves of creep rupture in a double logarithmic system of coordinates $\log\sigma$ and $\log t$ (Leckie, 1977). The applied stress σ leads to a rupture after a time t_R after load application. The curve of creep rupture is composed of two asymptotes, which are almost straight lines—Fig. 3.2.10. Similarly, the curves of constant strains caused by creep processes can be drawn.

For steady state creep, when no failure is involved, we have

$$\log\left(\frac{\dot{\varepsilon}}{\dot{\varepsilon}_0}\right) = n\log\left(\frac{\sigma}{\sigma_0}\right) \tag{3.2.22}$$

To account for the history of deformation, e.g., in cyclic loading, the following rupture summation criterion is used:

$$\sum \frac{t_{Ri}}{t_R} = 1 \tag{3.2.23}$$

where t_{Ri} is the duration of loading and t_R is the time of rupture corresponding to a constant stress for a given value of the load applied.

3.3 VISCOSITY MODELS

It was shown when discussing creep phenomena that effective viscosity is a very important parameter controlling the internal dynamics and thermal convection in the Earth. It will be shown that this parameter relates also to the attenuation of elastic waves and free oscillations of our globe. The response to surface stress loading, like that relating to ice-caps for example, can also be explained by viscosity effects. Recently a mechanism of stress propagation based on the elasto-viscous response has been proposed (Elsasser, 1967; Melosh, 1976).

The apparent viscosity of the upper mantle is commonly estimated from the isostatic behaviour of the Earth's crust under an applied or removed surface loading, such as e.g., an ice-cap or a large lake. In such estimations it is usually assumed that the lithosphere is an elastic layer $70 \div 150$ km thick overlying the asthenosphere, which behaves like a Newtonian viscous fluid. If the linear extension of a loaded area exceeds about 700 km, it is usually assumed that the lithosphere is deformed passively by slow elastic bending and that the rate of approaching an isostatic equilibrium depends exclusively on the viscosity of the asthenosphere, considered here as part of the upper mantle. When the loading is known, the rate of this process during a known period of time after its application (or removal) permits to estimate the viscosity of the astenosphere. For this purpose the solution of the Navier–Stokes equation for a specified initial distribution of stresses at the surface is needed. The uplift of Fennoscandia after the melting of the pleistocene ice cover is used for the determination of the upper mantle viscosity, as well as the very similar uplift of the southern part of Hudson Bay in Northern America. Recently the uplift caused by the last drying up of Lake Bonneville in the west of Utah, USA, was also used for viscosity determination (Cathles III, 1975). Calculations have shown that a relatively thin layer of low viscosity could also be present in the mantle (Artyushkov, 1967; Anderson and O'Connell, 1967). But determinations of mantle viscosity are mainly based on observations of the uplift of postglacial areas. In spite of the fact that the glacial epoch ended more than ten thousand years ago, the uplift of areas loaded by the glacier is going on with a rate of more than ten centimetres per century. The best known area is Fennoscandia, but uplift is observed in all the postglacial areas of Canada, Siberia and the islands of the Arctic Archipelago.

The fact that the shape of the Earth differs from that of a rotating spherical liquid body in hydrostatic equilibrium can be partially explained by the melting of the polar ice-caps (Wang, 1966), and also by the slowing down of the rotation of the Earth, caused by tidal friction (McDonald, 1960) and interpreted by using a certain average viscosity of the mantle. The viscous properties of the crust and mantle can also be estimated from observations relating to the attenuation of seismic waves (Anderson, 1967a, b), though under the doubtful assumption that attenuation and creep processes are due to the same mechanism.

The energy of deformation is progressively transformed into heat as a result of deviations from elasticity; such deviations are usually collectively called *solid friction.* This effect is known as *anelasticity.* The phenomenon occurs in all solids and is observed at infinitely small strains. Anelasticity causes the attenuation of the Earth's free oscillations and influences forced oscillations such as terrestrial tides, producing deformation delay in relation to the stress. Anelastic properties of the Earth's interior are thus of particular importance in studies of the attenuation of seismic waves, terrestrial tides and Chandler's wobbles. These problems are considered in more detail in another volume of this series.

Cathles III (1975) developed a general approach to the determination of the response effects caused by surface loads of different orders, expanding them into a series of spherical harmonics. Thus the response depends on the order of harmonics and this corresponds to the load size. Different models of viscosity distribution can be compared with observations. Data related to crust movements, both recent and those found from geomorphological and geological evidence, are used to check the adopted models and the predicted response behaviour in loaded areas. In his calculations Cathles III (1975) adopted the matrix form for the elastic and viscous types of the equations of motion.

3.3.1 Elastic and viscous equations of motion

Elastic and viscous equations of motion differ in the way in which the displacements control the equilibrium. The continuity equation and the balance of momentum for the case of equilibrium can be written as follows:

$$\frac{\partial \varrho}{\partial t} + \nabla(\varrho \mathbf{v}) = 0 \tag{3.3.1}$$

$$\nabla \boldsymbol{T} + \mathbf{F} = \mathbf{0} \tag{3.3.2}$$

where ϱ is the density, $\mathbf{v}$ is the velocity field, $\boldsymbol{T}$ is the stress tensor, and $\mathbf{F}$ are the body forces.

The following prestress model is assumed

$$\boldsymbol{T} = -p_0 \boldsymbol{1} + \boldsymbol{\sigma} \tag{3.3.3}$$

where $\boldsymbol{\sigma}$ are the perturbation stresses. In the Lagrangian description an element in motion drags its pressure $p_0(X, t)$ along, and thus

$$\frac{dp_0}{dt} = 0$$

while in the Eulerian description $p_0(x(X, t), t)$ we have

$$\frac{\partial p_0}{\partial t} = -\mathbf{v}\nabla p_0$$

For an elastic body the pressure is related to the material points:

$$p_0(X) = p_0(x-u) \approx p_0(x) - \mathbf{u}\nabla p_0 \tag{3.3.4}$$

where $x = X+u$ represents the space coordinates, after deformation, related to the material description X before deformation, $\mathbf{u}$ is the vector of displacement. It is assumed that the state of equilibrium is controlled by elastic adjustments. This is called the *elastic equation of motion*; from (3.3.2), (3.3.3) and (3.3.4) it follows that

$$\nabla\boldsymbol{\sigma} - \nabla p_0 + \nabla(\mathbf{u}\nabla p_0) + \mathbf{F} = \mathbf{0} \tag{3.3.5}$$

For continuous displacements the equilibrium controlled by the viscous flow is

$$\nabla\boldsymbol{\sigma}-\nabla p_0+\mathbf{F} = \mathbf{0} \tag{3.3.6}$$

This is the viscous type of equation, which can also be applied to a visco-elastic medium.

For the self-gravitating model of the Earth, consisting of a number of spherical layers, eqns. (3.3.5) and (3.3.6) can be solved by the approximation method. The zero and first order equations can be obtained by assuming the following relations: $\varrho = \varrho_0+\varrho_1$ for density, $\boldsymbol{T}_0 = -p_0\boldsymbol{1}$, $\boldsymbol{T}_1 = \boldsymbol{\sigma}$ for stresses and $\Phi = \Phi_0+\Phi_1$ for gravity potential. Now the zero order equations are

$$\nabla p_0-\mathbf{F}_0 = \mathbf{0} \quad \text{or} \quad \nabla p_0+\varrho_0\nabla\Phi_0 = \mathbf{0} \tag{3.3.7}$$

$$-\nabla\Phi_0 = \mathbf{g}_0 = -g_0\hat{\mathbf{r}} \tag{3.3.8}$$

$$\nabla^2\Phi_0 = 4\pi G\varrho_0 \tag{3.3.9}$$

where $\hat{\mathbf{r}}$ is the unit vector normal to a sphere.

The first order equations (where $\dot{\mathbf{u}} = \mathbf{v}$ is taken to integrate the continuity equation) become

$$\varrho_1+\nabla(\varrho_0\mathbf{u}) = 0 \tag{3.3.10}$$

$$\nabla^2\Phi_1 = 4\pi G\varrho_1 \tag{3.3.11}$$

$$\nabla\boldsymbol{\sigma}+\nabla(\mathbf{u}\nabla p_0)-\varrho_0\nabla\Phi_1-\varrho_1\nabla\Phi_0 = \mathbf{0} \quad \text{(elastic type)} \tag{3.3.12}$$

$$\nabla\boldsymbol{\sigma}-\varrho_0\nabla\Phi_1-\varrho_1\nabla\Phi_0 = \mathbf{0} \quad \text{(viscous type)} \tag{3.3.13}$$

These equations must be supplemented by boundary conditions at each layer boundary. For the viscous type of material response a drop of the normal stress vector should compensate the weight of material displaced at the boundary (buoyant force), while in the elastic case this term is cancelled by the pressure dragged along by the material elements (the second term in eqn. (3.3.12)). Thus finally the boundary conditions are

$$\hat{\mathbf{r}}[\boldsymbol{\sigma}]_-^+ = 0 \quad \text{elastic response} \tag{3.3.14}$$

$$\hat{\mathbf{r}}[\boldsymbol{\sigma}+\mathbf{u}\varrho_0\nabla\Phi_0]_-^+ = 0 \quad \text{viscous response (buoyant force)} \tag{3.3.15}$$

For a layered medium, mixed boundary conditions can also be used. For a model of the Earth containing several thin layers the divergence of the buoyant force might be added to the viscous type of equilibrium equations:

$$\nabla\boldsymbol{\sigma}-\varrho_0\nabla\Phi_1-\varrho_1\nabla\Phi_0-(\hat{\mathbf{r}}\mathbf{u})\mathbf{g}_0\left(\frac{\partial\varrho_0}{\partial r}\right)_{\mathrm{NA}} = \mathbf{0} \tag{3.3.16}$$

The last term, however, would appear only for the nonadiabatic gradient (NA) of density (the difference between the existing gradient in the Earth's interior and the adiabatic gradient which arises solely from changes of pressure with depth). For the adiabatic change of density no buoyant force is present in the equilibrium equation.

For a stratified model of the Earth composed of thin layers matrix techniques can be used (Gilbert and Backus, 1966). These techniques will be described in detail in another volume of this series, and only some main points of the method are briefly mentioned here. The exact solution of the linear differential equation

$$\frac{\partial}{\partial z}\mathbf{u}(z) = A(z)\mathbf{u}(z) \tag{3.3.17}$$

is given by the formula

$$\mathbf{u}(z) = \mathbf{u}(z_0)\exp\left[\int_{z_0}^{z} A(\xi)\mathrm{d}\xi\right] \tag{3.3.18}$$

This can also be applied to matrix equations. For a thin stratified medium the solution is

$$\mathbf{u}(z) = P(z, z_0)\mathbf{u}(z_0), \qquad P(z, z_0) = \prod_i \mathrm{e}^{A(\bar{z}_i)d_i} \tag{3.3.19}$$

where $P(z, z_0)$ is called the *propagator matrix*, $\bar{z}_i$ is the mean value of z between z_{i-1} and z_i, and d_i are the thicknesses of the corresponding layers (Fig. 3.3.1).

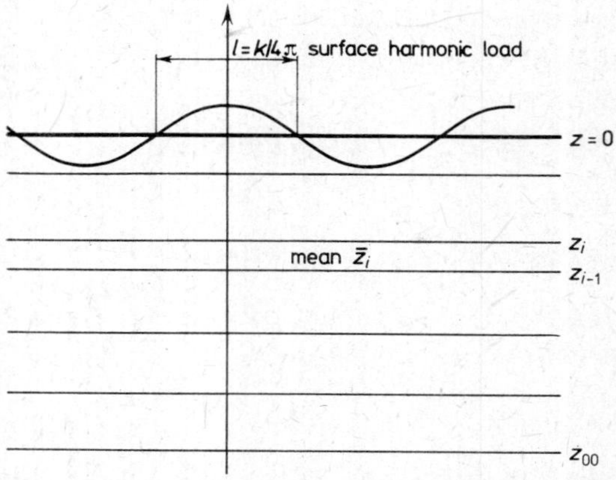

Fig. 3.3.1. Surface load, the case of a layered Earth's model.

The more complex type of differential equation

$$\frac{\partial}{\partial z}\mathbf{u}(z) = A(z)\mathbf{u}(z)+\mathbf{b}(z) \tag{3.3.20}$$

can formally be solved as follows:

$$\mathbf{u}(z) = \int_{z_0}^{z} P(z, \xi)\mathbf{b}(\xi)\mathrm{d}\xi + P(z, z_0)\mathbf{u}(z_0)$$

or for a layered structure, approximately,

$$\mathbf{u}(z) = \sum_i P(z, \bar{z}_i)\mathbf{b}(\bar{z}_i)d_i + P(z, z_0)\mathbf{u}(z_0) \tag{3.3.21}$$

The propagation matrix $P(z, z_0) = f(A)$ can be found by using the eigenvalue method. If λ_i are the roots of the determinant $A-\lambda I| = 0$, then the representation is

$$f(A) = \sum_n \prod_{\substack{i \\ i \neq n}} \frac{(A-\lambda I)}{(\lambda-\lambda_i)} f(\lambda_n) \tag{3.3.22}$$

where I is the identity matrix. This representation is valid only for unique roots. In the case of multiple roots a more complex formula must be used. For example, in the case of two roots with multiplicity two we have

$$f(A) = \sum_{\substack{i,j=1,2 \\ (i \neq j)}} (A-\lambda_j I)^2 \left[\frac{f(\lambda_i)}{(\lambda_i-\lambda_j)^2} + \left(\frac{-2f(\lambda_i)}{(\lambda_i-\lambda_j)^3} + \frac{f'(\lambda_i)}{(\lambda_i-\lambda_j)^2} \right) \right] \tag{3.3.23}$$

3.3.2 *Flat model of the Earth* (*constant gravity*)

The flat layered model of the Earth is the simplest model, but deserves special consideration. The material properties depend on depth. Let us consider the case of propagation in direction x. To all quantities $\sigma(x, z)$, $\mathbf{u}(x, z)$ a Fourier transform of the following type can be applied:

$$\bar{u}(k_x, z) = \int_{-\infty}^{\infty} u(x, z)\mathrm{e}^{\mathrm{i}k_x x}\mathrm{d}x$$

$$\frac{\partial}{\partial x}\bar{\mathbf{u}} = \mathrm{i}k_x\bar{\mathbf{u}} \tag{3.3.24}$$

The constitutive equations

$$\sigma_{is} = \lambda u_{n,n}\delta_{is} + \mu(u_{i,s}+u_{s,i})$$

for an elastic medium can now be expressed by the following set ($\Lambda = \lambda + 2\mu$):

$$\frac{\partial}{\partial z}\bar{u}_x = -ik_x\bar{u}_z + \mu^{-1}\bar{\sigma}_{xz}$$

$$\frac{\partial}{\partial z}\bar{u}_y = \mu^{-1}\bar{\sigma}_{yz} \tag{3.3.25}$$

$$\frac{\partial}{\partial z}\bar{u}_z = -\Lambda^{-1}\lambda ik_x\bar{u}_x + \Lambda^{-1}\bar{\sigma}_{zz}$$

and equation of motion (3.3.12) (with $\nabla\Phi_0 = g_0\hat{\mathbf{r}}$, $\nabla\Phi_1 = 0$) can be transformed into the form

$$\frac{\partial}{\partial z}\bar{\sigma}_{xz} = -ik_x\bar{\sigma}_{xx} + \varrho_0 g_0 ik_x\bar{u}_z$$

$$\frac{\partial}{\partial z}\bar{\sigma}_{yz} = -ik_x\bar{\sigma}_{xy} \tag{3.3.26}$$

$$\frac{\partial}{\partial z}\bar{\sigma}_{zz} = -ik_x\bar{\sigma}_{xz} - \varrho_0 g_0 ik_x\bar{u}_x$$

Equations (3.3.25) and (3.3.26) can now be written in matrix form (eliminating $\bar{\sigma}_{xx}$ and $\bar{\sigma}_{xy}$ from the other constitutive relations: $\bar{\sigma}_{xx} = \Lambda ik_x\bar{u}_x + \lambda(\partial u_z/\partial z)$, $\bar{\sigma}_{xy} = \mu ik_x\bar{u}_y$), which can be presented separately by the 4×4 matrix related to the propagation plane and by the 2×2 matrix on the perpendicular plane (the so-called divergent and non-divergent equations):

$$\frac{\partial}{\partial z}\begin{bmatrix}\bar{u}_x\\ \bar{u}_z\\ \bar{\sigma}_{xz}\\ \bar{\sigma}_{zz}\end{bmatrix} = \begin{bmatrix} 0 & -ik_x & \mu^{-1} & 0\\ -\Lambda^{-1}k_x\lambda & 0 & 0 & \Lambda^{-1}\\ 4\mu\Lambda^{-1}(\lambda+\mu)k_x^2 & \varrho_0 g_0 ik_x & 0 & -\lambda\Lambda^{-1}ik_x\\ -\varrho_0 g_0 ik_x & 0 & -ik_x & 0\end{bmatrix}\begin{bmatrix}\bar{u}_x\\ \bar{u}_z\\ \bar{\sigma}_{xz}\\ \bar{\sigma}_{zz}\end{bmatrix} \tag{3.3.27}$$

$$\frac{\partial}{\partial z}\begin{bmatrix}\bar{u}_y\\ \bar{\sigma}_{yz}\end{bmatrix} = \begin{bmatrix}0 & \mu^{-1}\\ k_x^2\mu & 0\end{bmatrix}\begin{bmatrix}\bar{u}_y\\ \bar{\sigma}_{yz}\end{bmatrix} \tag{3.3.28}$$

Equations (3.3.28) correspond to transversal propagation (SH) and are not important when we consider the response to surface harmonic loads only in direction x. For the viscous response the matrix form of linear differential equations is obtained, in a similar manner, using the stress–strain relation for an incompressible fluid

$$\sigma_{is} = 2\nu\hat{\dot{\varepsilon}}_{is}, \quad \dot{u}_{i,i} = v_{i,i} = 0 \tag{3.3.29}$$

where ν is the coefficient of viscosity, and v_x, v_y, v_z, are the velocities

$$\frac{\partial}{\partial z}\begin{bmatrix}\bar{v}_x\\ \bar{v}_z\\ \bar{\sigma}_{xz}\\ \bar{\sigma}_{zz}\end{bmatrix} = \begin{bmatrix}0 & -ik_x & \nu^{-1} & 0\\ -ik_x & 0 & 0 & 0\\ 4\nu k_x^2 & 0 & 0 & -ik_x\\ 0 & 0 & -ik_x & 0\end{bmatrix}\begin{bmatrix}\bar{v}_x\\ \bar{v}_z\\ \bar{\sigma}_{xz}\\ \bar{\sigma}_{zz}\end{bmatrix} + \bar{u}_z\begin{bmatrix}0\\ 0\\ 0\\ -g_0\left(\frac{\partial}{\partial z}\varrho_0\right)_{\mathrm{NA}}\end{bmatrix} \tag{3.3.30}$$

where $\bar{u}_z$ is the total viscous displacement at the moment t

$$\bar{u}_z = \int_0^t \bar{v}_z \mathrm{d}t \tag{3.3.31}$$

The elastic incompressible case without gravity can be reduced to the following form

$$\frac{\partial}{\partial z}\begin{bmatrix}2\mu^* ik_x\bar{u}_x\\ 2\mu^* k_x\bar{u}_z\\ i\bar{\sigma}_{xz}\\ \bar{\sigma}_{zz}\end{bmatrix} = k_x^4\begin{bmatrix}0 & 1 & 2\bar{\mu}^{-1}\\ -1 & 0 & 0\\ 2\bar{\mu} & 0 & 0\\ 0 & 0 & -1\end{bmatrix}\begin{bmatrix}0 & 2\mu^* ik_x\bar{u}_x\\ 0 & 2\mu^* k_x\bar{u}_z\\ 1 & i\bar{\sigma}_{xz}\\ 0 & \bar{\sigma}_{zz}\end{bmatrix} \tag{3.3.32}$$

where $\bar{\mu} = \mu/\mu^*$ and μ^* is the rigidity of the bottom layer. For a half-space problem $\mu^* = \mu$, $\bar{\mu} = 1$; the eigenvalues of the matrix A are given by the double roots k_x, $-k_x$.

Using formula (3.3.23), the propagation matrix is obtained

$$\mathrm{e}^{Az} = G^+ - G^-$$

$$G^+ = \tfrac{1}{2}\mathrm{e}^{k_x z}\left\{\begin{bmatrix}1 & 0 & 1 & 0\\ 0 & 1 & 0 & 1\\ 1 & 0 & 1 & 0\\ 0 & 1 & 0 & 1\end{bmatrix} + k_x z\begin{bmatrix}1 & 1 & 1 & 1\\ -1 & -1 & -1 & -1\\ 1 & 1 & 1 & 1\\ -1 & -1 & -1 & -1\end{bmatrix}\right\} \tag{3.3.33}$$

$$G^- = -\tfrac{1}{2}\mathrm{e}^{-k_x z}\left\{\begin{bmatrix}1 & 0 & -1 & 0\\ 0 & 1 & 0 & -1\\ -1 & 0 & 1 & 0\\ 0 & -1 & 0 & 1\end{bmatrix} + k_x z\begin{bmatrix}-1 & 1 & 1 & -1\\ -1 & 1 & 1 & -1\\ 1 & -1 & -1 & 1\\ 1 & -1 & -1 & 1\end{bmatrix}\right\}$$

Because of the exponential factor, $G^- u(z_{00})$ must vanish (here $z \leqslant 0$, see Fig. 3.3.1). Thus by calculating the product of corresponding matrices acting on $u(z_{00})$ the following condition is obtained:

$$G^{-}u(z_{00}) = G^{-}\begin{bmatrix} 2\mu i k_x \bar{u}_x(z_{00}) \\ 2\mu k_x \bar{u}(z_{00}) \\ i\bar{\sigma}_{xz}(z_{00}) \\ \bar{\sigma}_{zz}(z_{00}) \end{bmatrix} = 0 \tag{3.3.34}$$

which leads to the relations

$$2\mu k_x \bar{u}_x = \bar{\sigma}_{xz}, \quad 2\mu k_x \bar{u}_z = \bar{\sigma}_{zz} \tag{3.3.35}$$

which are also valid for boundary conditions.

For the vertical harmonic load at the surface $z = 0$ we put

$$\bar{\sigma}_{zz} = -\varrho g h, \quad \bar{\sigma}_{xz} = 0 \tag{3.3.36}$$

and thus we have

$$\bar{u}_z = -\frac{\varrho g h}{2\mu k_x}, \quad \bar{u}_x = 0 \tag{3.3.37}$$

Similar boundary conditions can be found in the case of an incompressible viscous half-space. For the vertical component of velocity at the surface $z = 0$ we will get

$$\bar{v}_z = -\frac{\varrho g h}{2\mu k_x} \tag{3.3.38}$$

where $h(t)$ expresses the quasi-static change of load:

$$\bar{v}_z = \frac{\mathrm{d}h}{\mathrm{d}t} \tag{3.3.39}$$

By comparing these relations we find the solution for the surface deflection $h(t)$ forced by the applied or removed load at the moment $t = 0$:

$$h = h_0 \mathrm{e}^{-t/\tau} \quad \text{or} \quad h = h_0(1 - \mathrm{e}^{-t/\tau}) \tag{3.3.40}$$

where

$$\tau = \frac{3\nu k_x}{\varrho g} \tag{3.3.41}$$

is the decay or the adjustment time.

3.3.3 Self-gravitating spherical models

Self-gravitating spherical models of the layered Earth can be processed like those in the previous section, but instead of a Fourier transform we should use an expansion into spherical harmonics. Thus instead of dealing with a response caused by the $k = 2\pi/\lambda$ harmonic load we are to deal with the order of harmonics n; for a given spherical surface of radius r this evidently corresponds to the wavelength of the load $n \propto kr$.

For the application of a spherical harmonic analysis the quantities appearing in matrix equations must be expressed by scalars. For this purpose we can use the representation theorem for vectors and tensors on a sphere in the general form presented by Backus (1967). For a vector field we have

$$\mathbf{u} = \hat{\mathbf{r}}U + \nabla_1 V - (\hat{\mathbf{r}} \times \nabla_1)W \tag{3.3.42}$$

where $\hat{\mathbf{r}}$ is the unit vector normal to a sphere, ∇_1 is the surface gradient, and U, V and W are the scalar fields (radial, spheroidal, and torroidal).

Similarly, for the normal vector of stress we have

$$\hat{\mathbf{r}}\boldsymbol{\sigma} = \hat{\mathbf{r}}P + \nabla_1 Q - (\hat{\mathbf{r}} \times \nabla_1)R \tag{3.3.43}$$

The scalar fields can now be expanded into surface spherical harmonics. Then the corresponding equations of motion can be presented in a matrix form, and the matrix equations split into two systems. Only the matrices corresponding to divergent equations (3.3.27), relevant for further discussion of the surface load effects, will be considered. The respective coefficients of spherical expansion of the order n: U_n, V_n, P_n, Q_n appear here, while W_n and R_n appear in the nondivergent matrix equations.

The system of divergent equations will be further supplemented by two first order equations taking into account gravitational effects (3.3.11), and thus two additional scalars, Φ_1 and $g_1 = (1/r)(\partial \Phi_1/\partial r)$, will appear. Thus the full divergent system is obtained in the form

$$\frac{\partial}{\partial r}\begin{bmatrix} U_n \\ V_n \\ P_n \\ Q_n \\ (\Phi)_n \\ (g_1)_n \end{bmatrix} = A \begin{bmatrix} U_n \\ V_n \\ P_n \\ Q_n \\ (\Phi)_n \\ (g_1)_n \end{bmatrix} \tag{3.3.44}$$

where $(\Phi)_n$, $(g_1)_n$ represent the coefficients in the respective expansions into surface spherical harmonics. Matrices A (3.3.17) for the elastic type of equations and also for the viscous type can easily be found in work by Cathles III (1975).

The models of mantle viscosity considered by Cathles III (1975) relate to the response effects to loads of different orders, where the order of a load corresponds to size of loaded area. The models can be verified by observations of the crustal vertical movements in various regions (loaded for example by ice-caps of given dimensions). The response is expressed by a decay constant τ, as in the case of a flat Earth model (3.3.41), but of course it now depends on the order n of a spherical load. To calculate the response to harmonic loads Cathles III (1975) used visco-elastic equations of motion which are formed by the sums of elastic and viscous equations and in which the variables are also formed by suitable sums. The models are especially sensitive to the presence of layers with nonadiabatic gradients—see eqn. (3.3.16). The review of the parameters of viscosity and nonadiabatic density gradients used by Cathles III (1975) in his various models is given in Fig. 3.3.2.

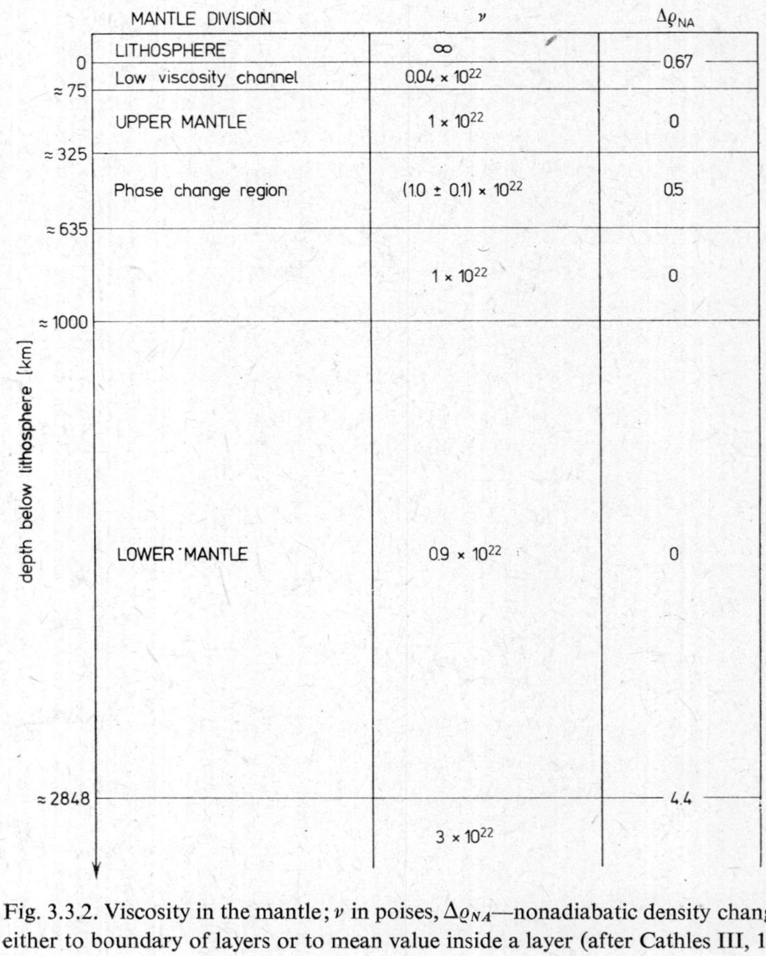

Fig. 3.3.2. Viscosity in the mantle; ν in poises, $\Delta\varrho_{NA}$—nonadiabatic density change related either to boundary of layers or to mean value inside a layer (after Cathles III, 1975).

The decay spectra for the surface vertical displacement can now be computed for different models. The decay spectra for a constant viscosity model with a nonadiabatic gradient in a range of depths between 335 and 635 km is shown in Fig. 3.3.3 (after Cathles III, 1975), where the order

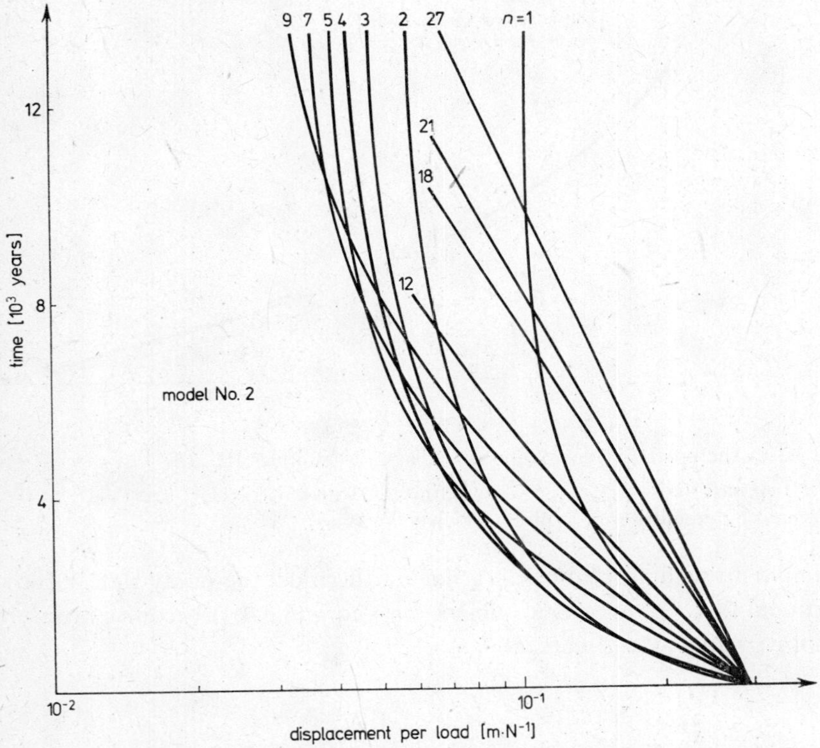

Fig. 3.3.3. Viscous decay spectra—decay time versus displacements per load. (From L. M. Cathles III, *The Viscosity of the Earth's Mantle* (copyright © 1975 by Princeton University Press), Fig. III-20, p. 108. Reprinted by permission of Princeton University Press).

n of harmonics is the parameter and the starting point for the curves is the value of the elastic response. The comparison of this model and the model of uniform viscosity and zero nonadiabatic gradient with the observed uplift of Fennoscandia, found from geological evidence, is shown in Fig. 3.3.4.

More realistic models should account for the elastic crust and for the possible existence of a low viscosity channel. It is believed that such a channel should correspond to the low-velocity channel beneath the lithosphere. The elastic crust and low viscosity channel would substantially influence the decay spectra of the response caused by surface harmonic loads. For

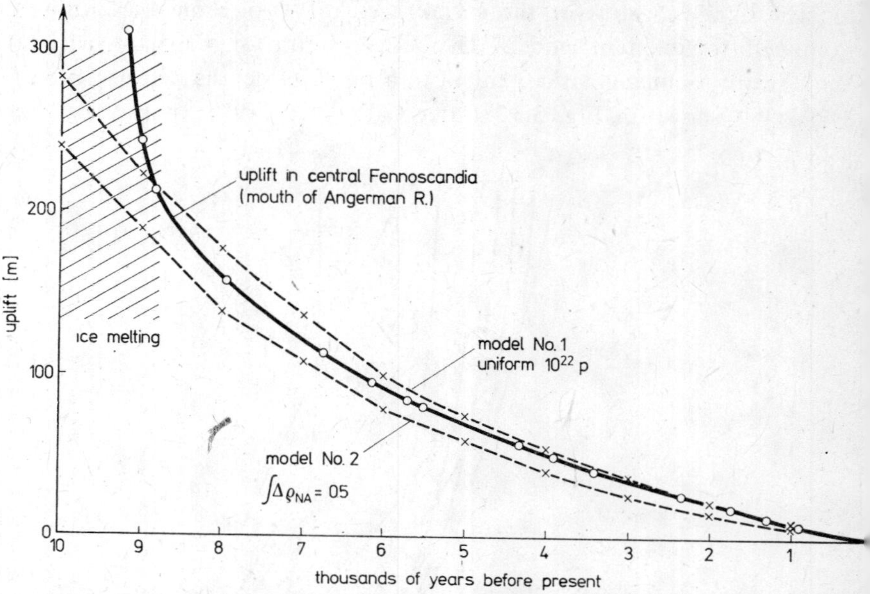

Fig. 3.3.4. The uplift of Fennoscandia. (From L. M. Cathles III, *The Viscosity of the Earth's Mantle* (copyright © 1975 by Princeton University Press), Fig. IV-37, p. 195. Reprinted by permission of Princeton University Press).

example, for a fluid channel in a flat Earth model the decay time τ (proportional to k_x for an elastic substratum—eqn. (3.3.41)) becomes inversely proportional to the square of k_x

$$\tau = \frac{3\nu}{\varrho g k_x^2 H^3} = \frac{1}{k_x^2 D} \tag{3.3.45}$$

where D is the diffusion constant of a fluid layer

$$D = \frac{\varrho g H^3}{3\nu} \tag{3.3.46}$$

On the other hand, for the elastic crust overlying a fluid substratum the adjustment (decay) time, for a flat Earth model, is

$$\tau = \frac{2\nu k_x}{\varrho g + k_x^4 D'} \tag{3.3.47}$$

where D' is another constant, called the flexural rigidity of the crust

$$D' = \frac{EH^3}{12(1-\eta^2)} = \frac{\mu(\lambda+\mu)}{3(\lambda+2\mu)} H^3 \tag{3.3.48}$$

where E is the Young modulus, and η is the Poisson ratio. In both cases

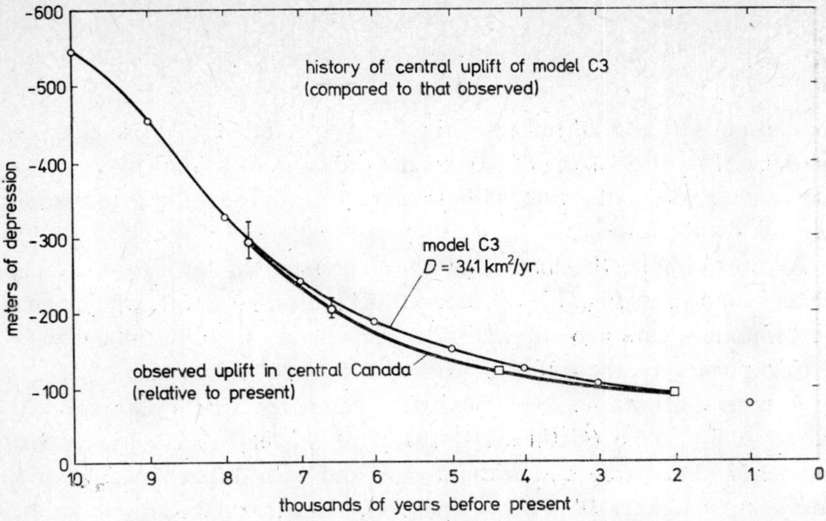

Fig. 3.3.5. Central uplift for Canada. (From L. M. Cathles III, *The Viscosity of the Earth's Mantle* (copyright © 1975 by Princeton University Press), Fig. IV-47, p. 212. Reprinted by permission of Princeton University Press).

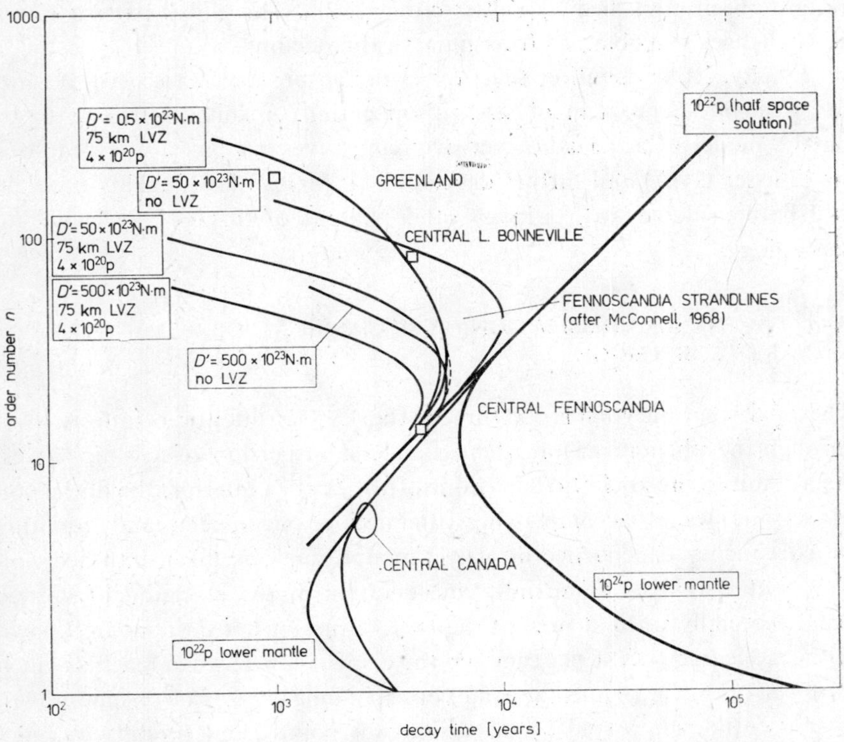

Fig. 3.3.6. The decay spectra: comparison of different models with geological evidence. It is also indicated whether a low-viscosity zone LVZ is taken into account in the respective solution or not. (From L. M. Cathles III, *The Viscosity of the Earth's Mantle* (copyright © 1975 by Princeton University Press), Fig. IV-36, p. 181. Reprinted by permission of Princeton University Press; simplified).

the constants D and D' influence the decay spectra. Both these constants, the flexural rigidity of the crust and the diffusion of a low-viscosity layer, play similar roles—the rate of horizontal spreading of the load response depends on their values.

As an example, the history of the depression of the central Canada region is shown in Fig. 3.3.5. Observations indicating uplift, which started ten thousand years ago, fit well into the viscous response model with $\nu = 10^{19}$ poise and the diffusion constant $D = 34\ \text{km}^2/\text{year}$.

A review of various geological data compared with decay spectra is shown in Fig. 3.3.6 (Cathles III, 1975) for models with a low-viscosity channel (75 km thick) and without one, and with different values of the crust flexural rigidity D'. This comparison, and other data as well, strongly suggest the presence of a low-viscosity layer in the mantle, this layer would certainly play an important role in global dynamics.

A review of isostatic adjustment as the response to deglatiation has recently been published by Peltier *et al.* (1978). The global pattern of sea level changes was obtained from numerical modelling.

Finally, let us mention that, for dynamic processes, viscosity permits to consider the problem of stress propagation. A suitable theory, based on the model of an elastic layer overlying a viscous layer, was proposed by Elsasser (1967) and further developed by Melosh (1976). This problem will be considered further in an other volume of this series, devoted to seismology.

3.4 *Q* FACTOR AND DISSIPATION OF MECHANICAL ENERGY IN THE EARTH'S INTERIOR

The anelastic and complex nature of the Earth's interior manifests itself through the phenomena of attenuation and dispersion of seismic waves. As a result of the anelastic behaviour of the Earth's interior, the dispersion of seismic waves accompanying attenuation phenomena and resulting in a frequency dependence of elastic moduli must be taken into account. This fact should be taken into consideration in the inversion of seismic data, especially since at present the travel times of body waves are known with an accuracy of 0.1 per cent and the dispersion effect can reach as much as 1 per cent at a frequency band corresponding to that of seismic waves. The essential role played by dispersion was pointed out several years ago by Jeffreys (1967), but quantitative relations describing dispersion and attenuation phenomena in the Earth's interior were formulated only recently (see e.g., Randall, 1976; Liu *et. al.*, 1976; Anderson *et al.*, 1977). The anelastic behaviour of the material in the Earth's interior is interpreted as being due to relaxation phenomena with mechanisms accounting for slippage on grain boundaries, partial melting of the material, phase transformations and thermoelasticity.

Taking into account the factors affecting the dissipation of mechanical energy, the deviations from the ideal elastic behaviour can be divided into the groups listed below. We are interested in the attenuation of seismic waves, tidal movements and the free oscillations of the Earth.

1. Anelasticity resulting from interactions between the mechanical energy flux and nonmechanical flows, such as heat flow, electric or magnetic energy flux, and diffusion of particle flow.

2. Viscoelastic effects related to viscous flow of solids. Several linear models of viscoelastic equations are usually considered. A liquid flow, through the rock skeleton in two-component media is also an effect belonging to this group. Melting effects also contribute to viscoelasticity.

3. Dissipation of mechanical energy caused by internal friction along grain boundaries and fractures, and by damped vibrations of dislocation segments.

4. Scattering of elastic waves on various heterogeneities such as grain or block boundaries, cracks and dislocations.

To account for these physical interactions, various models of solids are adopted, described by constitutive equations combining stress and strain or strain rate measures.

In a nondispersive system a simple relation between the quality factor Q and the attenuation coefficients is introduced. The nondimensional measure of attenuation, the factor Q, is expressed by the ratio of the dissipated energy to the average vibration energy W_{AV} per cycle or to its maximum value W_M

$$4\pi Q^{-1} = \frac{\Delta W}{W_{AV}}, \qquad 2\pi Q^{-1} = \frac{\Delta W}{W_M} \tag{3.4.1}$$

Both formulas are equivalent in a homogeneous medium. The factor Q constitutes an important property of the internal mechanical behaviour of the Earth, characterizing the degree of rock homogeneity. In heterogeneous rocks the values of Q are usually lower than in the homogeneous rocks.

Knopoff (1965) introduced an additional term into the differential equation of motion to account for attenuation effects. He introduced the nondimensional constant Q, which is related to the space ($e^{-\alpha x}$) and time

($e^{-\gamma t}$) attenuation coefficients as follows

$$\alpha = \frac{\omega}{2Qc}, \quad \gamma = \frac{\omega}{2Q} \tag{3.4.2}$$

where c is the phase velocity of the plane wave motion under consideration. Recently O'Connell and Budiansky (1978) derived exact relation based on definition (3.4.1)

$$Q = \frac{1}{2}\left(\frac{\omega}{\alpha c} - \frac{\alpha c}{\omega}\right) \tag{3.4.3}$$

which differs from relation (3.4.2) for small values of ω, that is for long waves and free oscillations of the Earth.

Attenuation also distorts dispersion properties. Futterman (1962) pointed out that physical dispersion must accompany wave attenuation to preserve the causality principle. In a medium with a constant Q the correction to the dispersion of surface and body waves is approximately expressed by the following relations (Kanamori and Anderson, 1977; Liu *et al.*, 1976):

$$\frac{v(\omega)}{v(\omega_0)} \approx 1 + \frac{1}{\pi Q} \ln \frac{\omega}{\omega_0}$$
$$\frac{v(\omega)}{v(\omega_0)} = \frac{1}{1 - \frac{1}{\pi Q} \ln \frac{\omega}{\omega_0}} \tag{3.4.4}$$

The latter formula remains correct for a whole spectrum, including $\omega = 0$. Here ω_0 shall be chosen as the upper cut-off frequency. In general Q depends on ω; the corresponding formula can be obtained by introducing a suitable dispersion relation $c(\omega)$ into eqn. (3.4.3).

Anelastic relaxation phenomena (recoverable processes) account for frequency dependent internal friction. Anelastic creep is small in comparison to elastic strain, and energy losses caused by anelasticity are mainly related to heat generation. Denoting by τ_n the relaxation time for different attenuation mechanisms, anelastic creep can be described as follows

$$\varepsilon = \sum \varepsilon_n (1 - e^{-t/\tau_n})$$

while damping is described by

$$Q^{-1} = \sum \varepsilon_n \frac{f\tau_n}{1 + f^2\tau^2}$$

where $f = \omega/2\pi$. An important mechanism here is the damped resonance of a pined dislocation segment. The quality factor Q for frequencies $\omega \ll \omega_0$, where ω_0 is the frequency of resonance, according to van Bueren (1961) is given by

$$Q^{-1} = \frac{\omega^2 \beta}{\omega_0^2 m}$$

where β and m are coefficients in the oscillatory equation of vibrating dislocation, β is responsible for the damping and m for the effective mass of dislocation.

An important role in the internal friction mechanism can be played by partial melting of rock material (Walsh, 1968, 1969—two models), when a fractional concentration of the melted phase influences the viscous response of the medium.

In polycrystals the temperature-activated relaxation caused by stress relaxation at grain boundaries should be considered.

An important role in the energy dissipation can be played by rheological changes of the material induced by irreversible processes, such as crack processes. In this case the energy losses can be studied by considering the hystheresis loops. Dilatation processes in earthquake focus areas should be included to this group of phenomena.

According to the Boltzmann theory, energy dissipation is caused by a delay in strain adjustment to the stress changes. A kind of memory connects the stress and strain histories (O'Connell and Budiansky, 1978)—see eqn. (3.1.8).

For a simple Maxwell body (3.1.5) the following expression for the quality factor Q can be obtained from relation (3.4.1)

$$Q = \frac{\omega \nu}{\mu} \tag{3.4.5}$$

while for a Voigt–Kelvin body the factor Q is

$$Q = \frac{\mu}{\omega \nu} \tag{3.4.6}$$

Both relations (3.4.5) and (3.4.6) fail either at $\omega \to \infty$ or at $\omega \to 0$, and therefore another linear model of a viscoelastic body is often used, preserving a finite value of Q at low and high frequencies.

The major result of the recent studies of seismic wave attenuation in the upper mantle and crust is a finding that Q is not dependent on frequency within an interval from 15 to 1200 Hz (see e.g., Knopoff, 1963; Scheidegger, 1971).

A similar independence of Q from frequency was observed in laboratory experiments on rocks and metals. The observed values of Q for rocks are between 50 and 1000 (Bradley and Fort, 1965). In laboratory experiments both on rocks and on metals it was found that Q is almost independent of frequency, strain is of the order of 10^{-6}–10^{-5}, and also almost independent of temperature.

For porous rocks Q increases with increasing confining pressure, and this implies that at low pressures pores could play an important role in the mechanisms of anelasticity. Apparently, under laboratory conditions the most dominating anelastic process is the slippage along grain boundaries, causing energy losses as a result of friction. This process is most probably eliminated by the pressure at depths greater than a few kilometers below the Earth's surface. This in turn would imply that laboratory results are probably not representative of the anelasticity mechanisms in the Earth's mantle.

Thus, for short-period body and surface waves, we should assume in accordance with laboratory data and seismic wave observations, a somewhat different model (but situated between the Maxwell (3.4.5) and the Voigt–Kelvin (3.4.6) relations), with Q remaining independent of the frequency (Anderson and Hart, 1978). This gives a possibility to construct a distribution of Q versus depth. For longer periods a slight increase of Q^{-1} for Love waves and a decrease for Rayleigh waves is observed (Knopoff, 1964). Free oscillations of the Earth (standing waves) are characterized by very long wavelengths, and a full spectral analysis of these waves should be made to obtain the factor Q for each mode of vibration.

The dissipation of wave motion can be considered separately for shear and for volume deformations, the respective Q factors being denoted by Q_μ and Q_K. The Q values related to P and S waves (Q_P and Q_S) are connected with Q_μ and Q_K through the following obvious formula:

$$Q_S = Q_\mu, \quad Q_P^{-1} = LQ_\mu^{-1} + (1-L)Q_K^{-1} \tag{3.4.7}$$

or

$$Q_K = \frac{(1-L)Q_P Q_S}{Q_S - LQ_P}$$

where $L = \frac{4}{3}(v_P/v_S)^2$.

Since $L > 1$ and Q should be positive by definition, the following condition is valid:

$$\frac{Q_S}{Q_P} \leqslant L$$

This ratio is larger in areas where phase transformations occur.

A number of studies of seismic wave attenuation have been made, mainly based on spectral analyses of body and surface waves. As an example a paper by Kanamori (1967) relating to P wave attenuation in the upper and lower mantle can be quoted. For body waves it is important that the effects of crustal layering be removed. In a broad frequency band the values of Q are frequency independent.

Brune (1977) estimated the values of Q_S from the correlation of S and SS amplitudes. He found that the average value of Q_S is about 131 and that it increases slightly with distance—from a value close to 100 at 30° to 250 at 70°.

According to observational data, a model of Q distribution (for small wavelength) with depth should take into account the following layers (Anderson and Hart, 1978):

lithosphere—high Q,
low Q channel (possibly),
upper mantle—low Q,
lower mantle—high Q,
mantle–core transition zone—low Q (possibly),
outer core—nonabsorbing liquid $Q_K = \infty$, $Q_S = 0$,
inner core—with rather low attenuation.

The model of Anderson and Hart (1978) is shown in Fig. 3.4.1. In all layers with $Q_K^{-1} = 0$ (except the outer core) energy losses are caused by shearing.

For strongly dispersive media such as the Earth the attenuation modes are influenced by dispersion caused by the heterogeneous structure of the substratum. A suitable theory for the layered Earth was developed by Anderson and Archambeau (1964). In an homogeneous layer the quality factor Q can be expressed by the complex wave number $\bar{k} = k + ik^*$ or by the complex velocity $\bar{v} = v + iv^*$. From the Knopoff relation (3.4.2) it follows that ($\bar{v} = \omega/\bar{k}$)

$$Q_P^{-1} = \frac{2k_P^*}{k_P} = \frac{2v^*}{v_P}, \quad Q_S^{-1} = \frac{2k_S^*}{k_S} = \frac{2v_S^*}{v_S} \tag{3.4.8}$$

Taking into account the layered structure and velocity dispersion, we obtain the following formula for the quality factor Q for a wavetrain whose group velocity is U:

$$Q = \frac{ck}{2Uk^*} \tag{3.4.9}$$

where c is the phase velocity.

The relation between the wave number k and the velocities v_{Pl} and v_{Sl} in the layer l is

$$k = \frac{\omega}{c} = k(h_l, v_{Pl}, v_{Sl}, \varrho_l)$$

where h_l and ϱ_l are the layer thickness and the density. This relation leads to a formula for a complex wave number (Anderson and Archambeau, 1964):

$$\bar{k} = k + \mathrm{i} \sum_i \left[\left(\frac{\partial k}{\partial v_P} \right)_l v_{Pl}^* + \left(\frac{\partial k}{\partial v_S} \right)_l v_{Sl}^* \right] \tag{3.4.10}$$

where according to (3.4.8) the relation $v_l^* = \frac{1}{2} v_l Q^{-1}$ can be used for each layer. Thus, the attenuation coefficients are connected with the properties of each layer.

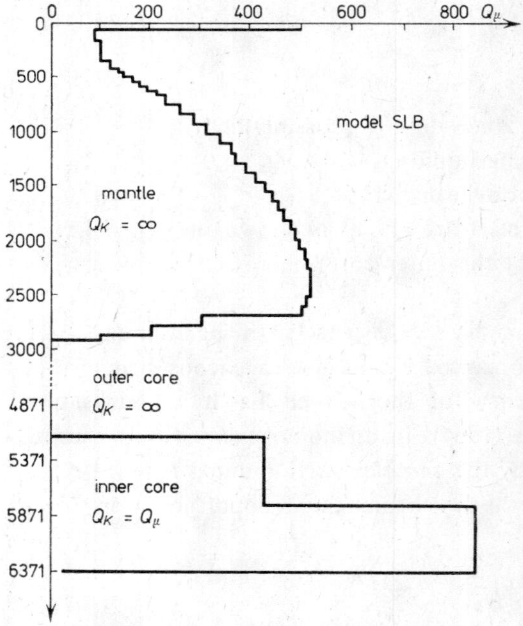

Fig. 3.4.1. Q model (after Anderson and Hart, 1978).

For very long periods (e.g., free oscillations of the Earth) Q depends on frequency and according to (3.4.3) Q shall be expressed by the formula $Q = \frac{1}{2}(v/v^* - v^*/v)$. Brune (1962) and later Knopoff *et al.* (1964) noted that there are some discrepancies for Q obtained from the body waves and that from free oscillation modes. For propagation wavetrains the attenu ation with distance, $\exp(-\pi x/cTQ)$, and for standing waves the attenuation at fixed point, $\exp(-\pi t/TQ)$, are joined by the relation (comp. (3.4.2))

$$UQ_T = cQ_x$$

The factor Q should be connected with the modes of vibrations, which can be done using the vector representation theorem (3.3.42) for a sphere. Dispersion effects for spheroidal and toroidal modes can be expressed by equations similar to formulas (3.4.10). For a spheroidal mode we have

$$Q_{SP}^{-1} = \sum_l \left[\frac{v_{Pl}}{c_{SP}} \left(\frac{\partial c_{SP}}{\partial v_{Pl}} \right)_{K,\varrho,v_S} Q_{Pl}^{-1} + \frac{v_{Sl}}{c_{SP}} \left(\frac{\partial c_{SP}}{\partial v_{Sl}} \right)_{K,\varrho,v_P} Q_{Sl}^{-1} \right] \quad (3.4.11)$$

and similarly for a toroidal mode it follows that

$$Q_{TR}^{-1} = \sum_l \left[\frac{v_{Sl}}{c_{TR}} \left(\frac{\partial c_{TR}}{\partial v_{Sl}} \right)_{K,\varrho,v_P} Q_{Sl}^{-1} \right] \quad (3.4.12)$$

where c_{SP} and c_{TR} are the spheroidal and the toroidal phase velocities. The spheroidal and toroidal modes are divided into harmonic tones ${}_jS_n$ and ${}_jT_n$. The respective values, ${}_j(Q_{SP}^{-1})_n$ and ${}_j(Q_{TR}^{-1})_n$, present a full pattern of attenuation for free oscillations of the Earth.

The corresponding Q values are determined from the width of spectral lines and from the decrease of the observed amplitudes of free oscillations with time. The bulk modulus K is practically not subject to relaxation and therefore it can be assumed that $Q_K = \infty$. The problem lies in such a choice of Q_μ as a function of depth that the calculated values of the attenuation of spheroidal and toroidal oscillations agree with the observed values. The values of Q_{SP} and Q_{TR} are related to attenuation distributions and the layered structure of the corresponding model of the Earth by formulas (3.4.11) and (3.4.12).

The amplitudes of spheroidal ${}_jS_n$ and toroidal ${}_jT_n$ oscillations display an important feature: for a fixed j and an increasing n the more and more external layers of the Earth undergo oscillations, while for a fixed n and an increasing j the oscillations descend deeper and deeper into the Earth's interior. Thus the attenuation of different oscillations provides information about Q or about the dissipation function Q^{-1}. The most recent review of the attenuation modes was published by Dziewonski (1979).

Summing up, it can be concluded that though there are several uncertainties as regards our knowledge about the distribution of Q with depth, the following main facts have been ascertained:

— The distribution of Q has two substantial minima of the order of 100. The first minimum corresponds to the low-velocity zone at depths of 50 to 300 km. The second minimum is observed in the lower mantle near the mantle-core boundary at depths of 2600 to 2900 km. Between these two areas, at depths of 1000 to 2200 km, the values of Q are one order of magnitude higher ($Q \approx 1000$) and the medium differs little from an ideally elastic one.

— In the transition zone between 300 and 1000 km the values of Q increase from 100 to about 1000, while in the lower mantle, at depths of 2200 to 2600 km, they decrease from 1000 to about 100.

— The distribution of Q in the core is not well known as yet. In the liquid external core the values of Q are distinctly higher than in the mantle ($Q \gg 1000$), while in the inner core, at depths of 5100 to 6471 km, they are between 100 and 300.

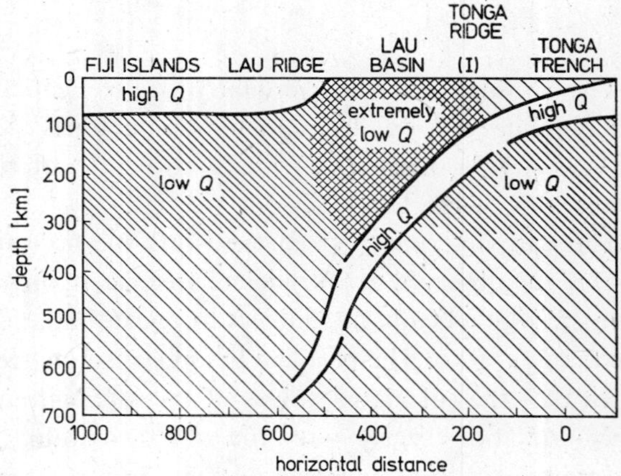

Fig. 3.4.2. Q at the subduction zone. (From M. Barazangi and B. Isacks, *J. Geophys. Res.*, **76**, 8493 (1971), copyrighted by the American Geophysical Union).

All these values were found from free oscillations and in general they agree with the model of Anderson and Hart (1978) described before.

Lateral variations are also observed in the Q distribution. They reflect the global pattern of internal dynamics. The theory of plate tectonics should be mentioned here, especially the submergence processes of plates into deeper parts of the mantle. These areas, known as the Benioff seismic zones, are characterized by low attenuation, i.e., high Q values.

From Fig. 3.4.2 it can be seen that to the left of the Benioff zone, in the area below the volcanic zone, a region of very high attenuation (extremely low Q) is distinctly marked. The high heat flow values and distinct gravity anomaly in that area indicate that the processes in the upper mantle are responsible for volcanic activity.

REFERENCES

Akulov N. S. (1964), "On dislocation kinetics", *Acta Met.*, **12**, 1195–1196.

Anderson D. L. (1967a), "The anelasticity of the mantle", *Geophys. J. R. Astron. Soc.*, **14**, 135–164.

Anderson D. L. (1967b), "Latest information from seismic observations", in: *The Earth's Mantle*, T. F. Gaskell (ed.), Academic Press, London, 355–420.

Anderson D. L. and Archambeau C. B. (1964), "The anelasticity of the Earth", *J. Geophys. Res.*, **69**, 2071–2085.

Anderson D. L. and O'Connell R. (1967), "Viscosity of the Earth", *Geophys. J. R. Astron. Soc.*, **14**, 287–296.

Anderson D. L. and Hart R. S. (1978), "Attenuation models of the Earth", *Phys. Earth and Planet. Inter.*, **16**, 289–306.

Anderson D. L., Hart R. S. and Liu H. (1977), "The Earth as a seismic absorption band", *Science*, **196**, 1104–1106.

Artyushkov E. V. (1967), "On the isostatic equation of the Earth's crust", *Geophys. J. R. Astron. Soc.*, **14**, 251–260.

Ashby M. F. and Verall R. A. (1977), "Micromechanism of flow and fracture and their relevance to the rheology of the upper mantle", *Philos. Trans. R. Soc. Lond.*, **A288**, 59–95.

Backus G. E. (1967), "Converting vector and tensor equations to scalar equations in spherical coordinates", *Geophys. J. R. Astron. Soc.*, **13**, 71–101.

Barazangi M. and Isacks B. (1971), "Lateral variations of seismic-wave attenuation in the upper mantle above the inclined earthquake zone of the Tonga Island arc; deep anomaly in the upper mantle", *J. Geophys. Res.*, **76**, 8493–8516.

Bland D. R. (1960), *The Theory of Linear Viscoelasticity*, Pergamon Press, New York.

Brace W., Paulding W. F. and Scholz C. (1966), "Dilatancy in the fracture of crystalline rocks", *J. Geophys. Res.*, **71**, 3939–3953.

Bradley J. J. and Fort A. N. (1965), "Internal friction in rocks", in: *Handbook of Physical Constants* (revised edition), S. P. Clark(Jr) (ed.), *Mem. Geol. Soc. Am.*, **97**, 175–193.

Brune J. N. (1962), "Attenuation of dispersed wave trains", *Bull. Seismol. Soc. Am.*, **52**, 109–112.

Brune J. N. (1977), "*Q* of shear waves estimated from *S-SS* spectral ratios", *Geophys. Res. Lett.*, **5**, 179–181.

Bueren van H. G. (1961), *Imperfections in Crystals*, North-Holland, Amsterdam.

Byerlee J. D. (1968), "Brittle-ductile transition in rocks", *J. Geophys. Res.*, **73**, 4741–4750.

Cable R. L. (1962), "A model for boundary diffusion controlled creep in polycrystalline materials", *J. Appl. Phys.*, **34**, 1679–1682.

Cathles III L. M. (1975), *The Viscosity of the Earth's Mantle*, Princeton University Press, Princeton.

Cruden D. M. (1970), "A theory of brittle creep in rock under uniaxial compression", *J. Geophys. Res.*, **75**, 3434–3442.

Domany E., Gubernatis J. E. and Krumshal J. A. (1975), "The elasticity of polycrystals and rocks", *J. Geophys. Res.*, **80**, 4851–4856.

Dziewonski A. M. (1979), "Elastic and anelastic structure of the Earth", *Rev. Geophys. Space Phys.*, **17**, 303–312.

Elsasser W. M. (1967), "Convection and stress propagation in the upper mantle", in: *The Application of Modern Physics to the Earth and Planetary Interiors*, S. K. Runcorn (ed.), John Wiley, New York, 223–246.

Futterman W. I. (1962), "Dispersive body waves", *J. Geophys. Res.*, **67**, 5279–5291.

Garofalo F. (1966), *Fundamentals of Creep and Creep-Rupture in Metals*, Macmillan, New York, Collier-Macmillan, London.

Gilbert F. and Backus G. E. (1966), "Propagator matrices in elastic wave and vibration problems", *Geophysics*, **31**, 326–332.

Griffith A. A. (1971), "The phenomena of rupture and flow in solids, *Philos. Trans. R. Soc. Lond.*, A**221**, 163–198.

Griggs D. T. (1967), "Hydrolytic weakening of quartz and other silicates", *Geophys. J. R. Astron. Soc.*, **14**, 19–32.

Griggs D. T. and Handin J. (1960), "Observations on fracture and a hypothesis of earthquakes", in: *Rock Deformation*, D. T. Griggs and J. Handin (eds.), *Mem. Geol. Soc. Am.*, **79**, 347–364.

Holcomb D. J. (1977), "A quantitative model of dilatancy in dry rock and its application to westerly granite", University of Colorado, NOAA, Boulder, Colorado (preprint), pp. 34.

Jaeger J. C. and Cook N. G. W. (1971), *Fundamentals of Rock Mechanics*, Chapman and Hall, London.

Jeffreys H. (1967), "The damping of *S* waves", *Geophys. J. R. Astron. Soc.*, **14**, 189–190.

Kanamori H. (1967), "Attenuation of *P* waves in the upper and lower mantle", *Bull. Earth Res. Inst.*, **45**, 299–312.

Kanamori H. and Anderson D. L. (1977), "Importance of physical dispersion in surface wave and free oscillation problems", *Rev. Geophys. Space Phys.*, **15**, 105–112.

Knopoff L. (1956), "The seismic pulse in materials possesing solid friction", *Bull. Seismol. Soc. Am.*, **46**, 175–183.

Knopoff L. (1963), "Equations of state at moderately high pressures" in: *High Pressure Physics and Chemistry*, R. S. Bradley (ed.), Vol. 2, Academic Press, New York, 227–246.

Knopoff L. (1964), "*Q*", *Rev. Geophys.*, **2**, 625–660.

Knopoff L., Aki K., Archambeau C. B., Ben-Menahem A. and Hudson J. A. (1964), "Attenuation of dispersed waves", *J. Geophys. Res.*, **69**, 1655–1657.

Leckie F. A. (1977), "The constitutive equations of continuum creep damage mechanics", *Philos. Trans. R. Soc. Lond.*, **A288**, 27–47.

Li J. C. M. (1963), "A dislocation mechanism of transient creep", *Acta Met.*, **11**, 1269–1270.

Liebovitz H. (1968), *Fracture. An Advanced Treatise*, Academic Press, New York.

Liu H., Anderson D. and Kanamori H. (1976), "Velocity dispersion due to anelasticity; implication for seismology and mantle composition", *Geophys. J. R. Astron. Soc.*, **47**, 41–58.

McDonald G. J. F. (1960), "Orientation of anisotropic minerals in a stress field", in: *Rock Deformation*, D. T. Griggs and J. Handin (eds.), *Mem. Geolog. Soc. Am.*, **79**, 1–8.

Melosh H. J. (1976), "Nonlinear stress propagation in the Earth's upper mantle", *J. Geophys. Res.*, **81**, 5621–5632.

Misra A. K. and Murrell S. A. F. (1965), "An experimental study of the effect of temperature and stress on the creep of rocks", *Geophys. J. R. Astron. Soc.*, **9**, 509–535.

Mott N. F. (1953), "Theory of work hardening of metals. Flow without slip-line", *Philos. Mag.*, **44**, 742–765.

Murrell S. A. F. (1965), "The effect of triaxial stress systems on the strength of rocks", *Geophys. J. R. Astron. Soc.*, **10**, 231–281.

Murrell S. A. F. (1967), "An experimental study of the effect of temperature and stress on the creep of rocks with a discussion of Earth tide damping, isostasy and mantle convection", *Geophys. J. R. Astron. Soc.*, **14**, 51–56.

Nadai A. (1938), "The influence of time upon creep", in: *Timoshenko Anniversary Volume*, Van Nostrand, New York, 155–170.

O'Connell R. J. and Budiansky B. (1978), "Measure of dissipation in viscoelastic media", *Geophys. Res. Lett.*, **5**, 5–8.

Ohnami M. (1972), "Study of plasticity laws of polycrystallinic metals at elevated temperatures", in: *Mechanical Behavior of Materials*, Vol. III, Kyoto.

Orowan E. (1960), "Mechanism of seismic faulting", in: *Rock Deformation*, D. T. Griggs and J. Handin (eds.), *Mem. Geol. Soc. Am.*, **79**, 323–345.

Orowan E. (1965), "Convection in a non-Newtonian mantle, continental drift and mountain building", in: "A Symposium on Continental Drift", *Philos. Trans. R. Soc. Lond.*, **A258**, 284–313.

Orowan E. (1967), "Seismic damping and creep in the mantle", *Geophys. J. R. Astron. Soc.*, **14**, 191–218.

Peltier W. R., Farrel W. E. and Clark J. A. (1978), "Glacial isostasy and relative sea level; a global finite element model", *Tectonophysics*, **59**, 81–110.

Pratt P. L. (1967), "Strength and deformation of ionic materials", *Geophys. J. R. Astron. Soc.*, **14**, 5–12.

Raleigh C. B. (1967), "Tectonic implications of serpentinite weaking", *Geophys. J R. Astron. Soc.*, **14**, 113–118.

Raleigh C. B. and Paterson M. S. (1965), "Experimental deformation of serpentinite and its tectonic implications", *J. Geophys. Res.*, **70**, 3365–3985.

Randall M. J. (1976), "Attenuative dispersion and frequency shifts of the Earth free oscillations", *Phys. Earth Planet. Inter.*, **12**, 1–4.

Robertson E. C. (1964), "Viscoelasticity of rocks", in: *State of Stress in the Earth's Crust*, N. R. Judd (ed.), Elsevier, New York, 181–233.

Rogan J. and Alexander J. M. (1966), "An investigation into room temperature creep of steels and copper", *J. Inst. Metals*, **94**, 61–69.

Scheidegger A. E. (1970a), "The rheology of the Earth in the intermediate time range", *Ann. Geofis.*, **23**, 27–43.

Scheidegger A. E. (1970b), "The rheology of the Earth in the long-time range", *Ann. Geofis.*, **23**, 325–346.

Scheidegger A. E. (1971), "Rheology of the tectonosphere in the short-time range", *Ann. Geofis.*, **24**, 311–324.

Stacey F. D. (1963), "Theory of creep in rocks and the problem of convection in the Earth's mantle", *Icarus*, **1**, 304–313.

Stacey F. D. (1967), "Consequences of energy dissipation by creep in the mantle", *Geophys. J. R. Astron. Soc.*, **14**, 433–438.

Stroh A. N. (1957), "Theory of fracture of metals", *Adv. Phys.*, **6**, 418–465.

Teisseyre R. (1970), "Crack formation and energy release caused by the concentration of dislocations along fault planes", *Tectonophysics*, **9**, 547–557.

Thomsen L. (1972), "Elasticity of polycrystals and rocks", *J. Geophys. Res.*, **77**, 315–327.

Walsh J. B. (1968), "Attenuation in partially melted material", *J. Geophys. Res.*, **73**, 2209–2216.

Walsh J. B. (1969), "New analysis of attenuation in partially melted rock", *J. Geophys. Res.*, **74**, 4333–4337.

Wang C. Y. (1966), "Earth's zonal deformations", *J. Geophys. Res.*, **71**, 1713–1720.

Weertman J. (1957), "Steady-state creep of crystals", *J. Appl. Phys.*, **28**, 1185–1189.

Weertman J. and Weertman J. R. (1975), "High temperature creep of rock and mantle viscosity", in: *Annual Review of Earth and Planetary Sciences*, F. A. Donath, F. G. Stehli and G. W. Wetherill (eds.), Vol. 3, Annual Reviews Inc., Palo Alto, California, 293–315.

Weertman J. R. (1977), "Creep laws for the mantle of the Earth", *Philos. Trans. R. Soc. Lond.*, **A288**, 59–95.

4. Thermal Energy and Heat Flow in the Earth's Interior

It is well known that the average surface heat flow of the Earth is about 63 $mW \cdot m^{-2}$ (1.5 $\mu cal \cdot cm^{-2} \cdot s^{-1}$). This means that at the present time the terrestrial globe loses about 10^{13} $J \cdot s^{-1}$ of its internal energy resources through thermal conductivity, and the annual loss of energy is of the order of 10^{21} J. For comparison (Verhoogen *et al.*, 1970), the products of volcanic eruptions transport energy of the order of $10^{18} \div 10^{19}$ $J \cdot year^{-1}$ outside the globe, contemporary epeirogenic movements use energy of the order of 10^{17} $J \cdot year^{-1}$,* and the total energy released by earthquakes of different intensities (about one million earthquakes per year) is of the order of 10^{19} $J \cdot year^{-1}$.

Thus, these exoenergetic processes are at least two orders of magnitude less effective than the phenomenon of global heat flow. The energy outflow forced by the thermal gradient is most probably the principal factor determining the direction and rate of evolution of the Earth, and thus also the present physical state of its interior.

Since the heat flow is equal to the product of the effective thermal conductivity and the temperature gradient, this chapter is devoted to various types of thermal conductivities in the crust and mantle, as well as to the possibilities of estimation of the present thermal gradient distribution in the Earth's interior.

4.1 HEAT FLOW OF THE EARTH

4.1.1 Heat flow in the crust

Heat flow, or, more properly, heat flow density, is the amount of energy passing through a surface unit during a time unit under the influence of the temperature gradient. The basic unit of heat flow is 1 $W \cdot m^{-2}$, but 1 $cal \cdot cm^{-2} \cdot s^{-1}$ (1 $W \cdot m^{-2}$ = 23.9 $\mu cal \cdot cm^{-2} \cdot s^{-1}$) is also used. In geo-

---

* An estimation made under the assumption that the rate of continent uplift balances the rate of their denudation (the rate is of the order of 10^{-3} $cm \cdot year^{-1}$) and the average height of the continents above sea level never exceeds 1 km (Verhoogen *et al.*, 1970).

thermics the commonly used unit is 1 HFU (Heat Flow Unit) = 1 $\mu cal \cdot cm^{-2} \cdot s^{-1} = 41.9\ mW \cdot m^{-2}$.

Between the heat flow, which is a vectorial quantity (**j**), and the temperature gradient there is the well-known relation $\mathbf{j} = -\mathcal{K}\ \mathrm{grad}\ T$, where $\mathcal{K}$ is the effective thermal conductivity of a solid medium (in general, $\mathcal{K}$ is a symmetric tensor of the second order).

In solids the thermal conductivity $\mathcal{K}$ is the sum of the conductivities corresponding to various types of energy carriers, e.g., phonons (quanta of elastic vibration energy), photons of temperature radiation, free electrons and holes, excitons, etc. In general, these conductivities depend on various physical properties of the solid, and explicitly on temperature.

In the range of temperatures prevailing in the Earth's crust the only conductivity of importance is the phonon thermal conductivity ($\varkappa$), often called the lattice conductivity or ordinary *thermal conductivity*.

The heat flow of the Earth is determined indirectly, i.e., in the ground the vertical temperature gradient ($\Delta T/\Delta z$), called the *geothermal gradient*, and the thermal conductivity ($\mathcal{K}$) of the medium are measured, and then the value of heat flow, or more properly, its vertical component, is calculated from the relation $J = \mathcal{K}(\Delta T/\Delta z)$. Problems connected with heat flow measurements have been considered in several papers (e.g., Beck, 1965; Langseth, 1965).

The geothermal gradient on the continents varies from about 8 to 40 $K \cdot km^{-1}$, while in sediments of the oceanic floor it is slightly lower. The thermal conductivity of rocks of the continental crust varies in general from 1.7 to 4.2 $W \cdot m^{-1} \cdot K^{-1}$ ($\approx 4 \div 10\ mcal \cdot cm^{-1} \cdot s^{-1} \cdot K^{-1}$), it depends on the mineral composition, structure and texture of the rocks, the water content in the pores, and also on pressure and temperature. Oceanic sediments show slightly lower conductivity, corresponding to the total saturation of rocks by water.

An extraordinary characteristic of the Earth's heat flow is practically the same average value of surface flows, determined separately for the continental and for the oceanic crust; this value is 63 $mW \cdot m^{-2} \approx 1.5$ HFU. The surface distribution of heat flow, on the other hand, is highly diversified (see Grim, 1976), and the value of flow can vary from 0 to 420 $mW \cdot m^{-2}$ (from 0 to 10 HFU). Certain geothermal regions can be distinguished, corresponding to large structural units of the crust and most probably also of the mantle (see Table 4.1.1, after Jacobs, 1974). The lowest values of heat flow on the continents were found on old Precambrian platforms (e.g., the Canadian, Australian and East-European Platforms). In the oceanic crust their "counterparts" correspond to the cold areas of the Pacific

TABLE 4.1.1
Heat flow in various tectonic regions (according to Jacobs, 1974)

Region	Number of observa- tions	Average heat flow and standard deviation	
		$mW \cdot m^{-2}$	HFU
Precambrian shields	214	41 ± 10	0.98 ± 0.24
Non-orogenic post-Precambrian areas	96	62 ± 17	1.49 ± 0.41
Orogenic Paleozoic areas	88	60 ± 17	1.43 ± 0.40
Orogenic Mesozoic-Cenozoic areas	159	74 ± 24	1.76 ± 0.58
Ocean basins	683	53 ± 22	1.27 ± 0.53
Midoceanic ridges	1065	79 ± 62	1.90 ± 1.48
Ocean trenches	78	49 ± 29	1.16 ± 0.70
Continental margins	642	75 ± 39	1.80 ± 0.93

trenches. High values of heat flow are observed in regions of young orogenic movements in the continental crust and in rift areas of the oceanic crust (the mid-Atlantic ridge, the Red Sea, the Gulf of California, etc.). Of course, higher values of heat flow are found in regions of volcanic activity. In geothermal regions in which heat transport is connected with water circulation (hot springs, geysers, fumaroles, etc.), the heat flow can exceed the average value even by a factor of 100. Geothermal regions, however, cover less than 1% of the entire surface of the continents, and their contribution to the whole of the Earth's heat flow is thus negligible (Verhoogen *et al.*, 1970).

The heat flow at the Earth's surface can be considered conventionally as a flow composed of the heat coming from the mantle interior and the heat generated in the Earth's crust. The heat source is the decay of natural radioactive elements. These are the elements forming the uranium series (^{238}U), actinium series (^{235}U) and thorium series (^{232}Th), and the radioactive isotope potassium ^{40}K. The parent elements of the radioactive series and potassium ^{40}K are long-lived isotopes; their half-lives are of the order of $10^9 \div 10^{10}$ years—see Table 4.1.2 (after Friedlender *et al.*, 1964; York and Farquhar, 1971; Rybach, 1976), where the verified efficiencies of radiogenic heat generation by these isotopes (elements) are also given. It should be noted that the radiogenic heat associated with ^{238}U, ^{235}U and ^{232}Th is the heat produced by all the isotopes of a given series in a state of radioactive equilibrium.

Because of its small thickness (on the average about 35 km) and its

TABLE 4.1.2

Radiogenic heat generated by uranium, thorium and potassium (data after Friedlender *et al.* 1964; York and Farquhar, 1971; Rybach, 1976)

Isotope (Element)	Decay constant [10^{-10} year^{-1}]	Half-life [10^9 years]	Radiogenic heat production [10^{-7} W $\cdot$ g^{-1}]	Isotopic ratio %
^{238}U	1.54	4.51	0.918	99.28
^{235}U	9.72	0.713	5.74	0.711
^{232}Th	0.499	13.90	0.256	100
^{40}K	5.31	1.31	0.295	0.0118
U	—	—	0.951	—
Th	—	—	0.256	—
K	—	—	3.47×10^{-5}	—

old age, the Earth's crust can be modelled simply by a plane parallel layer with a steady heat flow. In this case the heat flow at the depth z is

$$j = J - \int_0^z Q(\zeta)\,d\zeta \tag{4.1.1}$$

where $Q(\zeta)$ is the vertical distribution of the radiogenic heat generation per unit of volume and J is the heat flow at the layer surface. The maximum thickness z_{max} of the layer must fulfil the condition

$$\int_0^{z_{max}} Q(\zeta)\,d\zeta \leqslant J$$

resulting from the assumption that the heat flow is steady.

In the case of a multi-layer (n-layer) model of the crust, to which corresponds a "stepwise" distribution of the function $Q(z)$, it can be written

$$j_m(z) = J - \sum_{i=1}^{m-1} Q_i \Delta z_i - Q_m \left(z - \sum_{i=1}^{m-1} \Delta z_i\right); \quad m \leqslant n \tag{4.1.1a}$$

where Q_i = const and Q_m = const is the radiogenic heat in the layers i and m, respectively, Δz_i is the thickness of the layer i, and $j_m(z)$ is the heat flow inside the layer m.

Approximating the continental crust by a two-layered granite–basaltic model with the thickness of the granitic layer $\Delta z_G = 25$ km and of the basaltic layer $\Delta z_B = 10$ km, and accepting as representative the values of radiogenic heat for granite $Q_G \approx 2.1\ \mu W \cdot m^{-3}$ (5 HGU)* and for basalt

* 1 HGU (Heat Generation Unit) $\equiv 10^{-13}$ cal $\cdot$ cm^{-3} $\cdot$ s^{-1} is used in geothermics as a unit of the efficiency of radiogenic heat generation; 1 HGU $\approx 0.419\ \mu W \cdot m^{-3}$.

$Q_B \approx 0.42\ \mu W \cdot m^{-3}$ (1 HGU), it is readily found (see (4.1.1a)) that the heat flow at the Moho boundary ($z \approx 35$ km) is $J_{cM} = J - Q_G \Delta z_G - Q_B \Delta z_B \approx 10\ mW \cdot m^{-2}$ (0.24 HFU). In the case of the oceanic crust (without sediments) a one-layer basaltic model with the thickness $\Delta z'_B = 7$ km gives $J_{oM} = J - Q_B \Delta z'_B \approx 59\ mW \cdot m^{-2} \approx 1.4$ HFU ($J = 63\ mW \cdot m^{-2} \approx 1.5$ HFU for models of both the continental and the oceanic crust). Since the continents (including the shelves) cover only about 40% of the total surface of the terrestrial globe, the average global heat flow at the Moho is $J_M = 0.4 J_{cM} + 0.6 J_{oM} \approx 42\ mW \cdot m^{-2}$ (1 HFU). This is about 70% of the average value of heat flow at the Earth's surface. It is also obvious that the heat from the mantle interior flows mainly out through the oceanic crust.

In various regions of the continental crust, "heat flow provinces", or "geothermal provinces", are distinguished, where the surface heat flow (J) is a linear function of the radiogenic heat (Q) produced in a crystalline ground

$$J = DQ + j_0 \tag{4.1.2}$$

where D and j_0 are the empirical coefficients characterizing individual provinces (Roy *et al.*, 1968; Čermak and Jessop, 1971; Swanberg *et al.*, 1974; Rao and Jessop, 1975; Rao *et al.*, 1976; Hyndman *et al.*, 1979; and others).

To relation (4.1.2) corresponds, for example, a simple model of the upper crust in the form of a plane parallel layer with the thickness D but composed of vertical blocks with different radiogenic heat generation Q (for a given block Q = const) and heated at floor by a constant heat flow j_0 (see (4.1.1)).

In general, relation (4.1.2) can be obtained from eqn. (4.1.1) for each distribution of radiogenic heat of the type $Q = Q_0 f(D, z)$, where Q_0 is the surface radiogenic heat, if the following condition is fulfilled

$$D^{-1} \int_0^{z_m} f(D, z) \mathrm{d}z = 1 \tag{4.1.2a}$$

Thus, for example, $f(D, z) = 1$ and $z_m = D$, $f(D, z) = 1 - (z/kD)^{1/(k-1)}$ at $k > 1$ and $z_m = kD$, $f(D, z) = \cos^2(\frac{1}{4}\pi z/D)$ and $z_m = 2D$, $f(D, z) = e^{-z/D}$ and $z_m \to \infty$, etc. (see also Lachenbruch, 1970; Albarede, 1975).

A convenient model of the continental crust (or its parts) is the model with an exponential decreases in the concentrations of natural radioactive elements (and thus of radiogenic heat as well) with depth. In this case $Q = Q_0 e^{-z/D}$, where the parameter D, which is a logarithmic decrement,

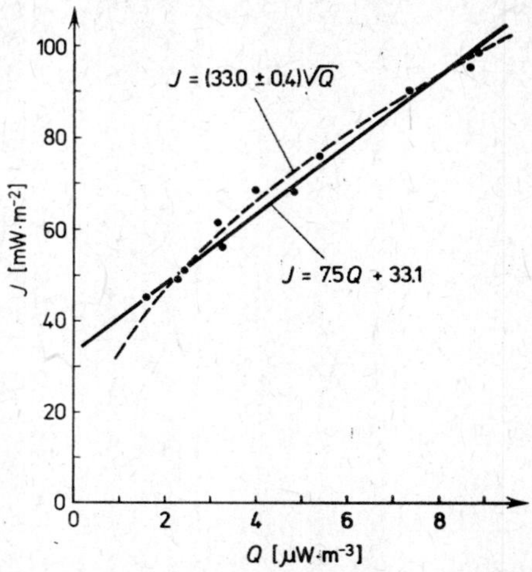

Fig. 4.1.1. Surface heat flow (J) versus radiogenic heat production (Q) of crystalline ground for the Eastern US geothermal province, according to Roy *et al.* (1968). The solid and dashed lines show the approximation of empirical data by linear relation (4.1.2) and by relation (4.1.4) in the form of $J = A'\sqrt{Q}$, respectively.

determines the "mean H steepness" of the vertical distribution of heat sources, and then (see (4.1.1))

$$j = J - Q_0 D(1 - e^{-z/D})$$

The best known and often considered as, to some extent, "standard geothermal provinces" are the three geothermal provinces in the United States distinguished by Roy *et al.* (1968): Sierra Nevada (SN), Eastern US (EUS) (see Fig. 4.1.1) and Basin and Range (BR). The values of the parameters D and j_0 are 10.1 km and 17 $mW \cdot m^{-2}$ (0.40 HFU) for province SN, 7.5 km and 33 $mW \cdot m^{-2}$ (0.79 HFU) for province EUS, and 9.4 km and 59 $mW \cdot m^{-2}$ (1.4 HFU) for province BR, respectively (Roy *et al.*, 1968)*. Thus, the values of the parameter D in all the three standard provinces are similar to each other, being of the order of 10 km, while the values of the parameter j_0, often called the reduced heat flow, are distinctly different. This may indicate a thermal differentiation of both the lower crust and the upper mantle.

For various geothermal provinces Pollack and Chapman (1977a,b) found the general empirical relation

$$j_0 \approx 0.6\langle J \rangle \tag{4.1.3}$$

where $\langle J \rangle$ is the average value of heat flow for a given province, and j_0 is the reduced heat flow (Fig. 4.1.2).

* New empirical data on the heat flow and radiogenic heat generation suggest several various linear relationships for "Basin and Range" geothermal provinces.

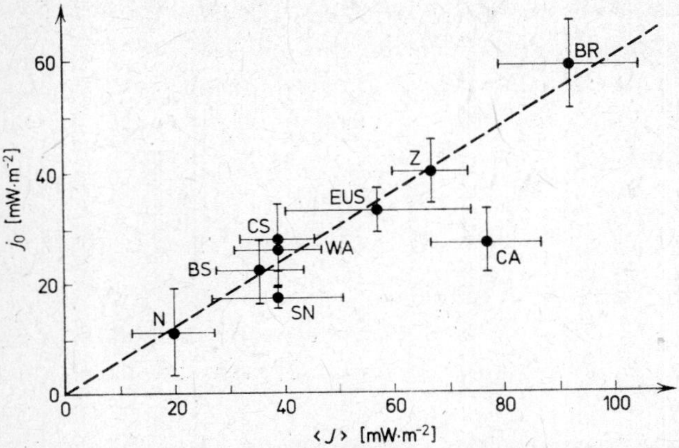

Fig. 4.1.2. Reduced heat flow (j_0) versus mean surface heat flow ($\langle J \rangle$) in geothermal provinces: Sierra Nevada (SN), Eastern US (EUS), Basin and Range (BR), Canadian Shield (CS), Baltic Shield (BS), Central Australia (CA), Western Australia (WA), Niger (N), and Zambia (Z). The horizontal and vertical bars represent the standard deviation of the mean heat flow and the uncertainty of the reduced heat flow estimation; the dashed line is linear approximation (4.1.3) (after Pollack and Chapman, 1977a).

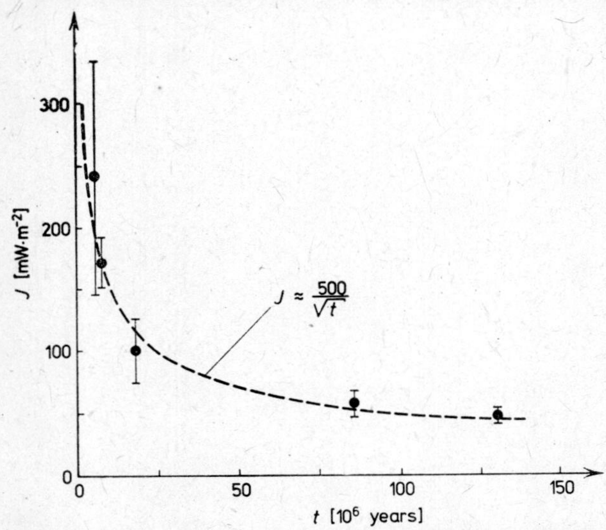

Fig. 4.1.3. Heat flow versus age of the floor for the Pacific Ocean. The points represent the mean values for selected provinces of the floor, the vertical bars show the standard deviations of the mean values, and the dashed line is an approximation. (From "The flow of heat from the Earth's interior" by H. N. Pollack and D. S. Chapman, copyright © 1977 by Scientific American Inc.).

The establishment of the Roy–Blackwell–Birch relation (4.1.2) stimulated the search for other relations between the surface heat flow, radiogenic heat and various chemical and physical properties of the rocks forming the Earth's crust. It was found, for example, that the value of heat flow decreases with the geological age (t) of the ground in both the oceanic (see Fig. 4.1.3) and the continental crust. Lister (see Hamza, 1979) noted that the heat flow in the oceanic crust (J), undisturbed by hydrothermal circulation, seems to fulfil the relation

$$J = \frac{500}{\sqrt{t}}$$

where J is in $mW \cdot m^{-2}$ and t is in millions of years.

The dependence of heat flow on the thickness of sediments and the state of their consolidation was also found (see e.g., Crough and Thompson, 1976).

Rybach (1976, 1979) has shown that it is possible to relate, through the medium density, the values of radiogenic heat generation (Q) in magmatic rocks to the velocity of longitudinal elastic waves (v_P). The radiogenic heat generated in a crystalline ground decreases with increasing velocity of the P wave as follows

$$Q = \tilde{q} \exp(-b v_P)$$

where $\tilde{q}$ and b are empirical coefficients dependent on the pressure (see Fig. 4.1.4).

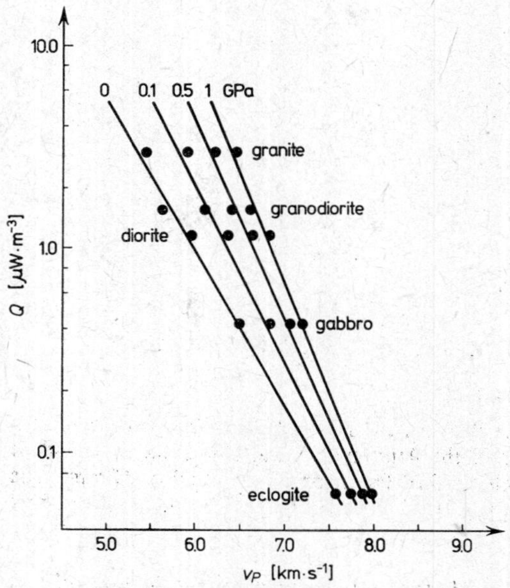

Fig. 4.1.4. Empirical relationship between the compressional wave velocity (v_P) and the radiogenic heat production (Q) in magmatic rocks for different pressures, according to Rybach (1979).

Maj (1980) suggested that in standard geothermal provinces, besides relation (4.1.2), there is another relation between the heat flow J and radiogenic heat Q of the type

$$J^2 = AQ \tag{4.1.4}$$

where the thermal parameter A is inversely proportional to the temperature coefficient of phonon thermal resistivity (β) of a crystalline ground (see Fig. 4.1.1). The values of the parameter A are 846 $W \cdot m^{-1}$ (2.02 $cal \cdot cm^{-1} \cdot s^{-1}$) for province SN, 1090 $W \cdot m^{-1}$ (2.60 $cal \cdot cm^{-1} \cdot s^{-1}$) for province EUS, and 2450 $W \cdot m^{-1}$ (5.85 $cal \cdot cm^{-1} \cdot s^{-1}$) for province BR (the data for calculations were taken from Roy *et al.* (1968), Lachenbruch (1968) and Lachenbruch *et al.* (1976)). From the fact that the distribution $J(Q)$ can be approximated by a relation both of type (4.1.2) and of type (4.1.4), it follows that the parameters D, j_0 and A in a given geothermal province should fulfil the condition $Dj_0 A^{-1} \leqslant 0.25$, and that the reduced heat flow j_0 and the average heat flow $\langle J \rangle$ of a province might be connected by a relation of the form $j_0 = 0.5 \langle J \rangle$, i.e., by a relation very similar to that of Pollack and Chapman (4.1.3).

Pollack and Chapman (1977b) carried out a spherical-harmonic analysis of the continental and oceanic values of heat flow, reduced to the Moho boundary, i.e., at the surface of the Earth's upper mantle. To determine the continental values of heat flow underneath the crust, a two-layer model of the crust was used. In the upper layer the concentration of radiogenic heat sources decreased exponentially with depth, while in the lower layer radiogenic heat generation was constant and equal to 0.21 $\mu W \cdot m^{-3}$ (0.5 HGU). The relations (4.1.2) and (4.1.3) were also used. For the oceanic crust the authors applied a simple model of a coolling homogeneous basaltic layer with a thickness of 6.5 km and constant radiogenic heat generation $Q = 0.5$ $\mu W \cdot m^{-3}$ (1.2 HGU), which had started to form in a rift zone t years ago.

The results of the spherical-harmonic analysis suggest the following average values of heat flow: 28 $mW \cdot m^{-2}$ (0.67 HFU) for the surface of the continental mantle, 57 $mW \cdot m^{-2}$ (1.36 HFU) for the surface of the oceanic mantle, and 48 $mW \cdot m^{-2}$ (1.15 HFU) for the surface of the entire mantle on a global scale.*

It was also found that the horizontal distribution of heat flow at the

* The adequate average values of heat flow at the Earth's surface are: 53 $mW \cdot m^{-2}$ (1.27 HFU) for the continental crust, 62 $mW \cdot m^{-2}$ (1.48 HFU) for the oceanic crust, and 59 $mW \cdot m^{-2}$ (1.4 HFU) for the whole globe.

Moho boundary shows a direct correlation with the geopotential and an inverse correlation with the topography of the terrestrial globe.

Hamza (1979) estimated the value of heat flow from the continental mantle interior for various geothermal provinces (assuming that their thickness is $2D$—see (4.1.2) and (4.1.2a)) using the values of the reduced heat flow (j_0) and taking into account the radiogenic heat generated in the lower crust and in the lithosphere underneath the crust. For the 20 km thick lower crust $Q = 0.4\ \mu W \cdot m^{-3}$ ($\approx$ 1 HGU) was assumed, and for the continental lithosphere the thickness $\Delta z = 9.4\sqrt{t} - 40$ (in km), where t is the geological age (in millions of years) of the last thermal phenomenon, and the radiogenic heat is $Q = 0.02\ \mu W \cdot m^{-3}$ ($\approx$ 0.05 HGU), corresponding to the present radioactivity of peridotites.

A similar procedure was used for the oceanic crust. It was assumed that its thickness is 5 km and its radiogenic heat is that of the "average basalt" ($\approx 0.4\ \mu W \cdot m^{-3} \approx 1$ HGU). For the oceanic lithosphere with a thickness $\Delta z' = (9.4\sqrt{t} - 5)$ km (t in million years), the radiogenic heat $Q = 0.04\ \mu W \cdot m^{-3}$ ($\approx$ 0.1 HGU) was assumed.

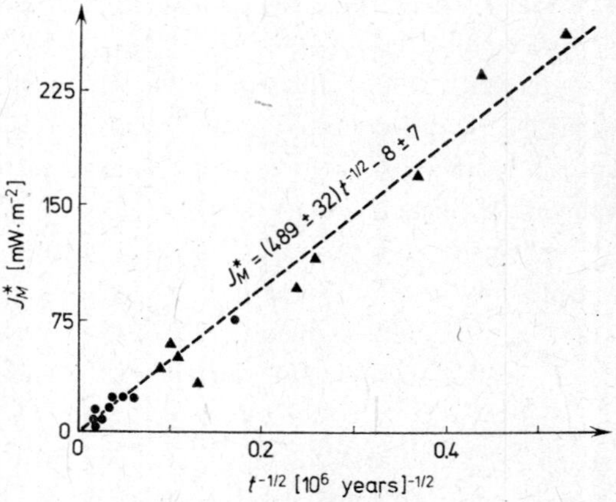

Fig. 4.1.5. Heat flow from the mantle J_M^* (corrected for radiogenic heat production in the lithosphere beneath the crust) versus $1/\sqrt{t}$, according to Hamza (1979). The solid circles and solid triangles represent continental and oceanic data, respectively; the dashed curve is a least square fit to the combined continental and oceanic data.

The calculation results suggest that the value of heat flow (J_M^*) from both the oceanic and the continental mantle, corrected for the radiogenic heat generated in the lithosphere, is a function of the geological age t (see Fig.

4.1.5), and can be approximated by

$$J_M^* = \frac{489 \pm 32}{\sqrt{t}} - (8 \pm 7),$$

where the flow J_M^* is in $\mathrm{mW \cdot m^{-2}}$ and the age t is in millions of years.

4.1.2 Radiogenic heat in the Earth's mantle

It seems that heat in the mantle is primarily produced by the decay of radioactive elements. Just as in the crust, radiogenic heat is produced by the radioactive series (uranium, actinium and thorium series) and by the radioactive isotope potassium ^{40}K.

Two principal geothermal models of the Earth's mantle can now be distinguished, which depend on the assumed present concentration of uranium, thorium and potassium in the mantle material. They are the "chondritic" and the "terrestrial" models.

In general, the chondritic model of the mantle has a special position in the large group of various thermal models of the Earth. It is a model with a chemical and mineral composition close to the average composition of chondritic meteorites,* especially with regard to the concentration of uranium, thorium and potassium (see Table 4.1.3, after MacDonald

TABLE 4.1.3
Chondritic model of the Earth's mantle (after MacDonald, 1963)

Isotope	Concentration $[(10^{-8}\ \mathrm{g}) \cdot \mathrm{g}^{-1}]$	Radiogenic heat production $[10^{-16}\ \mathrm{W \cdot g^{-1}}]$
^{238}U	1.092	10.0
^{235}U	0.0079	0.45
^{232}Th	4.4	11.3
^{40}K	9.52	28.0

(1963)). The results of studies on the abundance of elements, however, seem to limit the applicability of the chondritic model. It has been found that though the ratio Th/U is the same (Th/U $\approx$ 4) for terrestrial magmatic rocks, chondritic meteorites and lunar rocks, the ratio K/U differs

* Stony meteorites composed of silicate minerals (primarily of olivines) also present in the Earth. They have a specific structure characterized by the presence of "chondrules" or spherical aggregations of crystalline silicates.

substantially by one order of magnitude and is K/U $\approx 10^4$ in terrestrial rocks, $\approx 10^5$ in chondrites and $\approx 10^3$ in lunar rocks (Horai and Fujii, 1971). The ratio Rb/Sr is also $10 \div 20$ times as great in chondrites as it is in terrestrial magmatic rocks (Gast, 1972). It seems that terrestrial material is poorer in potassium in comparison with chondritic meteorites.

On the other hand, there is no doubt that the matter forming the Sun as well as the matter forming the planets, planetoids and meteorites originated from a primitive solar nebula. Recent results of studies on the abundance of elements in the photosphere of the Sun and in various types of meteorites, as well as theoretical results relating to nucleosynthesis of elements, imply that in the Solar System the abundance of potassium normalized to silicon (K atoms per 10^6 Si atoms) or the ratio K/Si $\approx 3200 \div 4200$. For uranium and thorium it is U/Si ≈ 0.0262 and Th/Si ≈ 0.058, respectively (Cameron, 1973). From this it follows that in the Solar System the ratio K/U $\approx 10^5$ and Th/U ≈ 2.2.

It is possible that the deficiency of potassium in the Earth is not real and on a global scale the Earth has the solar, thus also chondritic, ratio K/Si. If that was the case, the average content of potassium in the terrestrial globe would be $650 \div 960$ ppm (1 ppm $= (10^{-6}\ \mathrm{g}) \cdot \mathrm{g}^{-1}$). Thus the observed deviations would mean that potassium was concentrated in the lower mantle or in the core (about $80 \div 90\%$ of the total amount of potassium). It cannot be excluded therefore that the radioactive decay of isotope ^{40}K forms the required source of energy for electrodynamic processes in the Earth's core (Goettel, 1976). The observed differences in the ratio Rb/Sr and K/Rb can be explained by differentiation processes. Potassium and rubidium could have been separated from strontium at an early stage of the Earth's evolution and remained deeper in the Earth's interior.

In general, when using the chondritic model of the Earth's mantle it is very convenient to assume that the similarity between chondritic and mantle material is primarily reflected in the same present production of radiogenic heat per mass unit. From Table 4.1.3 it follows that the radiogenic heat generation is about $50 \times 10^{-16}\ \mathrm{W \cdot g^{-1}}$ ($\approx 12 \times 10^{-16}\ \mathrm{cal \cdot g^{-1} \cdot s^{-1}}$) for chondrite. Therefore the chondritic model of the Earth's mantle with a mass of about 4.08×10^{27} g produces almost 20×10^{12} W ($\approx 4.9 \times 10^{12}\ \mathrm{cal \cdot s^{-1}}$) or about 6.5×10^{20} J/year. Since the area of the external mantle surface (the area of a sphere with the radius $R_M = 6335$ km) is $5.04 \times 10^{14}\ \mathrm{m^2}$ and the average heat flow at the Moho boundary $J_M \approx 42\ \mathrm{mW \cdot m^{-2}} \approx 1$ HFU, the heat lost by the mantle would be about 21×10^{12} W ($5.04 \times \times 10^{12}\ \mathrm{cal \cdot s^{-1}}$) or $\approx 6.9 \times 10^{20}$ J/year. It is easy to notice that the chondritic model of the mantle, with steady heat flow, practically accounts for the total present loss of heat by the Earth's interior, and this has almost become a classic argument for the chondritic model of the mantle.

The terrestrial model of the mantle (see Table 4.1.4) proposed by Wasserburg *et al.* (1964) has the ratio K/U = 10^4 and Th/U = 3.7, which are characteristic for terrestrial basic magmatic rocks. The content of uranium in the terrestrial model is chosen so as to produce radiogenic heat of about 6.9×10^{20} J/year, equal to the heat outflowing from the mantle.

TABLE 4.1.4
Terrestrial model of the mantle (after Wasserburg *et al.*, 1964)

Isotope	Concentration $[(10^{-8}\ g) \cdot g^{-1}]$	Radiogenic heat production $[10^{-16}\ W \cdot g^{-1}]$
^{238}U	2.239	20.6
^{235}U	0.0162	0.93
^{232}Th	8.344	21.4
^{40}K	2.68	7.9

Thus the terrestrial model of the mantle contemporarily generates the same amount of radiogenic heat as that produced by the chondritic model. In many cases both thermal models of the mantle can be considered as equivalent models, except of course when the Earth's thermal evolution is considered.

The distribution of heat flow $j(z)$ is shown in Fig. 4.1.6 for three variants of a simple model of the Earth's upper mantle. The variants differ from one another only in the values of the generated radiogenic heat. It has been assumed that the model under consideration corresponds to a steady state

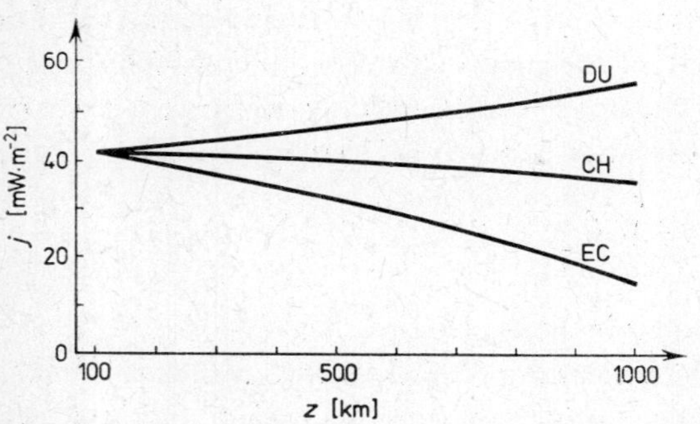

Fig. 4.1.6. Heat flow in a simple model of the Earth's upper mantle with natural radioactivity of dunite (DU), chondritic meteorite (CH) and eclogite (EC).

of heat flow and to the production of radiogenic heat per mass unit $Q_{\mathscr{M}} = \text{const}$. In this case the equation of transport is

$$\nabla_r^2 j(r) = Q_{\mathscr{M}} \varrho(r)$$

where ϱ is the density, $r = R_{\mathrm{Earth}} - z$ is the distance from the centre of the spherical globe, and ∇_r^2 is the radial part of the Laplace operator.

The calculations have been carried out for the following values of radiogenic heat $Q_{\mathscr{M}}$: 2.4, 50, and $106.3 \times 10^{-16}\ \mathrm{W \cdot g^{-1}}$, corresponding to the values of the present radiogenic heat of "average dunite", "average chondrite" and "average eclogite" (see e.g. MacDonald, 1964; Kaula, 1968), and with the boundary condition $j(R_M) \approx J_{\mathrm{Moho}} = 1\ \mathrm{HFU} \approx 42\ \mathrm{mW \cdot m^{-2}}$; the density distribution $\varrho(r)$ has been taken from Bullen (1963).

It is easy to see that the eclogitic variant (EC) of the upper mantle model could almost entirely explain the observed outflow of heat. An increase in the value of heat flow in the dunitic variant (DU) is a purely geometrical effect. The curves $j(z)$ from Fig. 4.1.6 suggest that the heat flow down to depths of a few hundred kilometres is not oversensitive to differences between the adopted values of radiogenic heat (and thus also of concentrations of natural radioactive elements) in the mantle material.

4.2 GEOTHERMAL GRADIENT WITHIN THE EARTH

4.2.1 Temperature gradient in the upper mantle

One of the peculiarities of the Earth's upper mantle is the Low-Velocity Layer (LVL), discovered by B. Gutenberg. This term is used for the roof layer of the mantle, where both positive and negative gradients of velocity of P and S waves are observed (see Fig. 4.2.1, according to Gutenberg (1958, 1959), Pekeris (1966) and Anderson (1967b)).

In general, the velocities v_P and v_S in the mantle depend on the chemical composition and phase state of the mantle material, and on the pressure p and temperature T. For an upper mantle model with a fixed chemical composition and phase (at least to a depth of 300 km) it can be written

$$\frac{\mathrm{d}v}{\mathrm{d}z} = \left(\frac{\partial v}{\partial p}\right)_T \frac{\mathrm{d}p}{\mathrm{d}z} + \left(\frac{\partial v}{\partial T}\right)_p \frac{\mathrm{d}T}{\mathrm{d}z}$$

where $v = v_P$ or $v = v_S$, and z is the depth.

It is well known that the velocities of elastic waves in silicate minerals (or, more generally, in chemically homogeneous solids) increase with pressure, and decrease with the growth of temperature (e.g., Soga *et al.*, 1966; Schreiber and Anderson, 1967; Birch, 1969; Liu *et al.*, 1975).

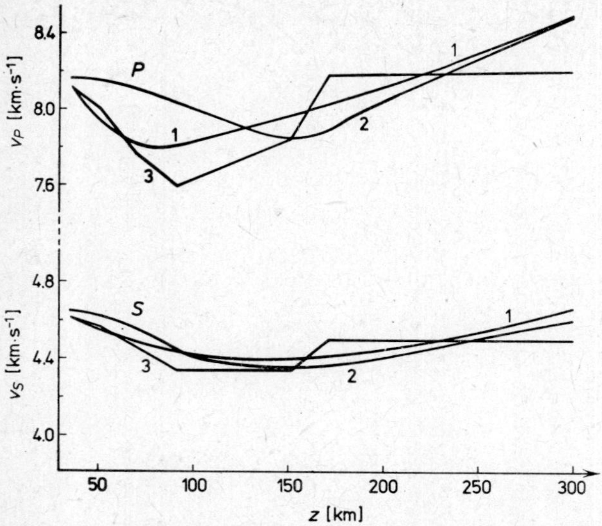

Fig. 4.2.1. Distribution of P and S wave velocities in the Low-Velocity Layer (LVL), according to Gutenberg (1958, 1959)—curves 1, Pekeris (1966)—curves 2 and Anderson (1967b)—curves 3 (model CIT 11 A).

If it is assumed that both factors, i.e., the positive gradients of pressure and of temperature, are responsible for the origin of the LVL, then the variation of P or S wave velocity in relation to depth can be described by the formula

$$\frac{\mathrm{d}v}{\mathrm{d}z} = \left(\frac{\partial v}{\partial p}\right)_T \frac{\mathrm{d}p}{\mathrm{d}z} - \left|\left(\frac{\partial v}{\partial T}\right)_p\right| \frac{\mathrm{d}T}{\mathrm{d}z}$$

since only $(\partial v/\partial T)_p < 0$.

Under conditions of hydrostatic equilibrium it follows that

$$\frac{\mathrm{d}p}{\mathrm{d}z} = \varrho g$$

where ϱ is the density and g is the gravity acceleration.

In this case

$$\frac{\mathrm{d}v}{\mathrm{d}z} = \left(\frac{\partial v}{\partial p}\right)_T \varrho g - \left|\left(\frac{\partial v}{\partial T}\right)_p\right| \frac{\mathrm{d}T}{\mathrm{d}z}$$

and therefore

$$\frac{\mathrm{d}T}{\mathrm{d}z} = \frac{\left(\frac{\partial v}{\partial p}\right)_T \varrho g - \frac{\mathrm{d}v}{\mathrm{d}z}}{\left|\left(\frac{\partial v}{\partial T}\right)_p\right|}$$

At the point $z_{\min}$ of the minimum velocity of a given P or S wave $\mathrm{d}v/\mathrm{d}z = 0$ and

$$\left.\frac{\mathrm{d}T}{\mathrm{d}z}\right|_{z_{\min}} = \frac{\left(\dfrac{\partial v}{\partial p}\right)_T}{\left|\left(\dfrac{\partial v}{\partial T}\right)_p\right|} \varrho g \tag{4.2.1}$$

From thermodynamic relations it follows that

$$\left(\frac{\partial v}{\partial p}\right)_T \left(\frac{\partial T}{\partial v}\right)_p = \left(\frac{\partial T}{\partial p}\right)_v$$

Formula (4.2.1) can thus be written as

$$\left.\frac{\mathrm{d}T}{\mathrm{d}z}\right|_{z_{\min}} = \left(\frac{\partial T}{\partial p}\right)_v \varrho g \tag{4.2.2}$$

When considering the thermal conditions leading to the origin of the LVL, Valle (1956) and MacDonald and Ness (1961) used the approximation

$$\left(\frac{\partial T}{\partial p}\right)_{v_S} \approx \left(\frac{\partial T}{\partial p}\right)_{\varrho}$$

Since

$$\left(\frac{\partial p}{\partial T}\right)_{\varrho} = \alpha K_T$$

where $\alpha \equiv -(1/\varrho)(\partial\varrho/\partial T)_p$ is the coefficient of thermal expansion and $K_T \equiv \varrho(\partial p/\partial\varrho)_T$ is the isothermal bulk modulus, relation (4.2.2) takes the form

$$\left.\frac{\mathrm{d}T}{\mathrm{d}z}\right|_{z_{\min}} = \frac{\varrho g}{\alpha K_T}$$

From the approximation

$$K_T \approx K_S = \varrho\phi$$

where K_S is the adiabatic bulk modulus and $\phi \equiv v_P^2 - \frac{4}{3}v_S^2$ is the seismic parameter, it follows that

$$\left.\frac{\mathrm{d}T}{\mathrm{d}z}\right|_{z_{\min}} = \frac{g}{\alpha}\phi^{-1} \tag{4.2.3}$$

Assuming that for the depth $z_{min} = 150$ km: $\phi = 36\ \mathrm{km}^2 \cdot \mathrm{s}^{-2}$, $g = 990\ \mathrm{cm} \cdot \mathrm{s}^{-2}$ and $\alpha = \alpha_0 \approx 4 \times 10^{-5}\ \mathrm{K}^{-1}$, we obtain

$$\left.\frac{\mathrm{d}T}{\mathrm{d}z}\right|_{150} \approx 6.9\ \mathrm{K} \cdot \mathrm{km}^{-1}$$

Formula (4.2.3) was slightly modified by MacDonald and Ness (1961), who took into account the dependence of the coefficient α on temperature and pressure. It is well known that the coefficient of thermal expansion of minerals (solids) increases with temperature and decreases with increasing pressure. At temperatures higher than the Debye temperature, however, the influence of temperature on the coefficient of thermal expansion becomes so small that it can be readily neglected. The influence of pressure on thermal expansion, on the other hand, can be described, according to MacDonald and Ness, by a semi-empirical relation (under an assumption that the pressure is not too large in comparison with the bulk modulus K_T)

$$\alpha = \alpha_0[1-\alpha'(p-p_0)K_T^{-1}]$$

which after the introduction of the approximation $K_T \approx K_S$ takes the form

$$\alpha = \alpha_0\left[1-\frac{\alpha'(p-p_0)}{\varrho\phi}\right] \tag{4.2.4}$$

where α_0 is the coefficient of thermal expansion under the pressure p_0, and α' is the empirical parameter. The value of this parameter is about 4 for such minerals as forsterite (Mg_2SiO_4) or periclase (MgO) (MacDonald and Ness, 1961).

Thus, assuming $\alpha_0 \approx 4 \times 10^{-5}\ \mathrm{K}^{-1}$, $\alpha' \approx 4$ and $p_0 \approx 1$ GPa (at the Moho boundary), the following value is obtained: $\alpha \approx 3.7 \times 10^{-5}\ \mathrm{K}^{-1}$ for $z_{min} = 150$ km, and

$$\left.\frac{\mathrm{d}T}{\mathrm{d}z}\right|_{150} \approx 7.4\ \mathrm{K} \cdot \mathrm{km}^{-1}$$

Thomsen (1967) has challenged the validity of the approximation $(\partial T/\partial p)_{v_S} \approx (\partial T/\partial p)_\varrho$, as laboratory experiments carried out by O. L. Anderson have shown that the following relation is valid for silicates and oxides:

$$\left(\frac{\partial T}{\partial p}\right)_\varrho > \left(\frac{\partial T}{\partial p}\right)_{v_S}$$

For the estimation of the thermal gradient in the LVL Thomsen proposed to use the well-known formula of Launay, connecting the Debye temperature θ with the elastic wave velocities v_P and v_S

$$\theta = \frac{h}{k}\left(\frac{9N\varrho}{4\pi\bar{A}}\right)^{1/3}\left(\frac{1}{v_P^3}+\frac{2}{v_S^3}\right)^{-1/3} \tag{4.2.5}$$

where h is the Planck constant, k is the Boltzmann constant, N is the Avogadro number, $\bar{A}$ is the mean atomic mass and ϱ is the medium density.

Since always $v_P > v_S$, the quotient v_S^3/v_P^3 is small enough to be neglected in formula (4.2.5), and thus after the calculation of the logarithmic derivative the following relation is obtained:

$$\left(\frac{\partial v_S}{\partial \varrho}\right)_T \approx \frac{v_S}{\varrho}\left(\gamma-\frac{1}{3}\right)$$

where $\gamma \equiv (\partial \ln\theta/\partial \ln\varrho)_T$ is the Grüneisen parameter. Since

$$\left(\frac{\partial v_S}{\partial \varrho}\right)_T = \left(\frac{\partial v_S}{\partial p}\right)_T \frac{K_T}{\varrho}$$

it can be written

$$\left(\frac{\partial v_S}{\partial p}\right)_T = \frac{v_S}{K_T}\left(\gamma-\frac{1}{3}\right)$$

and introducing the approximation $K_T \approx K_S = \varrho\phi$, we have

$$\left(\frac{\partial v_S}{\partial p}\right)_T = \frac{v_S}{\varrho\phi}\left(\gamma-\frac{1}{3}\right)$$

In this case the relation $(\partial v_S/\partial p)_T(\partial v_S/\partial T)_p^{-1} = (\partial T/\partial p)_{v_S}$ takes the following form:

$$\left(\frac{\partial T}{\partial p}\right)_{v_S} = \left|\left(\frac{\partial T}{\partial v_S}\right)_p\right| \frac{v_S}{\varrho\phi}\left(\gamma-\frac{1}{3}\right)$$

and therefore (see formula (4.2.1))

$$\left.\frac{\mathrm{d}T}{\mathrm{d}z}\right|_{z_{\min}} = \left|\left(\frac{\partial T}{\partial v_S}\right)_p\right| \frac{v_S}{\phi} g\left(\gamma-\frac{1}{3}\right)$$

Assuming for the depth $z_{\min} = 150$ km the values $-(\partial T/\partial v_S)_p \approx 3200$ $\mathrm{s \cdot K \cdot km^{-1}}$ (after Thomsen, 1967), $v_S = 4.35\ \mathrm{km \cdot s^{-1}}$, $\phi = 36\ \mathrm{km^2 \cdot s^{-2}}$, $g = 990\ \mathrm{cm \cdot s^{-2}}$ and $\gamma = 1.3$, the thermal gradient is obtained

$$\left.\frac{\mathrm{d}T}{\mathrm{d}z}\right|_{150} \approx 4\ \mathrm{K \cdot km^{-1}}$$

The thermal gradient in the LVL can also be estimated by extrapolation of laboratory results relating to the influence of pressure and temperature upon the velocities of elastic waves in basic rock samples or in silicate and oxide minerals. F. Birch was the first to estimate in this way the critical gradient* to be about 6.6 $K \cdot km^{-1}$ (see Ringwood, 1962). From laboratory data collected by Anderson and Sammis (1970) it follows that the values of the derivatives $(\partial T/\partial p)_{v_S}$ and $(\partial T/\partial p)_{v_P}$ are 84.5 and 251 $K \cdot GPa^{-1}$ for olivine, and 117 and 297 $K \cdot GPa^{-1}$ for pyroxene. According to the mantle model of Bullen (1963), at a depth of $z_{min} = 150$ km the density is $\varrho = 3.42$ $g \cdot cm^{-3}$ and the gravity acceleration is $g = 990$ $cm \cdot s^{-2}$. Consequently the value of the derivative $(\partial T/\partial p)_{v_S} \approx 85 \div 117$ $K \cdot GPa^{-1}$ implies, in an olivine–pyroxene model of the upper mantle, the critical temperature gradient

$$\left.\frac{dT}{dz}\right|_{150} = \left(\frac{\partial T}{\partial p}\right)_{v_S} \varrho g \approx 3 \div 4 \ K \cdot km^{-1}$$

Similarly, for a depth of $80 \div 90$ km (the density ≈ 3.38 $g \cdot cm^{-3}$), where a minimum velocity of the P wave occurs (see Fig. 4.2.1), the derivative $(\partial T/\partial p)_{v_P} \approx 251 \div 297$ $K \cdot GPa^{-1}$ gives the critical gradient as

$$\frac{dT}{dz} = \left(\frac{\partial T}{\partial p}\right)_{v_P} \varrho g \approx 8 \div 10 \ K \cdot km^{-1}$$

Similar values of the critical gradient are obtained from laboratory data relating to oxide minerals such as garnet, spinel, periclase (Anderson and Sammis, 1970; Schreiber and Anderson, 1966).**

The occurrence of the Low-Velocity Layer can also be connected with the existence of the Partial Melting Zone (PMZ) in the Earth's upper mantle. It is an area in which certain mineral components of the mantle are either in a molten phase or close to melting. A strong attenuation of seismic waves and the occurrence of velocity gradients would result from such a state.

The value of the temperature gradient in the PMZ should be close to the slope (gradient) value of the corresponding solidus curve (Magnitsky, 1971). However, the mineral composition of the upper mantle is not well known. Most silicates have such a slope coefficient dT_m/dp of the melting curve that at a pressure of 3 GPa (the depth ≈ 100 km) the value of the

* A temperature gradient neutralizing the influence of a pressure gradient on the velocity of P or S elastic waves.

** Laboratory data for dunite and eclogite under a pressure of $1 \div 3$ GPa suggest the critical gradient $dT/dz = (\partial T/\partial p)_{v_P} \varrho g \approx 6.3$ $K \cdot km^{-1}$ (Christensen, 1979).

corresponding gradient of the melting point dT_m/dz is $2 \div 5$ K · km^{-1}. For multi-component systems the slope of the melting curve is generally smaller. Yoder and Tilley (1962) found that for basalt $dT_m/dp \approx 65$ K · GPa^{-1} and for eclogite $dT_m/dp \approx 110$ K · GPa^{-1}. In general, volcanic rocks are characterized by a slope dT_m/dp of about 100 K · GPa^{-1}, which at the top of the upper mantle corresponds to a melting point gradient of about 3 K · km^{-1} (Anderson and Sammis, 1970). The possible presence of H_2O in the upper mantle is practically of no consequence for these results.

The observed rapid increase in the velocity of seismic waves at a depth of about 400 km is most probably caused by structural changes in silicates, which are the principal minerals of the Earth's mantle. Several possible phase transitions are usually considered, one of them being the transformation Mg_2SiO_4 (olivine structure) → Mg_2SiO_4 (spinel structure), associated with a density increase of about 10%.

According to Akimoto and Fujisawa (1968), an olivine composed in 90% of Mg_2SiO_4 and in 10% of Fe_2SiO_4 at a temperature of about 1470 K already begins to transform into a spinel structure under a pressure of about 13 GPa, corresponding to a depth of about 370 km below the Earth's surface. The pressure interval in which both phases coexist is about $2 \div 3$ GPa, according to the diagrams of phase transitions. A mantle layer with thickness $\Delta z \approx 50 \div 80$ km, which is in excellent agreement with the thickness of the so-called Seismic Transition Zone (Press, 1970), corresponds to such an interval of pressure. Magnitsky (1971) used the results of Akimoto and Fujisawa to estimate the thermal gradient at a depth of about 400 km, and obtained the following result:

$$\left.\frac{dT}{dz}\right|_{400} \approx 1.8 \div 3.6 \text{ K} \cdot \text{km}^{-1}$$

To estimate the temperature gradient in the Earth's mantle at a depth of 1000 km Maj (1978a) used the "seismic equation of state", introduced for silicates and oxides by Anderson (1967a),

$$\varrho = \Gamma \bar{A} \phi^{\zeta} \tag{4.2.6}$$

where ϱ is the density, $\bar{A}$ is the mean atomic mass, ϕ is the seismic parameter and Γ and ζ are empirical constants (see Chapter 2).

From eqn. (4.2.6), under the assumption that the coefficients Γ and ζ and $\bar{A}$ are constant, it follows that

$$\frac{d\varrho}{dz} = \varrho \zeta \phi^{-1} \frac{d\phi}{dz} \tag{4.2.7}$$

Since $\varrho = \varrho(p, T)$, we have on the other hand

$$\frac{d\varrho}{dz} = \left(\frac{\partial\varrho}{\partial p}\right)_T \frac{dp}{dz} + \left(\frac{\partial\varrho}{\partial T}\right)_p \frac{dT}{dz} = \frac{\varrho^2}{K_T} g - \varrho\alpha \frac{\partial T}{dz} \tag{4.2.8}$$

where K_T is the bulk modulus, α is the coefficient of thermal expansion and g is the gravity acceleration.

From eqns. (4.2.7) and (4.2.8), on introducing the approximation $K_T \approx K_S$, the following relation is obtained (for the case of hydrostatic pressure):

$$\frac{dT}{dz} = \frac{1}{\alpha}\left(g - \zeta\left|\frac{d\phi}{dz}\right|\right)\phi^{-1}$$

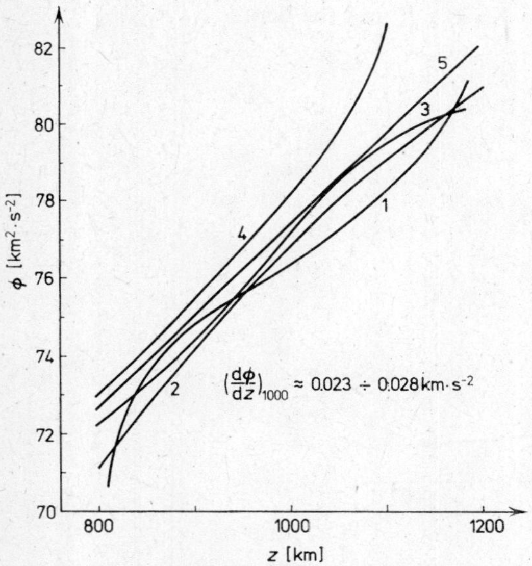

Fig. 4.2.2. Seismic parameter $\phi \equiv v_P^2 - \frac{4}{3}v_S^2$ in the Earth's mantle at a depth of 1000 km. P and S wave velocities were taken from: Gutenberg (1959)—1, Birch (Pekeris, 1966)—2, Jeffreys (Bullen, 1963)—3, Anderson (1967b)—4 and Dziewonski *et al.* (1975)—5.

The seismic parameter ϕ as a function of depth, for depths between 800 and 1200 km, is shown in Fig. 4.2.2 in several seismic models of the Earth's mantle (v_P and v_S after Gutenberg (1959), Birch (Pekeris, 1966), Jeffreys (Bullen, 1963), Anderson (1967b) and Dziewonski *et al.* (1975)). At a depth of 1000 km the average value of the seismic parameter is $\phi \approx 77$ km$^2 \cdot$ s^{-2} and of the gradient $d\phi/dz \approx 0.023 \div 0.028$ km $\cdot$ s^{-2}. Assuming that for the material of the mantle $\zeta = 0.3$ (Anderson, 1967a; Simmons and England, 1970), $g_{1000} = 995$ cm $\cdot$ s^{-2} (Bullen, 1963) and $\alpha_{1000} \approx 2\times$

$\times 10^{-5}$ K^{-1} (from the relation of MacDonald–Ness (4.2.4)), we obtain

$$\left.\frac{dT}{dz}\right|_{1000} \approx 1 \div 2 \text{ K} \cdot \text{km}^{-1}$$

The probable minimum values of the temperature gradient for depths of 150, 400 and 1000 km, equal to about 4, 1.8 and 1 K · km^{-1}, respectively, suggest a possible distribution of the thermal gradient in the mantle (see Fig. 4.2.3). The curve dT/dz from Fig. 4.2.3 can be approximated with high accuracy by the equation

$$\frac{dT}{dz} = \frac{530}{z} + 0.47 \tag{4.2.9}$$

where the gradient dT/dz is in K · km^{-1}, and the depth z is in km.

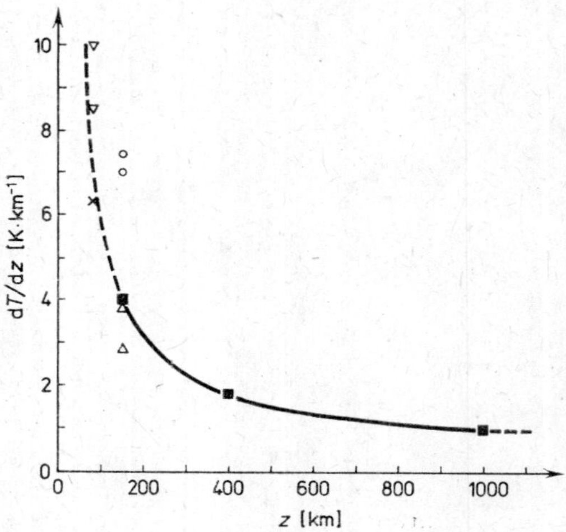

Fig. 4.2.3. Probable thermal gradient in the Earth's upper mantle (between 35 and 1000 km). The solid-dashed line represents eqn. (4.2.9); the solid squares denote the minimum values of the geothermal gradient at depths of 150, 400 and 1000 km, calculated by Thomsen (1967), Magnitsky (1971) and Maj (1978a); the open circles show data taken from Valle (1956) and MacDonald and Ness (1961); the open triangles represent the values calculated from laboratory data for olivine and pyroxene according to Anderson and Sammis (1970); the cross shows the value deduced from laboratory data for dunite and eclogite (Christensen, 1979).

From an extrapolation of eqn. (4.2.9) it follows that at a depth of $z = 35$ km, corresponding to the Moho boundary, under the continents the temperature gradient (on the side of the upper mantle) is about 16 K · km^{-1}. It is well known that the average global heat flow from the mantle has the value $J_M \approx 1$ HFU (≈ 42 mW · m^{-2}) (see e.g., Pollack and Chapman, 1977a). On the other hand, the phonon thermal conductivity at the top of a dunite–peridotite* model of the upper mantle is most probably $\varkappa' \approx 2.5\ \mathrm{W \cdot m^{-1} \cdot K^{-1}}$ or $\approx 6\ \mathrm{mcal \cdot cm^{-1} \cdot s^{-1} \cdot K^{-1}}$ (Horai and Simmons, 1970). In this case, since the flow $j(z = 35) \approx J_M$, we would have $\mathrm{d}T/\mathrm{d}z|_{35} = J_M/\varkappa' \approx 17\ \mathrm{K \cdot km^{-1}}$. Similarly, for the depth $z \approx 2800$ km eqn. (4.2.9) gives $\mathrm{d}T/\mathrm{d}z|_{2800} \approx 0.66\ \mathrm{K \cdot km^{-1}}$; thus the temperature gradient in the lower mantle is close to the adiabatic gradient.

The temperature distribution in the mantle corresponding to the thermal gradient (4.2.9) is as follows

$$T = 712 + 530 \ln(z/35) + 0.47(z - 35) \tag{4.2.10}$$

under the assumption that the temperature at a depth of 100 km below the Earth's surface is about 1300 K, according to volcanological and petrological data.

The Earth's mantle is known to be in a solid phase, and thus its temperature at any depth must be lower than the melting point T_m, although the shapes of the $T(z)$ and $T_m(z)$ curves may be very similar to each other (especially at greater depths). Of course, the melting point forms the upper limit of effective temperatures in the mantle. The lower limit is determined by the adiabatic temperature, which can be obtained from the adiabatic gradient.

4.2.2 *Adiabatic gradient in the lower mantle*

In general, the effective thermal gradient in the Earth's interior can be written in the form of a sum

$$\frac{\mathrm{d}T}{\mathrm{d}z} = \left(\frac{\mathrm{d}T}{\mathrm{d}z}\right)_{\mathrm{ad}} + \left(\frac{\mathrm{d}T}{\mathrm{d}z}\right)_{\mathrm{nad}}$$

where $(\mathrm{d}T/\mathrm{d}z)_{\mathrm{ad}}$ is the adiabatic gradient, i.e., the temperature gradient formed as a result of heating of the terrestrial globe in the process of gravitational self-compression, and $(\mathrm{d}T/\mathrm{d}z)_{\mathrm{nad}}$ is the nonadbiabatic** component or the temperature gradient related to other energy sources.

---

* According to Beck *et al.* (1978), the phonon thermal conductivity of dunites is equal to $2.5 \div 5\ \mathrm{W \cdot m^{-1} \cdot K^{-1}}$ at a temperature of about 600 K and under pressure of about 1 GPa.

** The term "superadiabatic gradient" is also used.

The adiabatic gradient is given by the following formula (see e.g., Jacobs, 1974):

$$\left(\frac{\mathrm{d}T}{\mathrm{d}z}\right)_{\mathrm{ad}} = \left(\frac{\partial T}{\partial p}\right)_S \frac{\mathrm{d}p}{\mathrm{d}z} = \left(\frac{\partial T}{\partial p}\right)_S \varrho g$$

where S is the entropy, ϱ is the density and g is the gravity acceleration.

From thermodynamic relations it follows that

$$\left(\frac{\partial T}{\partial p}\right)_S = \frac{\alpha}{\varrho}\frac{T}{c_p}$$

where α is the coefficient of thermal expansion and c_p is the specific heat under constant pressure. In this situation we have

$$\left(\frac{1}{T}\frac{\mathrm{d}T}{\mathrm{d}z}\right)_{\mathrm{ad}} = \frac{\alpha g}{c_p}$$

Using the Grüneisen law

$$\gamma = \frac{\alpha}{c_p}\frac{K_S}{\varrho} = \frac{\alpha}{c_p}\phi$$

where γ is the Grüneisen parameter, K_S is the adiabatic bulk modulus and ϕ is the seismic parameter, we obtain

$$\left(\frac{1}{T}\frac{\mathrm{d}T}{\mathrm{d}z}\right)_{\mathrm{ad}} = \frac{\gamma g}{\phi} \qquad (4.2.11)$$

The average adiabatic gradient in the lower mantle is about 0.33 $\mathrm{K \cdot km^{-1}}$ (Jacobs, 1974).

The results of seismic studies of the core–mantle boundary indicate that there is a local minimum in P and S wave velocities in the lowest layer of the lower mantle. It is possible that a decrease in v_P and v_S values resulted from the high thermal gradient in this thermal boundary layer of the lower mantle.

From the data listed in Sections 4.1.1 and 4.1.2, and relating to thermal balance in a chondritic model of the Earth's mantle, it follows that the difference between the heat emitted from the mantle (the total heat flow at the boundary) and the total radiogenic heat produced in the mantle at present is about $(1 \div 4) \times 10^{12}$ W. It is possible that this is the real energy generated in the core and transmitted to the lower mantle, and not an effect resulting from inaccurate estimations. In this case the heat flow at the core–mantle boundary ($j_{\mathrm{c-m}}$) would have a value between 7 and 26 $\mathrm{mW \cdot m^{-2}}$.

According to Loper (1978) the Earth's core should be assumed to have lost about 3.7×10^{12} $J \cdot s^{-1}$ to explain the existence of the inner core in a solid phase as well as its gradual growth during the Earth's history.

In a solid model of the lower mantle the phonon thermal conductivity ($\varkappa$) is the only significant element of thermal energy transfer. Assuming that in the lower mantle $\varkappa \leqslant 4$ $W \cdot m^{-1} \cdot K^{-1}$ (see e.g., Kieffer, 1976), the temperature gradient ($dT/dz = j_{c-m}/\varkappa$) in the thermal boundary layer could not be smaller than about $2 \div 7$ $K \cdot km^{-1}$. Thus the effective thermal gradient in the lowest mantle layer, which borders on the core–mantle boundary, might exceed the average adiabatic gradient in the lower mantle (≈ 0.3 $K \cdot m^{-1}$) by as much as one order of magnitude.

4.2.3 Melting point in the Earth's mantle and core

According to the elementary theory of Lindemann, the melting of a solid begins when the amplitude of vibrations of the crystalline lattice is large enough for direct interactions between neighbouring elements of the lattice, which are ions or molecules. For a simple space lattice with the melting point $T_m \gg \theta$ (Debye temperature) such a "thermal resonance" appears at vibration frequencies

$$\nu = \nu_L = c_L T_m^{1/2} \varrho^{1/3} \bar{A}^{-5/6} \tag{4.2.12}$$

where ϱ is the density, $\bar{A}$ is the mean atomic mass and c_L is a coefficient dependent on the lattice constant and on the size of the average effective sphere of molecule or ion activity. In the theory of Lindemann it is assumed that the coefficient c_L is constant for several different substances (see e.g., Ziman, 1962; Blackman, 1955; Knopoff and Shapiro, 1972). The frequency ν_L in formula (4.2.12) is often called the *Lindemann frequency*.

The theory of Lindemann was first used by Uffen (1952) for an estimation of the distribution of the relative melting point in a monoatomic ($A = 21$) model of the Earth's mantle (see also Gilvarry, 1957). He used the well-known condition

$$\nu_L = \nu_D$$

identifying the Lindemann frequency ν_L with the characteristic frequency of Debye ν_D. This frequency is connected with the elastic properties of a solid isotropic medium by the following relation:

$$\nu_D = \left(\frac{3N\varrho}{4\pi\bar{A}} \right)^{1/3} v_m \tag{4.2.13}$$

where N is the Avogadro number and v_m is the average velocity of sound defined as

$$v_m = \sqrt[3]{\frac{3}{v_P^{-3}+2v_S^{-3}}}$$

From relations (4.2.12) and (4.2.13) it follows that

$$T_m = \left(\frac{1}{c_L}\sqrt[3]{\frac{3N}{4\pi}}\right)^2 \bar{A}v_m^2$$

Thus for a model of the Earth's mantle with $\bar{A}$ = const and c_L = const we have

$$\frac{T_m}{T_{m_{100}}} = \left(\frac{v_m}{v_{m_{100}}}\right)^2 = \left[\frac{(v_S^3+2v_P^3)_{100}}{v_S^3+2v_P^3}\right]^{2/3}\left[\frac{v_P v_S}{(v_P v_S)_{100}}\right]^2 \tag{4.2.14}$$

where the index "100" denotes a depth of 100 km below the Earth's surface (Uffen, 1952).

The average velocity of sound can be defined in various ways. It can be defined, for example, as an arithmetical mean of P and S wave velocities $\langle v \rangle = \frac{1}{3}(v_P+2v_S)$ or as the square root of the seismic parameter ϕ

$$v_\phi = \sqrt{\phi} \equiv \sqrt{v_P^2-\tfrac{4}{3}v_S^2} = \sqrt{\frac{K_S}{\varrho}} \tag{4.2.15}$$

where K is the adiabatic bulk modulus. The velocity v is called the *hydrodynamic velocity of sound* (McQueen *et al.*, 1964).

Replacing v_m in formula (4.2.13) by the hydrodynamic velocity (4.2.15), the following relation is obtained:

$$\nu_D = \left(\frac{3N\varrho}{4\pi\bar{A}}\right)^{1/3} \phi^{1/2} \tag{4.2.16}$$

Since $\phi = K_S/\varrho$,

$$\nu_D = \left(\frac{3N}{4\pi}\right)^{1/3} \bar{A}^{-1/3}K_S^{1/2}\varrho^{-1/6}$$

This is the well-known *relation of Madelung–Einstein* (see e.g., Blackman, 1955; Meyer, 1968).

From eqns. (4.2.12) and (4.2.16) it follows that

$$T_m = \left(\frac{1}{c_L}\sqrt[3]{\frac{3N}{4\pi}}\right)^2 \bar{A}v_\phi^2$$

and for a mantle model with $\bar{A}$ = const and c_L = const it can be written

(Uffen, 1952)

$$\frac{T_m}{T_{m_{100}}} = \left(\frac{v_\phi}{v_{\phi_{100}}}\right)^2 = \frac{v_P^2 - \frac{4}{3}v_S^2}{(v_P^2 - \frac{4}{3}v_S^2)_{100}} \tag{4.2.17}$$

The distribution of the melting point in the Earth's mantle, found from relation (4.2.14) for P and S wave velocities corresponding to the model of Gutenberg–Bullen, is shown in Fig. 4.2.4.*

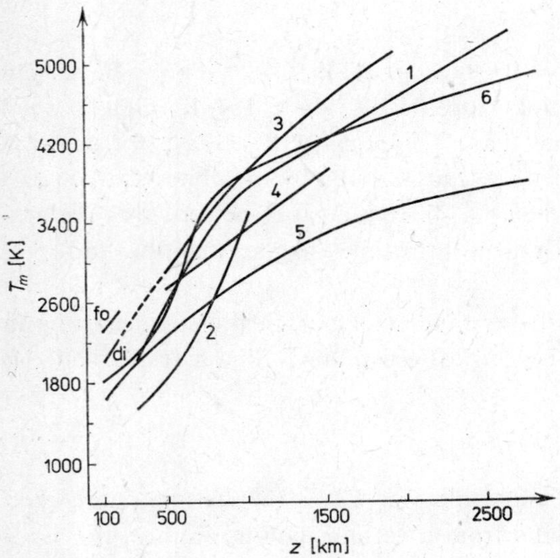

Fig. 4.2.4. Various distributions of the melting point (T_m) in the Earth's mantle: 1—the curve according to Uffen (eqn. (4.2.12) with v_S and v_P taken from Pekeris (1966)), 2—the branch of $T_m(z)$ curve according to Clark (1963), 3—the "diopside approximation" of the melting point (see text) according to MacDonald (1959), 4—the "enstatite approximation" (data from Boyd *et al.*, 1964), 5—the "basaltic approximation" (Jacobs, 1974), 6—eqn. (4.2.21a); di, fo—from laboratory data for diopside (Boyd and England, 1963) and forsterite (Davis and England, 1964).

Numerous laboratory experiments have been carried out under pressure of up to about 5 GPa to establish the melting curves, i.e., the relations $T_m(p)$, for the more important minerals of the Earth's mantle, such as forsterite (Mg_2SiO_4), enstatite ($MgSiO_3$) or diopside ($CaMgSi_2O_6$). Davis and England (1964) found that the relation describing the melting

* In general, since the velocities of seismic waves P and S depend on pressure and temperature, their values should be reduced to the same temperature, e.g., temperature $T_{m_{100}}$, when relations (4.2.14) and (4.2.17) are used.

of forsterite is of the form

$$T_m = 47.7\,p + 2171$$

where p is in GPa and T_m in K. Thus the gradient is $\mathrm{d}T_m/\mathrm{d}p = 47.7$ $\mathrm{K \cdot GPa^{-1}}$. For diopside Williams and Kennedy (1969) found the relation

$$T_m = 1652.1 + 131.94\,p + 63.57\,p^2$$

where p is in GPa. In this case the gradient is

$$\frac{\mathrm{d}T_m}{\mathrm{d}p} = 131.94 + 127.14\,p$$

thus under the pressure $p = 0$, $\mathrm{d}T_m/\mathrm{d}p \approx 131.9$ $\mathrm{K \cdot GPa^{-1}}$. Boyd and England (1963) found that for diopside $\mathrm{d}T_m/\mathrm{d}p \approx 154$ $\mathrm{K \cdot GPa^{-1}}$ when $p = 0$. Similarly, it was found that $\mathrm{d}T_m/\mathrm{d}p \approx 128$ $\mathrm{K \cdot GPa^{-1}}$ for enstatite (Boyd *et al.*, 1964) and $\mathrm{d}T_m/\mathrm{d}p \approx 140$ $\mathrm{K \cdot GPa^{-1}}$ for albite ($NaAlSi_3O_8$) (Boyd and England, 1963) when $p = 0$. It can easily be seen that the gradient $\mathrm{d}T_m/\mathrm{d}p$ for forsterite is almost three times smaller than that for other silicates.

Such laboratory data can be extrapolated to high pressures existing in the mantle by using the semi-empirical equation of Simon (Bullard, 1954)

$$p = \frac{a}{B}\left(\frac{T_m}{T_{m0}}\right)^B - 1 \tag{4.2.18}$$

where a and B are empirical constants and T_{m0} is the melting point when $p = 0$. Equation (4.2.18) results from a certain modification of the well-known Clausius–Clapeyron equation:

$$T_m\left(\frac{\mathrm{d}T_m}{\mathrm{d}p}\right)^{-1} = \frac{q}{\Delta V} \tag{4.2.19}$$

where q is the phase transition heat (heat of melting) and ΔV is the change of specific volume during a given transition. Assuming that $q/\Delta V$ is a linear function of pressure ($q/\Delta V = Bp + a$), eqn. (4.2.18) is obtained after the integration of relation (4.2.19).

The values of the parameters in the Simon equation are: $T_{m0} = 1830$ K, $a = 2.85$ GPa, $B = 5.01$ for enstatite; $T_{m0} = 1665$ K, $a = 2.33$ GPa, $B = 4.46$ for diopside; $T_{m0} = 1391$ K, $a = 1.95$ GPa, $B = 5.1$ for albite.

By correlating the values of $T_m(p)$ found from the Simon equation with the corresponding depths in the mantle, we obtain the curves $T_m(z)$, which may be considered as upper limits of the melting point in the real mantle (see Fig. 4.2.4). The Earth's mantle is a multi-component system and its melting point is always lower than that of the individual components.

It is well known that under given conditions an unequivocal melting point can be assigned only to a single solid phase. In mixed crystals, such as e.g., olivines or plagioclases, the melting process is related to a temperature interval determined by two curves—the solidus and the liquidus curves. Thus, for example, an olivine composed of forsterite (Mg_2SiO_4) in 90% and of fayalite (Fe_2SiO_4) in 10% needs an interval of temperatures of about 150 K (from about 1940 to about 2100 K) for complete melting; pure forsterite melts at a temperature of about 2160 K and pure fayalite at a temperature of 1480 K (Smith, 1963). In such cases the classic notion of the melting point cannot be used. It must be replaced by the solidus temperature for a given chemical composition.

A similar situation, though much more complex, exists in multi-component systems such as magmatic rocks. Of course, the melting points of such systems are lower than those of pure mineral components.

In general, a model of magmatic rock composed of n pure minerals has $(2^n-1)-n$ different melting points (not counting the n melting points of the individual mineral components). They are the temperatures corresponding to the eutectic points of the magmatic alloy. The temperature of the lowest eutectic point, i.e., the eutectic minimum,* will be considered as the melting point.

The eutectic minimum depends on the pressure. Laboratory data imply that in polymineral systems the eutectic minimum (T_m^{eut}) increases with the pressure more slowly than the melting point of the individual mineral components. A change of entropy (corresponding to the eutectic minimum) during the melting of a multi-phase system is given by

$$\Delta S = \sum_i \Delta S_i + \Delta S_M$$

where ΔS_i is the change of entropy during the melting of the i-th component and ΔS_M is the change of entropy caused by the mixing of melted components.

From the Clausius–Clapeyron equation (4.2.19) in the form $\mathrm{d}T_m/\mathrm{d}p = \Delta V/\Delta S$ it follows that the slope of the eutectic minimum curve $T_m^{\text{eut}}(p)$ should be smaller than slope of the melting curves $T_m(p)$ of the individual components in a given eutectic system. It should be noted that the change of entropy (ΔS_M) caused by the mixing of melted minerals, usually small when the chemical composition of the minerals is not changed, can reach

* The eutectic minimum is the temperature at which all mineral components of a crystallizing alloy are simultaneously disengaged.

a high value when the melting is accompanied by disintegration of the minerals into simpler compounds, e.g., oxides.

Eutectic mixtures seldom appear in magmatic rocks. During crystallization of the magma chemical reactions take place between the solid phase, i.e., the crystals already disengaged, and the liquid phase, i.e., the remaining magmatic alloy. As a result, minerals which can be systematized into the series called the *reaction series* are very often formed. Each cell in such a series is formed as a result of a chemical reaction between the primary cell and the magmatic alloy, provided that the crystals are not rapidly removed from the magma, e.g., under the influence of the gravity force. Two types of reaction series are distinguished—the continuous and discontinuous series. Plagioclases are an example of a continuous series, and the dark minerals of magmatic rocks (olivine → pyroxene → hornblende → biotite → ...) are an example of a discontinuous series.

Under normal pressure such acid magmatic rocks as granite and diorite begin to melt at temperatures of about 1100 and 1360 K respectively, and alkaline rocks such as peridotite and gabbro melt at temperatures of about 1025 and 1525 K (Smith, 1963). The melting point of basalt is about 1430 K and of dunite about 1670 K. Lherzolite (a variant of peridotite) begins to melt at a temperature of about 1370 K, but complete melting does not take place until the temperature is about 2170 K.

The first attempt to find the melting point distribution in the polymineral mantle was made by Clark (1963). He considered an olivine model of the mantle with the mean atomic mass $\bar{A} \approx 21$, the melting heat $Q_L \approx 419$ $\mathrm{J \cdot g^{-1}}$ (the average melting heat of silicates), the isothermal bulk modulus $K_{T0} = 125$ GPa, the melting point $T_{m0} = 1300$ K and the initial slope of the melting curve $T_m(p)$ the same as for basalt ($\mathrm{d}T_m/\mathrm{d}p = 65\ \mathrm{K \cdot GPa^{-1}}$ (Yoder and Tilley, 1962)). In his considerations Clark used the Simon equation (4.2.18), the Clausius–Clapeyron equation (4.2.19) and the Birch–Murnaghan equation of state which describes the relationship between the pressure and density of a solid. A fragment of the curve $T_m(z)$ after Clark is shown in Fig. 4.2.4.

To describe the melting process of solids under high pressures Kraut and Kennedy (1966) proposed the following empirical relation:

$$-\frac{\mathrm{d}T_m}{\mathrm{d}\left(\dfrac{\Delta\varrho}{\varrho}\right)} = \mathrm{const} \qquad (4.2.20)$$

where $\Delta\varrho = \varrho - \varrho_0$ is the density increase caused by the pressure (ϱ_0 is the density under the pressure $p = 0$). The diagram of $T_m(\Delta\varrho/\varrho)$ for iron under pressure of order of 100 GPa, after Higgins and Kennedy (1971), is shown in Fig 4.2.5.

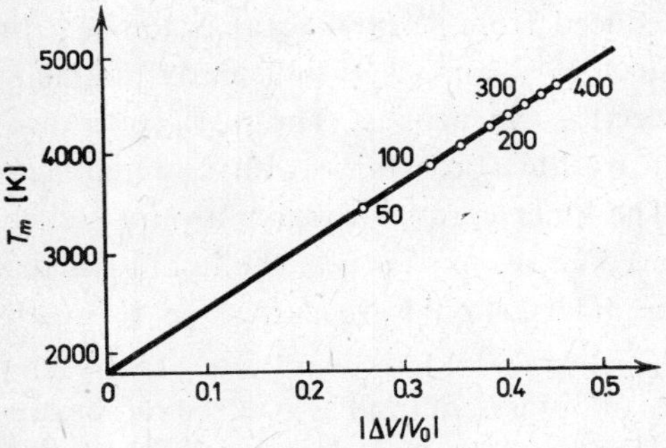

Fig. 4.2.5. Higgins–Kennedy linear relationship between the melting point (T_m) and the fractional change in volume ($|\Delta V/V_0|$) for iron under various pressures (marked on the curve in GPa) according to Jacobs (1974).

Relation (4.2.20) can be written in the form

$$-\frac{dT_m}{dp} = \text{const}\frac{d\left(\frac{\Delta\varrho}{\varrho}\right)}{dp} = \text{const}\frac{\varrho_0}{\varrho^2}\frac{d\varrho}{dp}$$

For low pressures

$$-\left(\frac{dT_m}{dp}\right)_0 = \text{const}\frac{1}{\varrho_0}\left(\frac{d\varrho}{dp}\right)_0$$

Thus

$$\frac{dT_m}{dp} = \left(\frac{dT_m}{dp}\right)_0 \varrho_0\left(\frac{dp}{d\varrho}\right)_0 \frac{d}{dp}\left(1-\frac{\varrho_0}{\varrho}\right)$$

and

$$T_m = T_m(\varrho_0)+\left(\frac{dT_m}{dp}\right)_0 \varrho_0\left(\frac{dp}{d\varrho}\right)_0\left(1-\frac{\varrho_0}{\varrho}\right) \tag{4.2.21}$$

For the mantle model of Clark (1963), in which $(dT_m/dp)_0 = 65$ $K \cdot GPa^{-1}$ and $K_T \approx \varrho_0(dp/d\varrho)_0 = 125$ GPa and $T_m(\varrho_0) \approx 1800$ K we have

$$T_m = 1800+8125\left(1-\frac{\varrho_0}{\varrho}\right) \tag{4.2.21a}$$

If $\varrho_0 \approx 3.4\ g \cdot cm^{-3}$ (at a depth of 100 km) is assumed, then the melting point at the lower mantle boundary ($\varrho \approx 5.5\ g \cdot cm^{-3}$) is $T_m \approx 4900$ K.

The distribution of melting point in the mantle with the density distribution of Bullen (model A1) is shown in Fig. 4.2.4.

It is well known that the liquid outer core cannot be composed entirely of pure iron since its density (deduced from seismic data) is too low by about 10%. Two elements—sulphur and silicon—are sufficiently abundant in nature to cause the relevant decrease in density. The outer core most probably contains sulphur in the troilite (FeS) form with a weight percentage of sulphur of 10÷15%. The inner core is probably a pure nickel–iron core which expands with time. During this process the liquid situated directly at the inner–outer core boundary should grow richer in FeS, thus becoming lighter.

On the other hand, the presence of potassium (and also of the radioactive isotope ^{40}K) in considerable quantities in the Earth's core cannot be excluded (see Section 4.1.2). Bukowinski (1976) believes that the thick, under normal conditions, potassium atom diminishes considerably already under pressure of 20÷30 GPa as a result of the valency electron transition from the large *s*-orbital into the much smaller *d*-orbital. Such modifications of atom (or ion) dimensions might considerably facilitate the penetration of potassium ions into crystalline structures linked with other elements formed under various conditions.

In some types of meteorites, such as enstatite chondrites and achondrites, potassium–iron sulphide (djerfisherite) was found (see Goettel, 1976), which implies that potassium can have distinct chalcophilit properties, e.g., the chemical affinity of sulphur. Data related to the stability of K_2S, metallurgy data relating to potassium in sulphide phases in metallurgical systems indicate that the penetration of potassium into Fe–FeS melts is highly probable. Similar conclusions can be drawn from experimental data on the partitioning of potassium into solid silicates and Fe–FeS melts (Goettel, 1976).

The melting point of pure iron under pressure of about 160 GPa, i.e., under the pressure at the core–mantle boundary, is about 5000 K, while at the inner–outer core boundary it probably reaches 7700 K (see Elsasser *et al.*, 1979).

Usselman (1975) experimentally studied the behaviour of the solution of FeS in Fe under pressures reaching up to 10 GPa. It was found that under 10 GPa eutectic Fe–FeS contains 23 per cent by weight of S, which is more than in the outer core matter (≈ 10÷15%). The results of measurements imply, however, that the eutectic state is displaced towards compositions with a smaller amount of sulphur as the pressure increases. The probable melting point of eutectic under pressure corresponding to the core–

mantle boundary is about 3100 K, while under pressure corresponding to the inner–outer core boundary it is about 4200 K.

4.2.4 *Temperature in the Earth's interior*

There are generally two methods of searching for the contemporary distribution of temperature in the Earth's interior.

One of these methods consists of typically modelling studies in which the well-known equation of heat transport is used, especially in the case of steady-state heat flow under assumed boundary conditions (and possibly initial conditions) and with arbitrarily adopted physical and chemical properties of the model of a given geosphere, such as e.g., types and distribution of internal heat sources.

The Earth's thermal models can be arbitrarily divided into two basic groups: solid thermal models (STM) and convective thermal models (CTM). Models in which individual geospheres are considered as ideal solids are included in the STM group. The energy flow is primarily determined by phonon thermal conductivity (the Earth's crust) or by phonon thermal conductivity and radiative thermal conductivity (the Earth's mantle), and in the case of a metallic core it is determined by thermal diffusion of electron gas (electron thermal conductivity). The CTM group is formed by the models of the mantle and core in which the heat flow is first of all connected with a simultaneous transport of mass.

An example of temperature distribution in a chondritic model of the mantle with steady-state heat flow is shown in Fig. 4.2.6 (MacDonald, 1959). In this model a temperature of 700 K and heat flow of 39.5 $mW \cdot m^{-2}$ (0.94 HFU), corresponding to the Moho boundary, were adopted as boundary conditions. The phonon thermal conductivity is constant and equal to 2.5 $W \cdot m^{-1} \cdot K^{-1}$ ($\approx$ 6 $mcal \cdot cm^{-1} \cdot s^{-1} \cdot K^{-1}$), while the radiative thermal conductivity is of the form (4.4.3). The medium opacity increases with temperature as a result of increasing carrier absorption* (see Section 4.4.3).

The second, very interesting and effective method of searching for the present temperature distribution (or geothermal gradient dT/dz) in the Earth's interior is based, as regards theory, on the physics of solids, on

* The following optical properties of the model were assumed: the refractive index of temperature radiation n = const $\approx$ 1.7, the opacity $\varepsilon = \varepsilon_0 + (120\pi/n)\sigma_i^* \exp(-E'/kT)$, where $\varepsilon_0 = 10\ cm^{-1}$ is the opacity under normal conditions, $E' = E_G/2 = 1.5$ eV is the activation energy of an intrinsic electrical conductivity, and $\sigma_i^* = 1 \div 2\ \Omega^{-1} \cdot cm^{-1}$ denotes the electrical conductivity of the medium at $T \to \infty$.

various data supplied by petrology, volcanology, seismology, geomagnetism and on laboratory experiments on minerals and rocks under high pressures and temperatures.

It is well known that seismic and laboratory data are also used in attempts to determine the possible distribution of melting points in the Earth's mantle or core (see Section 4.2.3). The distribution $T_m(z)$ after Uffen, for example, is well known, and the curves $T_m(z)$ obtained by an extrapolation of laboratory data by the Simon equation are adopted (see Fig. 4.2.4). Since the Earth's mantle is solid, the melting point is the upper limit of possible temperatures in the mantle, while the lower temperature limit is determined by the adiabatic temperature, whose distribution can also be found from seismic data.

From temperature measurements in alkaline lavas, taking into consideration possible heat losses during magma movements from an upper mantle area, and from petrological data, relating to crystallization conditions of specified minerals, it is possible to estimate that at a depth of 100 km temperatures reach values between 1300 and 1700 K.

Laboratory results relating to phase transitions in a pure binary system of the olivine–spinel type imply that at depths of 350÷400 km, where the upper boundary of a phase change area is situated, temperatures between about 1570 and 1720±120 K dominate (see Elsasser *et al.*, 1979). It should be noted that the presence of other elements, and not only of Mg and Fe, in the real material of the upper mantle could induce, at a given temperature, phase transformations of the type in question even under lower pressures. This would in turn imply that the estimated temperatures are lower than the real temperatures. Thus it is possible that at a depth of 400 km the temperature reaches 1800÷1920 K.

A probable distribution of temperature in the lower mantle was given by Wang (1972), who used high pressure shock-wave data relating to magmatic rocks. Wang used an empirical sound velocity–density–mean atomic mass relationship to estimate the type of material (corresponding to the mean atomic mass) of the lower mantle. He proposed the equation of state for a rock material with a specified atomic mass. This equation connects two independent variables—pressure and temperature—by an isentropic relation through the medium density. Assuming a specific model for the density distribution in the lower mantle, Wang concludes that at a depth of 1300 km the temperature reaches about 2800 K. It increases with depth almost linearly (the gradient $\mathrm{d}T/\mathrm{d}z \approx 0.33\ \mathrm{K \cdot km^{-1}}$) and at a depth of 2800 km reaches the value of 3300 K. The uncertainty of temperature estimation is $\Delta T = \pm 800$ K.

The possible distribution of the thermal gradient (eqn. (4.2.9)) within the upper 1000 km of the mantle is shown in Fig. 4.2.3, which was obtained from various empirical data by similar methods. The extrapolation of gradient (4.2.9) on the entire mantle leads to a temperature distribution of the form (4.2.10) if $T_{100} \approx 1300$ K at a depth of 100 km is assumed (see Fig. 4.2.6).

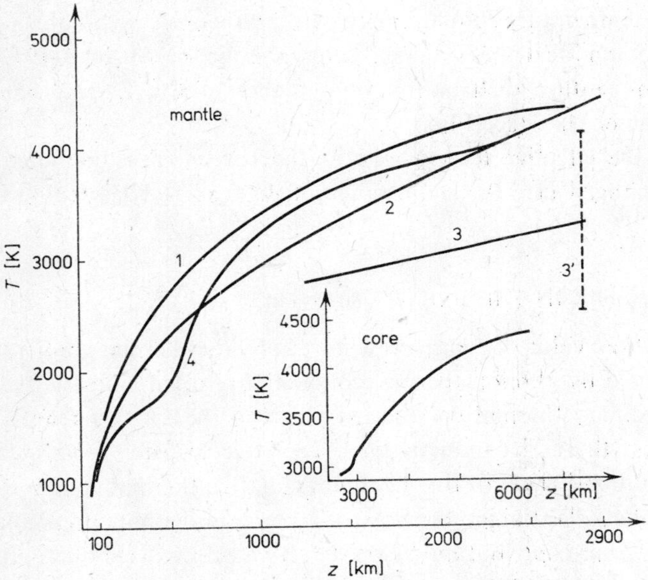

Fig. 4.2.6. Contemporary temperature distributions in the Earth. The mantle: 1—temperature calculated for a chondritic model of the mantle, according to MacDonald (1959), 2—eqn. (4.2.10), 3—according to Wang (1972) (the vertical bar 3′ denotes the error limit), 4—temperature estimated by Tozer (see e.g., Rikitake, 1966) from the distribution of electrical conductivity. The core: $T(z)$ curve represents the thermal outer-inner core model of Stacey (1977).

The effective temperatures in the mantle can be estimated from the distribution of electrical conductivity in the mantle interior. This distribution is found independently from analyses of long-period geomagnetic field variations. Such a procedure, however, demands the assumption of arbitrarily specific types of electrical conductivity, which in a given medium directly depend on temperature. The temperature distribution in an olivine mantle model in which intrinsic electrical conductivity dominates was determined by Tozer in this way (see e.g., Rikitake, 1966)—the result is shown in Fig. 4.2.6. The most recent laboratory results relating to silicate minerals imply, however, that the principal electrical conductivity in the mantle is rather the impurity electrical conductivity caused by deep donors or Fe ions (see Section 4.4.4).

It is convenient to use two extreme models of the Earth's core when its temperature distribution is estimated. In the first model the inner core is composed of pure iron, and in the second the outer core is an eutectic solution of FeS in Fe. In such a case the upper and lower bounds of possible temperatures can be determined. Since the outer core behaves like a liquid phase, its temperature should be higher than the melting point.

Using slightly modified data taken from Usselman, Stacey (1977) calculated the temperature distribution in a model of the Earth's core with an eutectic outer core (see Fig. 4.2.6).

It seems that the effective temperature at the core–mantle boundary, estimated by Verhoogen (1973), in the range of 4000 ± 500 K is real (see Elsasser *et al.*, 1979).

4.3 PHONON TRANSFER IN THE EARTH'S MANTLE

In the Earth's crust the value of energy flow, forced by the thermal gradient, is entirely determined by phonon thermal conductivity, often called lattice conductivity or ordinary thermal conductivity. But in the Earth's mantle, apart from phonon thermal conductivity, a radiative transfer connected with the temperature radiation of the medium (radiative thermal conductivity) may also be of substantial importance. In general, under high pressures and temperatures some contribution to the effective thermal conductivity of a dielectric-solid could well be expected from excitons and free carriers of electric charge: electrons and holes.

In the metallic core of the Earth, free electrons (electron gas) are the main energy carriers, and they determine the value and character of thermal conductivity.

4.3.1 Phonon thermal conductivity of silicate minerals

The heat transport with phonons as the energy carriers (quanta of elastic vibrations of the crystalline lattice) can be described by using the well-known equation for thermal conductivity adapted from the kinetic gas theory

$$\varkappa = \tfrac{1}{3} c v \Lambda \tag{4.3.1}$$

where c is the specific heat per unit of volume, v is the average speed and Λ is the mean free path of gas molecules. For phonon gas the specific heat of a solid (per unit of volume) may be used, and the average speed v may be replaced by the mean velocity of sound (see e.g., Ziman, 1962).

In pure ideal crystals the mean free path of a phonon is in general determined by the so-called "Umklapp" process (U-process) or by back scattering of phonons in phonon–phonon interactions (dynamic scattering). The significance of a U-process increases with temperature; at high temperatures Λ_U is inversely proportional to T. In real crystals the scattering of phonons on structural defects, called geometrical scattering, is also of importance.

In general, it can be written that

$$\frac{1}{\Lambda} = \frac{1}{\Lambda_U} + \frac{1}{\Lambda_{\text{geom}}}$$

According to Dudgale and MacDonald (1955)

$$\Lambda_U = \frac{d}{\alpha \gamma T}$$

where d is a lattice constant, α is the coefficient of thermal expansion and γ is the Grüneisen parameter. Making use of the Grüneisen relation

$$\gamma = \frac{\alpha}{c_V}\frac{K_T}{\varrho} = \frac{\alpha}{c_p}\frac{K_S}{\varrho}$$

we have in a more general form

$$\Lambda_U = \frac{Kd}{c\varrho\gamma^2 T} \tag{4.3.2}$$

where ϱ is the density, and c is the specific heat for a constant volume or pressure, and K is either the isothermal or the adiabatic bulk modulus, respectively.

For the description of phonon thermal conductivity at high temperatures Lawson (see Keyes, 1959) proposed the following formula (see (4.3.1), (4.3.2) and the definition of the hydrodynamic velocity of sound (4.2.15)):

$$\varkappa = \frac{K_S^{3/2} d}{3\gamma^2 \varrho^{1/2} T} \tag{4.3.2a}$$

Keyes (1959) used (4.3.2a) in his studies of the relationship between phonon thermal conductivity and the melting temperature T_m, based on the Lindemann theory of the melting of solids.

According to Klemens (1958), at high temperatures ($T \gg \theta$) the phonon thermal conductivity in pure crystals

$$\varkappa = B' \frac{\theta'^3}{\gamma^2 \varrho^{1/3} T} \tag{4.3.3}$$

where B' is an empirical coefficient dependent on the crystalline structure. Its value has been estimated only for a few simple structures (e.g., for a face centred cubic structure it is $\varkappa = (500\ \bar{A}\theta^3 d/\gamma^2 T)\ \mathrm{W \cdot m^{-1} \cdot K^{-1}}$ (Ziman, 1962)). In general, the Debye temperature (θ') in relation (4.3.3) is called the *acoustic temperature of Debye*

$$\theta' \equiv \theta^{\mathrm{ac}} = \frac{\theta}{\sqrt[3]{n}}$$

where θ is the dynamic Debye temperature calculated from the Launay formula (4.2.5), and n is the number of atoms in an unit cell of the crystal.

The results of laboratory investigations on the phonon thermal conductivity of rock-forming minerals under normal conditions ($\varkappa_N$) suggest that (Horai and Simmons, 1969, 1970; Horai, 1971):

(a) phonon conductivity $\varkappa_N$ of silicates is controlled by the structure of the silicon–oxygen network;

(b) in a series of silicates that forms a binary solid solution the phonon conductivity has a minimum at an intermediate composition;

(c) phonon conductivity is a linear function of density when the mean atomic mass of minerals is constant;

(d) in an isomorphous series of minerals phonon conductivity decreases as the mean atomic mass increases;

(e) phonon conductivity of silicates is linearly dependent on the velocity of P or S elastic waves.

For silicate minerals it is also possible to introduce (see Maj, 1976) empirical linear relations between the phonon thermal conductivity $\varkappa_N$ and the average velocity of sound defined as $v_m = \sqrt[3]{3/(v_P^{-3}+2v_S^{-3})}$ (Debye velocity), $\langle v \rangle = (v_P+2v_S)/3$ (mean arithmetic velocity) or $v_\Phi = \sqrt{v_P^2-\frac{4}{3}v_S^2}$ (hydrodynamic velocity). Relations of this type are listed in Table 4.3.1.

Using laboratory data for 14 silicate minerals (see Table 4.3.2) representing different crystalline structures, Horai and Simmons (1970) obtained the following interesting empirical relation:

$$\theta = (61.1 \pm 7.2)\varkappa_N + (385 \pm 28) \tag{4.3.4}$$

where θ is the dynamic Debye temperature (4.2.5) in K, and $\varkappa_N$ is the phonon thermal conductivity at room temperature in $\mathrm{W \cdot m^{-1} \cdot K^{-1}}$ ($1\ \mathrm{W \cdot m^{-1} \cdot K^{-1}} = 2.39\ \mathrm{mcal \cdot cm^{-1} \cdot s^{-1} \cdot K^{-1}}$) (see also Fig. 4.3.1). Relation (4.3.4) has been used in attempts to determine the phonon thermal conductivity distribution in the Earth's mantle. For this purpose the following

TABLE 4.3.1
Empirical relations between phonon thermal conductivity and various physical properties of silicates

Relation	References
$v_P = (0.41 \pm 0.05)\varkappa_N + 5.93 \pm 0.17$	Horai and Simmons (1969)
$v_S = (0.22 \pm 0.05)\varkappa_N + 3.31 \pm 0.16$	,,
$\theta = (61.1 \pm 7.2)\varkappa_N + 385 \pm 28$	Horai and Simmons (1970)
$\varkappa_N = (3.23 \pm 0.35)\langle v \rangle - 13.17 \pm 1.86$	Maj (1976)
$\varkappa_N = (2.80 \pm 0.03)v_\phi - 12.14 \pm 1.86$	,,
$\varkappa_N = (0.221 \pm 0.003)\phi - 3.44 \pm 0.88$	Maj (1978b)
$\varkappa_N = 16.3 - 1.36\bar{A} + 5.45\varrho$	Horai and Baldridge (1972)

where
$\varkappa_N$ is the phonon thermal conductivity under normal conditions, in $W \cdot m^{-1} \cdot K^{-1}$,
v_P is the velocity of longitudinal elastic wave, in $km \cdot s^{-1}$,
v_S is the velocity of transversal elastic wave, in $km \cdot s^{-1}$,
θ is the dynamic Debye temperature, in K,
$\langle v \rangle$ is the arithmetic mean sound velocity, in $km \cdot s^{-1}$,
v_ϕ is the hydrodynamic sound velocity, in $km \cdot s^{-1}$,
ϕ is the seismic parameter, in $km^2 \cdot s^{-2}$,
$\bar{A}$ is the mean atomic mass, in $g \cdot mole^{-1}$
ϱ is the density, in $g \cdot cm^{-3}$

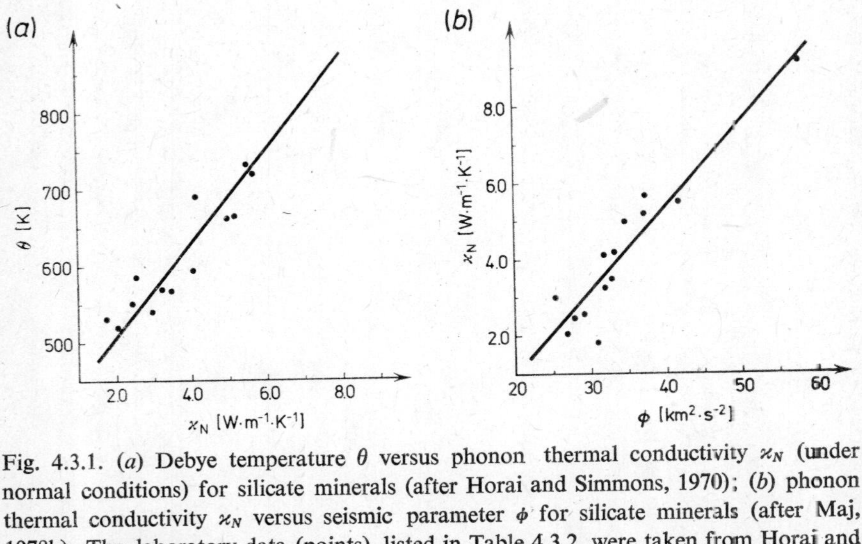

Fig. 4.3.1. (*a*) Debye temperature θ versus phonon thermal conductivity $\varkappa_N$ (under normal conditions) for silicate minerals (after Horai and Simmons, 1970); (*b*) phonon thermal conductivity $\varkappa_N$ versus seismic parameter ϕ for silicate minerals (after Maj, 1978b). The laboratory data (points), listed in Table 4.3.2, were taken from Horai and Simmons (1970); the solid lines represent regression equations (4.3.4) and (4.3.5).

TABLE 4.3.2
Thermal and elastic data on silicate minerals (after Horai and Simmons, 1970)

Silicate Ideal chemical formula	$\varkappa_N$ $mcal \cdot cm^{-1} \cdot s^{-1} \cdot K^{-1}$	$\varkappa_N$ $W \cdot m^{-1} \cdot K^{-1}$	$\overline{A}$ $g \cdot mole^{-1}$	ϱ $g \cdot cm^{-3}$	v_P $km \cdot s^{-1}$	v_S $km \cdot s^{-1}$	θ K	ϕ $km^2 \cdot s^{-2}$	$\langle C \rangle$* $J \cdot mole^{-1} \cdot K^{-1}$
Sillimanite Al_2SiO_5	21.73	9.11	20.26	3.179	9.60	5.09	772	57.6	20.0
Grossularite $Ca_3Al_2Si_3O_{12}$	13.05	5.47	22.52	3.525	8.55	4.86	732	41.6	18.9
Jadeite $NaAlSi_2O_6$	13.41	5.62	20.21	3.229	8.19	4.75	719	37.0	18.4
Bronzite $(Mg_{0.9}, Fe_{0.1})SiO_3$	9.84	4.12	20.71	3.284	7.64	4.59	691	30.3	19.7
Forsterite $(Mg_{0.93}, Fe_{0.07})_2SiO_4$	12.25	5.13	20.75	3.276	7.91	4.39	665	36.9	19.3
Diopside $CaMgSi_2O_6$	11.83	4.96	21.66	3.283	7.80	4.44	662	34.6	18.4
Wollastonite $CaSiO_3$	9.63	4.04	23.23	2.883	7.48	4.26	594	31.8	18.4
Idocrase $Ca_{10}Mg_2Al_4Si_9O_{34}(OH)_4$	6.03	2.53	22.57	3.198	7.11	4.02	586	29.0	19.1

TABLE 4.3.2 (continued)

Silicate Ideal chemical formula	$\varkappa_N$ $mcal \cdot cm^{-1} \cdot s^{-1} \cdot K^{-1}$	$\varkappa_N$ $W \cdot m^{-1} \cdot K^{-1}$	$\overline{A}$ $g \cdot mole^{-1}$	ϱ $g \cdot cm^{-3}$	v_P $km \cdot s^{-1}$	v_S $km \cdot s^{-1}$	θ K	ϕ $km^2 \cdot s^{-2}$	$\langle C \rangle$* $J \cdot mole^{-1} \cdot K^{-1}$
Monticellite $CaMgSiO_4$	7.61	3.19	22.33	3.016	7.26	3.96	570	31.8	19.3
Hortonolite $(Mg_{0.48}, Fe_{0.52})_2SiO_4$	8.23	3.45	24.79	3.755	7.21	3.79	568	32.8	19.3
Microcline $KAlSi_3O_8$	5.74	2.41	21.41	2.569	7.00	4.00	551	27.7	18.0
Serpentine $Mg_3Si_2O_5(OH)_4$	7.05	2.95	19.80	2.719	6.63	3.74	540	25.3	16.6
Anorthite $[(NaSi)_{0.18}, (CaAl)_{0.82}] \cdot$ $\cdot AlSi_2O_8$	4.17	1.75	21.18	2.754	7.02	3.72	530	30.8	18.6
Albite $[(NaSi)_{0.88}, (CaAl)_{0.12}] \cdot$ $\cdot AlSi_2O_8$	4.82	2.02	20.32	2.646	6.63	3.60	521	26.7	18.1

* From the Neumann–Kopp law.

empirical relation can also be used:

$$\varkappa_N = (0.221 \pm 0.003)\,\phi - (3.43 \pm 0.88) \tag{4.3.5}$$

where ϕ is the seismic parameter in $km^2 \cdot s^{-2}$ and $\varkappa_N$ is in $W \cdot m^{-1} \cdot K^{-1}$. This relation was found (Maj, 1978b) from the data listed in Table 4.3.2 (see Fig. 4.3.1).

Equation (4.3.1) is very useful in seeking relationships between phonon thermal conductivity and various physical properties of minerals under normal conditions.

The formula of Debye describing the atomic heat of solids is of the form

$$C = 3R\left(\frac{T}{\theta}\right)^3 \int_0^{\theta/T} \frac{\xi^4 e^\xi}{(e^\xi - 1)^2}\, d\xi$$

where θ is the Debye temperature and R is a gas constant (see e.g., Kittel, 1966). From Table 4.3.2 it follows that for silicate minerals the average value of θ is ≈ 600 K. Thus at room temperature (about 300 K) the value of the atomic heat of the "Debye model of silicate" would be $C \approx 21$ $J \cdot mole^{-1} \cdot K^{-1}$.

The Neumann–Koop empirical law indicates that the mole heat (C^m) of a solid chemical compound is the sum of the atomic heats (C^a) of its components. Thus

$$C^m = \sum_i C_i^a$$

and the following values of the atomic heat C^a (in $J \cdot mole^{-1} \cdot K^{-1}$) should be taken: 7.5 for carbon, 9.6 for hydrogen, 11 for boron, 16 for silicon, 17 for oxygen, 21 for fluorine, 22 for phosphorus and sulphur; for the other elements the value $C^a \approx 26$ $J \cdot mole^{-1} \cdot K^{-1}$ (6.2 $cal \cdot mole^{-1} \cdot K^{-1}$) is acceptable (see e.g., Weber, 1960).

Introducing the mean atomic heat as

$$\langle C \rangle = \frac{1}{n} \sum_i C_i^a$$

where n is the number of atoms in a molecule, the average value $\langle C \rangle \approx 19$ $J \cdot mole^{-1} \cdot K^{-1}$ for silicate minerals listed in Table 4.3.2 is obtained. Thus, under normal conditions, for the "monoatomic model of silicate" the specific heat per unit volume is

$$c \approx 19 \frac{\varrho}{A}\ J \cdot cm^{-3} \cdot K^{-1} \tag{4.3.6}$$

where ϱ is the density and $\bar{A}$ is the atomic mass.

The mean free path of a phonon is determined by Umklapp processes and by the scattering of phonons on geometrical defects. When impurity concentrations (x) are low, the changes of phonon thermal conductivity caused by impurities of the crystal can be described by the formula

$$\frac{\Delta\varkappa}{\varkappa} = x\frac{s}{d^3}\Lambda' \tag{4.3.7}$$

where d is a lattice constant, s is the phonon scattering cross-section of impurity, and Λ' is the mean free path of the phonon in a pure crystal. For many solids with dispersed impurities an approximation $s \approx d^2$ is reasonable. Horai (1971) estimated the mean free path Λ' at room temperature for a few silicates, using relation (4.3.7) and the changes $\Delta\varkappa$ resulting from the chemical composition. He obtained the following results (in units of the lattice constant): $\Lambda'_N = 1.1\,d$ for forsterite, $\Lambda'_N = 1.6\,d$ for enstatite and $\Lambda'_N = 2.7\,d$ for albite.

The principal element of the network of silicates is the silicon–oxygen tetrahedral $[SiO_4]^{4-}$ formed by one cation Si^{4+} and four anions of oxygen O^{2-}. Its size is estimated to be about 6×10^{-4} μm.*

It is very probable that under normal conditions the mean free path of a phonon in silicate minerals is

$$10^{-4} < \Lambda_N \leqslant 10^{-3} \quad \mu m$$

Using relations (4.3.1) and (4.3.6) and seismic equation of state (4.2.6) in the form (Anderson, 1967a)

$$\frac{\varrho}{\bar{A}} = 0.048\,\phi^{0.323}$$

where the seismic parameter ϕ is in $km^2 \cdot s^{-2}$ and the density ϱ in $g \cdot cm^{-3}$, Maj (1978b) obtained a semi-empirical relation between the phonon thermal conductivity $\varkappa_N$ and the seismic parameter ϕ for silicate minerals

$$\varkappa_N = 0.18\,\phi^{0.82} \quad W \cdot m^{-1} \cdot K^{-1}$$

where ϕ is in $km^2 \cdot s^{-2}$.

It is assumed that the mean free path of a phonon is comparable with the size of the tetrahedral $[SiO_4]^{4-}$ (or $\Lambda_N \approx 6\times10^{-4}$ μm), and the average velocity of a phonon is equal to the hydrodynamic velocity of sound ($v_\phi = \sqrt{\phi}$ (see (4.2.15)).

---

* The ionic radius Si^{4+} is about 0.4×10^{-4} μm, and the ionic radius O^{2-} is about 1.4×10^{-4} μm (Fyfe, 1964).

4.3.2 The Birch–Clark relationship

It is known that from about 300 K to the Debye temperature the phonon thermal conductivity decreases with temperature as T^{-1}. This phenomenon, however, can be described quantitatively only by empirical relations, e.g., in the form

$$\varkappa = \varkappa^* + \frac{B}{T} \tag{4.3.8}$$

where $\varkappa^*(p)$ and $B \equiv [\partial\varkappa/\partial(1/T)]_p$ are the empirical coefficients characterizing a given material (see e.g., Kittel, 1966).

In general the phonon thermal conductivity also depends on the pressure p. Under moderately high pressures and at a constant temperature the following linear relation is valid:

$$\varkappa = \varkappa_0 + \delta p \tag{4.3.8a}$$

where $\varkappa_0(T)$ is the phonon conductivity under the pressure $p = 0$ and $\delta = (\partial\varkappa/\partial p)_T$ is the empirical coefficient.

Beck *et al.* (1978) studied the influence of temperature and pressure on the coefficients B and δ (see (4.3.8) and (4.3.8a)), for quartz, olivine, polycrystalline coesite and a few samples of dunite within the ranges of $350 \div 650$ K and $2 \div 5.6$ GPa. In all the cases the coefficients B and δ were found to be positive.

Birch and Clark (1940) found empirically that for magmatic rocks the relation $\varkappa(T)$ can be expressed in the following very convenient form:

$$\varkappa^{-1} = \varkappa_N^{-1} + \beta(T-300) \tag{4.3.9}$$

where $\varkappa_N^{-1}$ is the reciprocal of the phonon thermal conductivity, i.e., a quantity called *phonon thermal resistivity*, under normal conditions, and

$$\beta \equiv \left(\frac{\partial\varkappa^{-1}}{\partial T}\right)_p$$

is the temperature coefficient of the phonon thermal resistivity (see also Schatz and Simmons, 1972).

The temperature coefficient of phonon thermal resistivity for crystalline rocks is $\beta \leqslant 0.07$ cm $\cdot$ W^{-1}—see Table 4.3.3 (the values of β are either those by Birch and Clark (1940) or calculated from the laboratory data of Kawada (1964)).

Under normal conditions the phonon thermal conductivity of granitic rocks is from 2.75 to 3.26 W $\cdot$ m^{-1} $\cdot$ K^{-1} and that of basalt is about 2.3 W $\cdot$ m^{-1} $\cdot$ K^{-1} (Horai and Baldridge, 1972). For peridotites (ultrabasic rocks) the conductivity $\varkappa_N$ is $\approx 4 \div 5$ $W \cdot m^{-1} \cdot K^{-1}$ (Kawada, 1964), while for dunites it varies from 3.7 to 8.7 $W \cdot m^{-1} \cdot K^{-1}$ (see e.g., Beck *et al.*, 1978).

TABLE 4.3.3
Temperature coefficient of phonon thermal resistivity of crystalline rocks

Rock	β [cm · W^{-1}]	References*
Granite	0.038	1
Diorite	0.018	2
Basalt	0.069	2
Gabbro	0.026	2
Peridotite	0.024÷0.036	2
Pyroxenite	0.031	1
Dunite	0.043÷0.048	1.2
Diabase	0.005	1
Albitite	0.012	1
Eclogite	0.033	2

* 1—after Birch and Clark (1940), 2—calculated from laboratory data of Kawada (1964).

Relation (4.3.9) can successfully be used to estimate the phonon thermal conductivity in the Earth's crust, especially if $\varkappa_N$ is related to the appropriate physical quantities, e.g., the velocity of P or S elastic waves, the seismic parameter ϕ or the density ϱ of the medium.

4.3.3 Phonon thermal conductivity in the mantle

The first attempt to estimate the phonon thermal conductivity ($\varkappa$) in the Earth's mantle was made by Uffen (1954). Using the Debye theory of heat conductivity in solids and seismic data, he calculated the distribution of the relavite conductivity $\varkappa/\varkappa_{100}$ (the index "100" denotes a depth of 100 km) without specifying, however, the temperature distribution in the mantle. Similarly Zharkov (see Lubimova, 1967), using relation (4.3.3), calculated the distribution of $\varkappa T/\varkappa_{100} T_{100}$ for a model of the mantle with Bullen's density $\varrho(z)$.

Horai and Simmons (1970) adjusted the Birch–Clark relationship (4.3.9) for calculations of the phonon thermal conductivity in the mantle by introducing the coefficient

$$\Psi \equiv \frac{\left(\dfrac{\gamma}{\gamma_0}\right)^2 \left(\dfrac{\varrho}{\varrho_0}\right)^{1/3}}{\left(\dfrac{\theta}{\theta_0}\right)^3}$$

where the index "zero" denotes specified physical conditions (see (4.3.3)), into (4.3.9).

The modified relation (4.3.9) is of the form

$$\varkappa^{-1} = \varkappa_N^{-1} + \Psi\beta(T-300)$$

where the phonon thermal resistivity $\varkappa_N^{-1}$ is a previously specified function of the Debye temperature θ or the seismic parameter ϕ (see relations (4.3.4) and (4.3.5)).

For the determination of the $\varkappa_N$ distribution in the mantle Horai and Simmons used the empirical relation (4.3.4). The Debye temperature θ was calculated from the Launay formula (4.2.5) for $\bar{A} \approx 21$ g·mole^{-1}. For the top of the upper mantle $\beta = \beta_0 \approx 0.048$ cm · W^{-1}, i.e., the value corresponding to dunites, was assumed. The Grüneisen parameter was calculated directly from its definition

$$\gamma \equiv \frac{\varrho}{\theta}\left(\frac{\partial\theta}{\partial\varrho}\right)_T \approx \frac{\varrho}{\theta}\frac{\Delta\theta}{\Delta\varrho} \tag{4.3.10}$$

Calculations were carried out for the mantle model with the distribution of P and S seismic wave velocities according to Jeffreys and density ϱ according to Bullen.

The distribution of $\varkappa_N$ (the phonon thermal conductivity referred to normal conditions) in the mantle can also be obtained from relation (4.3.5). It should be noted that the two distributions $\varkappa_N(z)$, obtained from (4.3.4) and (4.3.5) for the same distribution of P and S wave velocities in the mantle, are very similar, the difference between them not exceeding 7%. In the upper part of the mantle (down to depth of about 100 km), where $\phi \approx 38$ km^2 · s^{-2}, the conductivity from relation (4.3.5) is $\varkappa_N \approx 5$ W · m^{-1} · K^{-1}, which corresponds to the phonon thermal conductivity of dunites or peridotites at room temperature (Kawada, 1964; Horai and Baldridge, 1972; Beck *et al.*, 1978).

For calculations of the Grüneisen parameter distribution in the Earth's interior, besides relation (4.3.10), the well-known Grüneisen law can also be used

$$\gamma = \frac{\alpha}{c_p}\frac{K_S}{\varrho} = \frac{\alpha}{c_p}\phi$$

as well as the approximate relation

$$\gamma \approx \frac{1}{3} + \frac{\varrho}{v_S}\frac{\Delta v_S}{\Delta \varrho}$$

wihch can be derived from the Launay formula (4.2.5).

Changes of the Grüneisen parameter can also be estimated from the relation

$$\frac{\gamma}{\gamma_0} = \frac{\varrho_0}{\varrho}\frac{1}{1+\gamma_0\left(1-\frac{\varrho_0}{\varrho}\right)}$$

introduced by Rice under the assumption that the adiabatic bulk modulus of the medium depends only on the density ϱ (see e.g., Knopoff and Shapiro, 1969).

In the lower mantle the following equation is very useful (Anderson, 1979):

$$\gamma \varrho^q = \text{const}$$

where ϱ is the density and the exponent q is close to one (from seismic data it follows that $0.8 < q < 2.2$).

Kieffer (1976) estimated the phonon thermal conductivity in a mantle model composed of two parts; the upper mantle composed of olivine (to a depth of 400 km) and its spinel modification (from 400 to 750 km),* and the lower mantle composed of a mixture of two oxides: periclase (MgO) and stishovite (SiO_2). To estimate the conductivity of such a mixture composed of two parts the formulas of Hashin and Shtrikman (1962) were used, which give the upper and lower limits of thermal conductivity of a mixture in relation to the conductivities of its components and their fractional volume.

The influence of temperature on phonon thermal conductivity was accounted for by using the Birch–Clark relation (4.3.9), given by Schatz and Simmons (1972) for the pressure $p_0 = 0.1$ MPa

$$\varkappa^{-1}(p_0) = \varkappa_N^{-1} + \beta(T-300)$$

The following values were adopted for calculations: $\varkappa_N^{-1} = 21\ \text{cm} \cdot \text{K} \cdot \text{W}^{-1}$

---

* The phonon thermal conductivity (at room temperature) for pure forsterite (FO) was calculated first and then corrected for the content of fayalite (FA), i.e., reduced to a hypothetical composition of 85% FO and 15% FA (the coefficient of reduction 0.85).

and $\beta = 0.048$ cm · W^{-1} for ordinary forsterite, $\varkappa_N^{-1} = 13$ cm · K · W^{-1} and $\beta = 0.048$ cm · W^{-1} for forsterite with spinel structure, $\varkappa_N^{-1} = 1.8$ cm · K · W^{-1}, $\beta = 0.02$ cm · W^{-1} for periclase, and $\varkappa_N^{-1} = 4.5$ cm · K · W^{-1} and $\beta = 0.02$ cm · W^{-1} for stishovite.

The influence of pressure on phonon thermal conductivity was accounted for by using the following relation:

$$\frac{\varkappa}{\varkappa(p_0)} = 1 - \left(3\gamma_0 - \frac{1}{3} + 2\frac{\gamma_0'}{\gamma}\right)D + \ldots$$

where γ is the Grüneisen parameter, D is the dilatation defined as

$$D \equiv \frac{\varrho_0}{\varrho} - 1$$

and $\gamma_0' = (\mathrm{d}\gamma/\mathrm{d}D)_{D=0}$. The following values were taken: $\gamma_0 = 1.3$ and $\gamma_0' = 1.7$ for ordinary forsterite ($\varrho_0 = 3.21$ g · cm^{-3}), $\gamma_0 = 1.3$ and $\gamma_0' = 2.7$ for forsterite with spinel structure ($\varrho_0 = 3.55$ g · cm^{-3}), $\gamma_0 = 1.5$ and $\gamma_0' = 1.7$ for periclase ($\varrho_0 = 3.58$ g · cm^{-3}), and $\gamma_0 = 1.5$ and $\gamma_0' = 4.5 \div 9$ for stishovite ($\varrho_0 = 4.28$ g · cm^{-3}).

To calculate the dilatation distribution in the mantle it was assumed that the density distribution in the mantle is the same as in the Birch model (see Verhoogen *et al.*, 1970). For a mixture of periclase and stishovite (MgO and SiO_2) $\varrho_0 = 3.84$ g · cm^{-3} was assumed (the fractional volume of stishovite is about 0.4). Changes of the Grüneisen parameter were estimated by means of the formula

$$\gamma \approx \gamma_0 + \gamma_0' \Delta D$$

From the model considered by Kieffer an important conclusion follows, namely that the increase of phonon thermal conductivity in the mantle, caused by an increase in pressure, does not exceed 20% for 1 GPa or about 0.7% for each kilometre of depth. Down to a depth of 700 km, therefore, the influence of pressure on phonon thermal conductivity is certainly much smaller than that of temperature.

The attempts to estimate the phonon thermal conductivity made so far distinctly suggest regardless of some minor differences resulting e.g., from the temperatures adopted for given calculations, that the phonon transfer in the upper mantle is small and its value is about $1 \div 2$ W · m^{-1} · K^{-1} (or $3 \div 4$ mcal · cm^{-1} · s^{-1} · K^{-1}), while in the lower mantle its value most probably does not exceed 4.2 W · m^{-1} · K^{-1} (about 10 mcal · cm^{-1} · s^{-1} · K^{-1}).

4.4 RADIATIVE TRANSFER IN THE EARTH'S MANTLE

4.4.1 Radiative thermal conductivity

In the 300 to ≈ 500 K range of temperatures the effective thermal conductivity ($\mathscr{K}$) of silicate minerals decreases with temperature as T^{-1}. At higher temperatures, however, an increase, and not a decrease, of thermal conductivity with temperature has been observed (see Fig. 4.4.1,

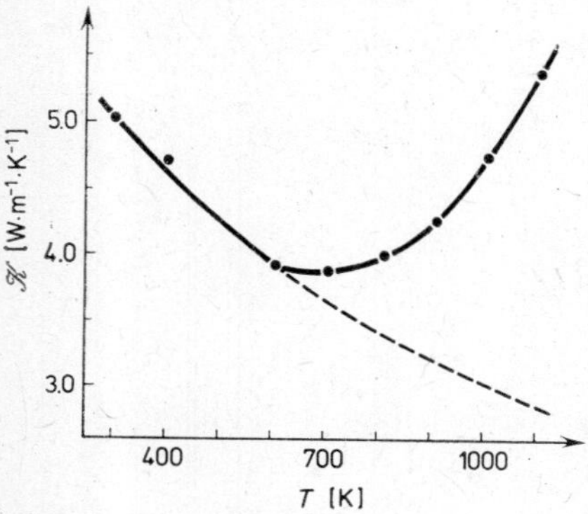

Fig. 4.4.1. Effective thermal conductivity ($\mathscr{K}$) versus temperature for single-crystal olivine ($(Mg_{0.82}, Fe_{0.18})_2SiO_4$, according to Kanamori *et al.* (1968). The dashed line, extrapolated from the left branch of the $\mathscr{K}(T)$ curve, shows the phonon thermal conductivity–temperature relationship.

after Kanamori *et. al.*, 1968). This effect can be explained by the increasing significance, at higher temperatures, of other energy carriers in solids than phonons, namely photons of temperature radiation (McQuarrie, 1954; Lee and Kingery, 1960; Stevels, 1962; Kawada, 1966; Hideyuki *et al.*, 1968).

The role possibly played by temperature radiation in the thermal state of the Earth's interior was pointed out by several authors (see e.g., Preston, 1956), but it was S. P. Clark Jr. (1956, 1957) who established the significance of radiative transfer in the physics of the Earth's interior and indirectly also in the physics of planetary interiors (Aronson *et al.*, 1967b).

The most convenient way to obtain a quantitative description of radiative transfer in solids is to introduce, on an analogy with the phonon thermal conductivity ($\varkappa$), a quantity called *radiative thermal conductivity* ($\varkappa_R$), since then the relation of the well-known form $\mathbf{j} = -(\varkappa + \varkappa_R) \operatorname{grad} T$ can be used.

In an ideally black body the radiative thermal conductivity is connected with temperature by the relation

$$\varkappa_R = \frac{4}{3}\int_0^\infty \frac{n^2(\lambda)}{\varepsilon(\lambda, T)}\frac{\partial J(\lambda,T)}{\partial T}\mathrm{d}\lambda \tag{4.4.1}$$

where $n(\lambda)$ and $\varepsilon(\lambda, T)$ are the refractive index and the opacity of the medium, respectively, for temperature radiation with the wavelength λ, and $J(\lambda, T)$ is the Planck function (see e.g., Clark, 1957; Fukao, 1969).

Assuming that the refractive index (n) of temperature radiation is not dependent on the wavelength and defining the so-called average opacity ($\langle\varepsilon\rangle$) as

$$\langle\varepsilon\rangle = \frac{\int_0^\infty \frac{\partial J(\lambda, T)}{\partial T}\mathrm{d}\lambda}{\int_0^\infty \frac{1}{\varepsilon(\lambda,T)}\frac{\partial J(\lambda, T)}{\partial T}\mathrm{d}\lambda} \tag{4.4.2}$$

the following relation is obtained from formula (4.4.1):

$$\varkappa_R = \frac{16}{3}\frac{n^2 s}{\langle\varepsilon\rangle}T^3 \tag{4.4.3}$$

where s is the Stefan–Boltzmann constant, which can be considered as the radiative thermal conductivity of a "grey body".

Relation (4.4.2) can be adopted for numerical calculations in the form

$$\frac{1}{\langle\varepsilon\rangle} = \frac{15}{4\pi^4}\sum_i \frac{1}{\varepsilon_i}\int_{x(\lambda_{i-1},T)}^{x(\lambda_i,T)} \frac{x^4 e^x}{(e^x-1)^2}\mathrm{d}x, \qquad x \equiv \frac{\hbar c}{kT} \tag{4.4.4}$$

where $\hbar$ is the "crossed" Planck constant, c is the light velocity in a vacuum and k is the Boltzmann constant.

From relation (4.4.3) it follows that the radiative thermal conductivity increases very fast with temperature, but the optical properties of the medium can play an essential role, especially the opacity $\langle\varepsilon\rangle$, which is a function of temperature and pressure.

4.4.2 Refraction of temperature radiation in the Earth's mantle

Lawson and Jamieson (1958) were the first to notice that for the determination of the refractive index of temperature radiation in the mantle the empirical Gladstone–Dale law $(\bar{n}-1)/\varrho = \text{const}$, sometimes called the specific refraction, can be used. This relation connects the mean refractive index of light in a crystal-dielectric ($\bar{n}$)* with its density (ϱ). According to Anderson and Schreiber (1965), the same aim can also be achieved by using the relation $(\bar{n}^2-1)/\varrho$ = const, known as the Drude law (molecular refraction). Of course, we assume in advance that both laws will also be valid at high pressures and temperatures.

Anderson and Schreiber (1965), using laboratory data relating to $\bar{n}$ and ϱ for a number of silicates and oxides with various crystalline structures, carried out a special control test to verify the usefulness of well-known relations between the light refractive index and density (the Lorentz–Lorenz law, the Drude law, the Gladstone–Dale law, the Allen law). The results of the test, though clearly giving preference to the Drude law, suggest that both the Drude law and the Gladstone–Dale law are only valid with sufficient accuracy for minerals with the mean atomic mass $\bar{A} \approx 21\,\text{g}\cdot\text{mole}^{-1}$ (Birch's law) containing large oxygen anions. The electric near field generated by anions, called the "overlap field", probably diminishes or even entirely neutralizes the influence of the Lorentz field inside a crystal.

The average value of molecular refraction (Drude's law) for silicate and oxide minerals fulfilling Birch's law is

$$\left\langle \frac{\bar{n}^2-1}{\varrho} \right\rangle \approx 0.52\,\text{cm}^3\cdot\text{g}^{-1} \tag{4.4.5}$$

and the value of specific refraction (the Gladstone–Dale law) is

$$\left\langle \frac{\bar{n}-1}{\varrho} \right\rangle \approx 0.20\,\text{cm}^3\cdot\text{g}^{-1}$$

according to Anderson and Schreiber (1965).

Let us consider a nonmagnetic and nonconductive medium composed of identical atoms, considering each of them as a classic elementary electron oscillator with the eigenfrequency ω_0. If interactions between the oscillators are neglected, then the following dispersion formula results from the electron theory of Lorentz (see e.g., Suffczyński, 1965):

$$n^2-1 = 4\pi \frac{e^2}{m} \frac{\mathcal{N}}{\omega_0^2-\omega^2} \tag{4.4.6}$$

where e is the electron charge, m is the electron mass, $\mathcal{N}$ is the number of oscillators (atoms) in a volume unit, ω is the frequency of electromagnetic radiation, and n is the refractive index of the isotropic medium.

* The arithmetic mean for three principal refractive indices of light of a given wavelength in a crystal.

Since

$$\mathcal{N} = \varrho \frac{N}{A}$$

where N is the Avogadro number, A is the atomic mass and ϱ is medium density, the following relation can be obtained from (4.4.6):

$$n^2 - 1 = 4\pi \frac{e^2 N}{mA} \frac{\varrho}{\omega_0^2 - \omega^2} \tag{4.4.7}$$

Relation (4.4.7) can be considered as Drude's law for radiation with a given frequency $\omega < \omega_0$.

In practice the refractive index is usually given for yellow light, or more precisely for the D line of sodium with $\lambda_D = 0.5893$ μm and $\omega_D = 3.2 \times \times 10^{15}$ s^{-1}. Thus we have

$$n^2 - 1 = 4\pi \frac{e^2 N}{mA} \frac{\varrho}{\omega_0^2 - \omega_D^2} \tag{4.4.8}$$

In the case of resonance, i.e., when $\omega \to \omega_0$, the increase of amplitude vibrations of an undamped oscillator is unlimited, and as a result disintegration of the oscillator composed of an electron and the attracting centre may occur. If the oscillator disintegration is identified with a single photoionization of an atom, then the eigenfrequency ω_0 can easily be estimated. It is well known that the kinetic energy E_{kin} of a photoelectron is related to ionizing radiation with the frequency ω as follows:

$$E_{\mathrm{kin}} = \hbar\omega - \mathcal{U}$$

where $\hbar$ is the "crossed" Planck constant and $\mathcal{U}$ is the ionization potential of a given atom. In other words, the photoionization phenomenon can occur only when the frequency ω fulfils the condition

$$\hbar\omega \geqslant \mathcal{U}$$

Thus it can be written that

$$\omega_0 = \frac{U}{\hbar} \tag{4.4.9}$$

where U is the potential of a single ionization of an atom.

The main components of the Earth's mantle are minerals composed of oxygen, silicon, magnesium, aluminium, iron, potassium, sodium, and calcium. Most of these minerals fulfil Birch's law $\bar{A} \approx 21$ g·mole^{-1}. Let us consider further on that the Earth's mantle is a monoatomic system, composed of atoms with the atomic mass

$$A = \bar{A} = \frac{\sum_i x_i A_i}{\sum_i x_i} \approx 21$$

where A_i and x_i are the atomic mass and the atomic abundance (e.g., in chondritic meteorites) of a given element, respectively. Nitsan and Shankland (1976a) determined from measurements of normal-incidence specular reflectance, in the spectral range $\approx 0.0885 \div 0.3100$ μm ($14 \div 4$ eV) at 300 K, the energy gap E_G in various silicate and oxide minerals (Table 4.4.1—data taken from Nitsan and Shankland 1976a, b).

TABLE 4.4.1.
Energy gap of silicate and oxide minerals under normal conditions, according to Nitsan and Shankland (1976a, b)

Mineral	Ideal chemical formula	E_G [eV]
Periclase	MgO	7.8
Spinel	$MgAl_2O_4$	8.3
Forsterite	Mg_2SiO_4	8.7
Pyrope	$Mg_3Al_2Si_3O_{12}$	8.7
Enstatite	$MgSiO_3$	8.8
Corundum	Al_2O_3	9.1
Andalusite	Al_2SiO_5	9.2
Kyanite	Al_2SiO_5	9.4
Quartz	SiO_2	10.2

It can be see in Table 4.4.1 that the width of energy gap (E_G) of such minerals as quartz, forsterite and enstatite is about $8 \div 10$ eV. Let us take the value $E_G \approx 9$ eV as the value representative for the most important silicates. If the energy gap E_G is now identified with the potential of a single ionization (U) of an atom, then it can be written that $U = E_G \approx 9$ eV, and according to (4.4.9) the corresponding eigenfrequency is $\omega_0 \approx 13.6 \times \times 10^{15}$ s^{-1} (1 eV $= 1.602 \times 10^{-19}$ J, $\hbar = 1.055 \times 10^{-34}$ J · s).

Substituting $A = \mathrm{A} \approx 21$ g · mole^{-1} and $\omega_0 = U/\hbar \approx 13.6 \times 10^{15}$ s^{-1} into relation (4.4.8), where $N = 6.023 \times 10^{23}$ atoms · mole^{-1}, $m = 9.11 \times 10^{-28}$ g and $e = 1.602 \times 10^{-19}$ C $\approx 4.803 \times 10^{-10}$ e.s.u., the molecular refraction (in cm^3 · g^{-1}) for a monoatomic model of the Earth's mantle is obtained:

$$\frac{n^2-1}{\varrho} = 0.516 \approx 0.52 \qquad (4.4.10)$$

It is easy to notice that this result is in very good agreement with eqn. (4.4.5).

4.4.3 Opacity of the Earth's mantle

It is well known that the opacity (ε)* of solids is determined by the absorption and scattering of radiation, both phenomena depending on the wavelength (λ).

In general, it can be written (Shankland *et al.*, 1979) that for radiation with the wavelength λ the opacity is

$$\varepsilon(\lambda) = \tilde{\alpha}(\lambda) + \tilde{\beta}(\lambda)\vartheta$$

where $\tilde{\alpha}(\lambda)$ and $\tilde{\beta}(\lambda)$ are the coefficients of absorption and of scattering, respectively, and ϑ is a parameter characterizing the spatial distribution of scattering directions. If the photons of temperature radiation are scattered with a forward-backward symmetry, then $\vartheta = 1$, while for a forward scattering $\vartheta = 0$.

Scattering of temperature radiation. Radiation scattering can play an important role in the opacity of fine-crystalline media, where the size of crystalline grains is smaller than 1 mm (Clark, 1957). When the scattering takes place along grain boundaries the coefficient $\tilde{\beta}$ is approximately equal to the reciprocal of grain size, and the parameter ϑ is about 0.02, which corresponds to strong forward scattering. Thus for grains of a size from 1 mm to 1 cm the contribution of this type of scattering to opacity would be $0.2 \div 0.02\ \mathrm{cm}^{-1}$ (Shankland *et al.*, 1979). It seems that the influence of scattering on the opacity of a fine-crystalline medium should be connected first of all with the refraction of radiation in individual grains, rather than with reflection from their boundary surfaces (Aronson *et al.*, 1967a).

The coefficient of scattering can be regarded with good approximation as independent of the wavelength λ. It is also highly probable that the coefficient is not very susceptible to a simultaneous increase in temperature and pressure. Thus it can be assumed that

$$\varepsilon = \varepsilon_0(\lambda) + \Delta\tilde{\alpha}(\lambda, p, T)$$

where $\varepsilon_0(\lambda) = \tilde{\alpha}_0(\lambda) + \tilde{\beta}_0$ is the opacity under normal conditions, and $\Delta\tilde{\alpha}(\lambda, p, T)$ is the change of absorption induced by an increase in temperature and pressure.

* The reciprocal of the mean free path of photons with a given frequency.

A very interesting example (see Fig. 4.4.2) of the role played by the scattering of temperature radiation in the opacity of solids is the comparison of thermal resistivities $\mathscr{K}^{-1}$ (the reciprocal of thermal conductivity) as functions of temperature in a synthetic polycrystal of olivine (Lee and Kingery, 1960) and in a natural monocrystal of olivine (Kanamori *et al.*, 1968). The phonon thermal resistivity in both samples is almost the same (Fig. 4.4.2). Since the chemical composition of both olivine samples is

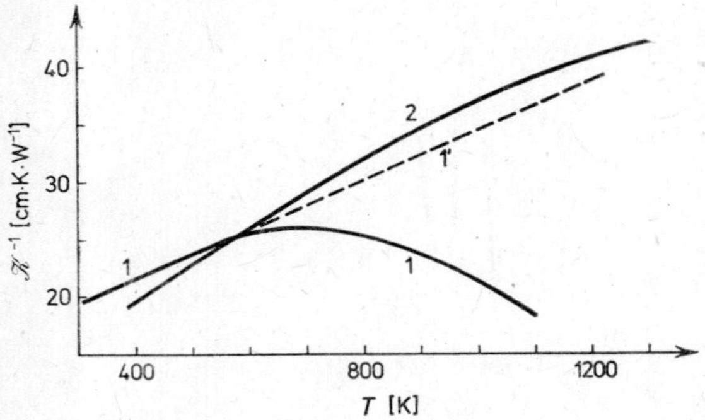

Fig. 4.4.2. Thermal resistivity ($\mathscr{K}^{-1}$) versus temperature for two olivine specimens: natural single-crystal (curve 1) and synthetic polycrystal (curve 2); the laboratory data were taken from Kanamori *et al.* (1968) and Lee and Kingery (1960). The dashed line (1′) shows the linear relation between the phonon thermal resistivity ($\varkappa^{-1}$) and temperature.

almost identical, such large differences between the further branches of the curves can only be explained by the high opacity of the polycrystal of olivine caused by a strong scattering of temperature radiation in a fine-crystalline structure (see also Fukao, 1969). As a result, the fast increase of radiative thermal conductivity with temperature is inhibited.

Absorption of temperature radiation. The optical pass band in the absorption spectrum of silicate and oxide minerals extends under normal conditions over near ultraviolet ($\lambda \lesssim 0.15$ μm), visible and near infrared radiation. The width of the pass band is determined by the edges of two absorption bands: the long-wave edge of basic absorption in ultraviolet and the short-wave edge of lattice absorption in infrared radiation. In real crystals there are absorption bands inside the pass band, originating from lattice defects (vacancies, shallow and deep impurities), excitons, and free carriers of electric charge.

A fragment of the absorption spectrum (the $\tilde{\alpha}(\lambda)$ curve) of a single-crystal of olivine $(Mg_{0.88}, Fe_{0.12})_2SiO_4$ at room temperature is shown in

Fig. 4.4.3 (Fukao *et al.*, 1968). Within the pass band, the region with zero opacity extends in the infrared range from about 1.8 μm to 3 μm. The absorption band at $\lambda \approx 1$ μm (with the corresponding local maximum $\tilde{\alpha} \approx 20$ cm^{-1}) was originated by an impurity of iron, i.e., the Fe^{++} ions (the absorption from electron transits between the *d* levels); the band is broadened by a local anisotropy of the crystal and phonon-assisted inter-

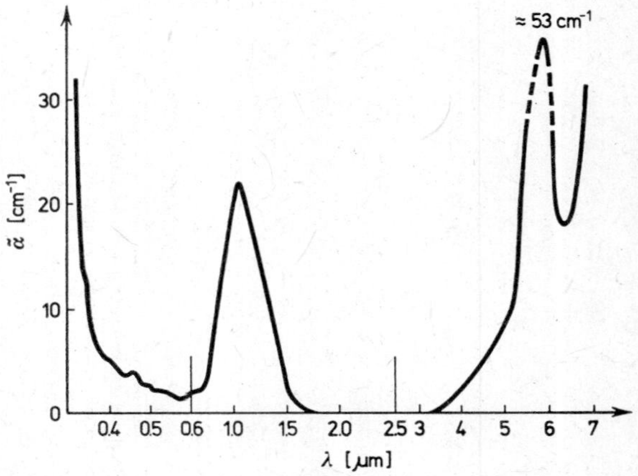

Fig. 4.4.3. Absorption spectrum of natural olivine $(Mg_{0.88}, Fe_{0.12})_2SiO_4$ at room temperature, according to Fukao *et al.* (1968).

actions. The strong absorption band at $\lambda \approx 5$ μm is the multiphonon band (overtone) of lattice vibrational bands (Shankland *et al.*, 1979). The absorption band corresponding to the wavelength $\lambda \approx 0.3$ μm is most probably connected with electron transitions to local levels related with deep impurities.

Shankland *et al.* (1979) studied the influence of temperature between 300 and 1700 K on the absorption spectrum (within the wavelength interval from 0.3 to 8 μm) of a natural monocrystal olivine $(Mg_{0.84}, Fe_{0.16})_2SiO_4$—see Fig. 4.4.4. Both the extreme absorption bands, i.e., the absorption band in ultraviolet and the lattice absorption band in infrared, gradually broaden as the temperature increases, simultaneously narrowing the pass band. At the same time absorption from the F^{++} impurity within the pass band is considerably strengthened. The average coefficient of absorption in the pass band increases with temperature from nearly zero at 300 K to ≈ 20 cm^{-1} at a temperature of 1700 K. Various mechanisms are responsible for the absorption of temperature radiation. A general model of energy structure in silicate minerals, very useful in

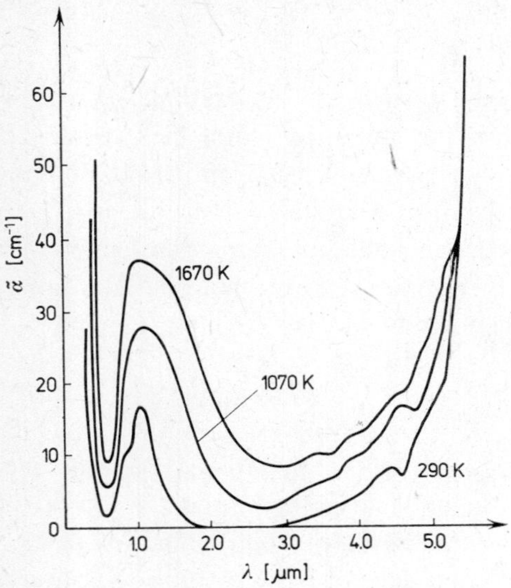

Fig. 4.4.4. Absorption spectra of natural single-crystal olivine $(Mg_{0.84}, Fe_{0.16})_2SiO_4$ at various temperatures. The beam is parallel to the α crystallographic axis. (According to T. J. Shankland, U. Nitsan and A. G. Duba, *J. Geophys. Res.*, **84**, no. 4 (1979), copyrighted by the American Geophysical Union).

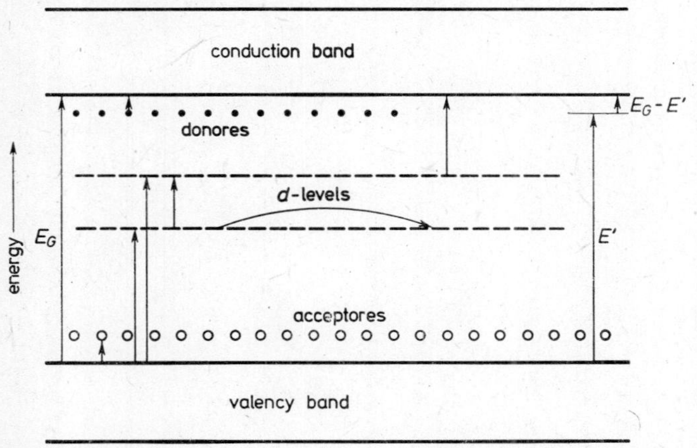

Fig. 4.4.5. Model of the electronic energy structure for silicates; E_G is the energy gap (valence–conduction level transition), E' is the exciton creation energy and $E_G—E'$ is the exciton binding energy. The vertical small arrows denote consecutive electronic transitions between different bands or levels; the semicircular arrow symbolizes transitions between *d*-states localized on different metal ions (deep impurities); the solid and open circles symbolize the levels from shallow impurities (donor and acceptor states) or structural defects. (According to U. Nitsan and T. J. Shankland, *Geophys. J. R. Astron. Soc.*, **45**, 81 (1976), copyrighted by the American Geophysical Union).

discussing these mechanisms, is shown in Fig. 4.4.5, after Nitsan and Shankland (1976a).

Basic absorption. In crystals-insulators, such as, for example, quartz or forsterite, basic absorption occurs for radiation with the wavelength $\lambda < 0.2$ µm, and is connected with electron transit from the basic band (valency band) to the conduction band of a crystal. In general, two types of transits are distinguished: straight and oblique. During straight transit the wave vector of the electron in a crystal is not changed. The lowest frequency ω_m at which such a transit is possible, is

$$\omega_m = \frac{E_G}{\hbar} \tag{4.4.11}$$

where E_G is the width of the energy gap in the crystal, i.e., the difference in energy between the bottom of the conduction band and the top of the valency band, and $\hbar$ is the "crossed" Planck constant. Relation (4.4.11) determines the position of the long-wave edge of basic absorption if diturbances from other transits are not present.

The width of the energy gap in a crystal is a function of temperature and pressure. In general it can be written that (e.g., Houghton and Smith, 1966)

$$\Delta E_G = \left(\frac{\partial E_G}{\partial T}\right)_p \Delta T + \left(\frac{\partial E_G}{\partial p}\right)_T \Delta p$$

From thermodynamic relations it follows that

$$\left(\frac{\partial E_G}{\partial T}\right)_p = \left(\frac{\partial E_G}{\partial \varrho}\right)_T \left(\frac{\partial p}{\partial \varrho}\right)_T \left(\frac{\partial \varrho}{\partial T}\right)_p$$

and since

$$\left(\frac{\partial p}{\partial \varrho}\right)_T \left(\frac{\partial \varrho}{\partial T}\right)_p = -K_T \alpha$$

where K_T is the isothermal bulk modulus and α is the coefficient of thermal expansion,

$$\Delta E_G = \left(\frac{\partial E_G}{\partial p}\right)_T (\Delta p - K_T \Delta T)$$

Values of the derivative $(\partial E_G/\partial T)_p$ are of an order of 10^{-4} eV · K^{-1} and of the derivative $(\partial E_G/\partial p)_T$ of 10^{-2} eV · GPa^{-1}, and the sign of the derivative $(\partial E_G/\partial p)_T$ can be either positive or negative. For silicate minerals both derivatives are negative.

From Table 4.4.1 (Nitsan and Shankland, 1976a, b) it follows that under normal conditions the energy gap E_G of important silicate and oxide minerals is of $\approx 8 \div 10$ eV (see also Shankland, 1968; Harlow and Shankland, 1974); they are therefore ideal insulators. It is clear from Fig. 4.4.6 that the width of the energy gap depends on the composition and crystalline structure of the minerals (Nitsan and Shankland, 1976a).

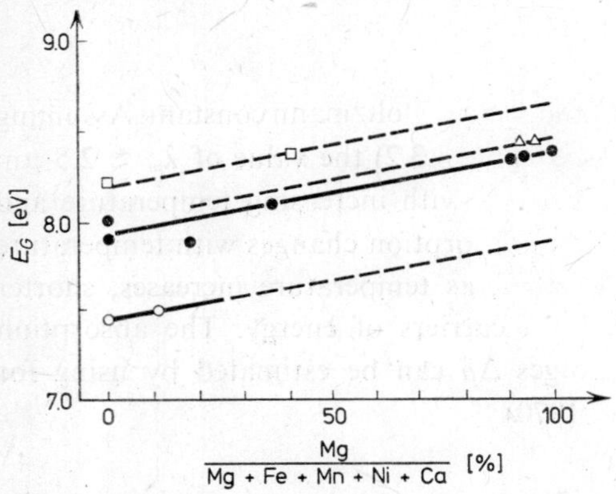

Fig. 4.4.6. Dependence of the energy gap (E_G) value on the chemical composition and crystal structure of silicates. The squares represent garnet, the triangles represent ortho-pyroxene, and the solid and open circles represent olivine and spinel structures, respectively. (According to U. Nitsan and T. J. Shankland, *Geophys. J. R. Astron. Soc.*, **45**, 76 (1976), copyrighted by the American Geophysical Union).

It is well known that olivine is considered to be the basic mineral of the Earth's mantle. Olivine is an isomorphic solid solution of orthosilicates of magnesium (Mg_2SiO_4) and iron (Fe_2SiO_4).* The lowest conduction band in olivine results from the "mixing" of the electron states 3*s* of magnesium and 4*s* of iron, while the highest valency band is formed by the electron states 2*p* of magnesium and 3*p* of iron. Pure forsterite, or $MgSiO_4$, has the energy gap $E_G \approx 8.8$ eV. The width of the energy gap in olivine decreases linearly with the increase of Fe/(Mg+Fe); for pure fayalite (Fe_2SiO_4) $E_G \approx 7.8$ eV (Nitsan and Shankland, 1976a).

Lattice absorption. Indirect interactions between photons of infrared radiation and phonons (interaction of the electromagnetic wave with electric dipoles of variable moment, arising in a crystalline lattice under the influence of elastic vibrations) result in a phenomenon called lattice

* Magnesium and iron appear in olivines as two-valent cations; ion radii for Mg^{++} and Fe^{++} are $\approx 0.65 \times 10^{-4}$ and 0.76×10^{-4} μm, respectively (Fyfe, 1964).

absorption. In general, three types of lattice absorption can be distinguished. They are: one-phonon absorption in the band of "Reststrahlen" (occurring only in ionic crystals), one-phonon absorption caused by defects of the crystalline lattice (e.g., by impurities), and multi-phonon absorption (Houghton and Smith, 1966).

The short-wave edge of lattice absorption is

$$\omega'_m = \frac{k\theta}{\hbar}$$

where θ is the Debye temperature, and k is the Boltzmann constant. Assuming for silicate minerals $\theta \approx 600$ K (see Table 4.3.2) the value of $\lambda_m \approx 2.5$ μm is obtained. Lattice absorption increases with increasing temperature and pressure. However, the significance of absorption changes with temperature for energy transport decreases because, as temperature increases, shorter and shorter waves become the main carriers of energy. The absorption changes induced by pressure changes Δp can be estimated by using for example the relation (Shankland, 1970)

$$\frac{\Delta\omega'_m}{\omega'_m} \approx \frac{\gamma}{K_T} \Delta p$$

where γ is the Grüneisen parameter, and K_T is the isothermal bulk modulus, which can readily be found from relation (4.3.10).

Exciton absorption. The excitation of electrons from the valency band to the conduction band of a crystal may be accompanied by the appearance of an electron–hole pair; such a pair, bound by electrostatic forces and electrically neutral, is called an *exciton.* Additional discrete energy levels originating from such a pair are situated below the bottom of the conduction band of the crystal. Electron transition to exciton states may occur at energies smaller than the energy gap (E_G) of a crystal. As a result of the appearance of excitons, additional absorption lines appear in the absorption spectrum, on both sides of the long-wave edge of the basic absorption. A number of characteristic properties of the exciton spectrum are satisfactorily explained by a hydrogen-like model of the exciton, called the Mott–Wannier exciton (see e.g., Dexter and Knox, 1965).

In general, the long-wave edge of exciton absorption is determined by the relation

$$\omega_g = \frac{E'_{\mathrm{ex}} - E^{\mathrm{m}}_{\mathrm{ph}}}{\hbar}$$

where E'_{ex} is the energy of exciton creation, and $E^{\mathrm{m}}_{\mathrm{ph}}$ is the maximum energy brought by a phonon during an oblique transit.

Impurity absorption. Radiation with an appropriate frequency ω may cause a series of optical transits: impurity centres → conduction band, valency band → impurity centres. As a result, the spectrum is enriched by new lines of absorption.

If δE is the energy difference between the bottom of the conduction band and the specific impurity level, then for photons with energy $\hbar\omega > \delta E$ electron transits from a given centre to the conduction band will occur. The same is true for the valency band at $\hbar\omega > \delta E'$, where $\delta E'$ is the energy difference between the impurity level and the top of the valency band; valency band → impurity centre electron transits may occur.

In general, two types of impurities are distinguished: shallow or deep donors and shallow or deep acceptors. If δE (or $\delta E'$) is of the order of kT (where k is the Boltzmann constant) the impurities are said to be shallow, and if $\delta E \gg kT$, the impurities are said to be deep.

Impurity centres are characterized by several excitation states, resembling, for example, the states in a hydrogen atom: the corresponding energies are

$$E_{n*} = -\frac{e^4 m_e^*}{2\varepsilon_d^2 \hbar^2 n^{*2}}$$

where n^* is the main quantum number, ε_d is the macroscopic dielectric constant of the medium, e is the electron charge, and m_e^* is its effective mass. Thus transits within the impurities themselves can occur and influence the absorption spectrum. There exist also electron transits between various impurities within a given band.

Shallow impurity states, situated close to the valency band (acceptors) or to the conduction band (donors), create absorption lines first of all in the far infrared. But transits from the valency band to empty donor levels or from populated acceptor levels to the conduction band can also occur. Absorptions of such types extend just behind the edge of the basic absorption, on its long-wave side. They are generally weak and usually masked by absorption connected with excitons and free carriers of charge (Kittel, 1966).

In silicates the most important impurity is iron, or more precisely the Fe^{++} ion.* It displays a characteristic maximum of absorption for waves with a length of about 1 μm (≈ 1.24 eV), related to the level d (see e.g., Fig. 4.4.3).

---

* In many cases ions of iron should be considered as a basic component of a crystal and at the same time its deep impurity.

The high absorption of ultraviolet radiation in olivines seems to be related primarily to the presence of iron.* Two processes can be distinguished here: oxygenation of defects, i.e., excitation of electrons from oxygen ions to the levels d of Fe^{++} ions (the bands of 4.6 and 6.0 eV), and electron transfer between neighbouring Fe^{++} ions in the 4.2 and 4.9 eV bands (Nitsan and Shankland, 1976a).

It can be stated in general that three types of intrinsic transits connected with the d states give the principal contribution to the absorption of the optical region of the pass band, i.e., at energies $\hbar\omega < E_G$ (the s–p gap): (1) levels $d \to$ conduction band, (2) valency band $\to$ levels d, and (3) transitions between d states localized on different Fe sites (see e.g., Nitsan and Shankland, 1976a).

Shankland (1970) studied the influence of pressure up to about 4 GPa on the Fe^{++} absorption band in olivine containing 7.7 per cent of Fe_2SiO_4, at a wavelength $\lambda \approx 1.06$ μm, and in periclase with an iron concentration Fe/Mg ≈ 0.004 at $\lambda \approx 1.02$ μm. It was found that in both crystals, as the pressure increases, the absorption bands broaden towards a "window" interior in the near infrared. The changes, however, are smaller than those observed during the increases of temperature. Aronson *et al.* (1970) measured radiation absorption in olivine, oligoclase and diopside (with various orientations of the crystals) at temperatures reaching 1723 K. The results confirmed the significance of absorption bands from Fe^{++} in the opacity of silicates.

Carrier absorption. Photons with energies $\hbar\omega \ll E_G$ may be absorbed as a result of electron transits inside a given band, e.g., the conduction band.

According to classical electrodynamics, the relation between the carrier absorption η and electric conductivity σ has, irrespective of the conductivity mechanism, the form

$$\eta = \frac{120}{n(\omega)}\sigma(\omega) \tag{4.4.12}$$

where $n(\omega)$ is the refractive index of radiation with frequency ω, and $\sigma(\omega)$ is in $(\Omega \cdot \mathrm{cm})^{-1}$ (see e.g., Suffczyński, 1965).

In general,

$$\sigma(\omega) = \sigma(0)\frac{\left\langle \dfrac{\tau}{1+(\omega\tau)^2} \right\rangle}{\langle \tau \rangle}$$

* Minerals of the olivine series may also contain small quantities of ions of nickel, manganese, calcium, aluminium and titanium.

where $\sigma(0)$ is the electrical conductivity at low frequencies ($\omega \to 0$) and τ is the relaxation time of charge carriers (of a given type).

4.4.4 Electrical conductivity in the Earth's mantle

It is well known that the electron–hole electrical conductivity of a solid is

$$\sigma(0) = e(n_e u_e + n_h u_h)$$

where n_e and n_h are the concentrations, and $u_e = e\langle\tau_e\rangle/m_e^*$ and $u_h = e\langle\tau_h\rangle/m_h^*$ are the mobilities of the electrons and the holes, m_e^* and m_h^* are the effective masses of the electrons and the holes, and $\langle\tau_e\rangle$ and $\langle\tau_h\rangle$ are the average relaxation times of electrons and holes, respectively.

In an ideal crystal the mobility of electric charge carriers (electrons) is determined by the scattering resulting from collisions with phonons. In real crystals some defects of the structure (vacancies, impurities) are always present, and they also cause scattering of carriers. At high temperatures, however, the dominant role is played by vibrations of the crystalline lattice.

In crystals without impurity atoms we have $n_e = n_h$; this case is called *intrinsic conductivity* (σ_i). The following relation is then valid:

$$\sigma_i(0) = 2e(u_e + u_h)\left(\frac{kT}{2\pi\hbar^2}\right)^{3/2}(m_e^* m_h^*)^{3/4}\exp\left(-\frac{E_G}{2kT}\right)$$

where $\hbar$ is the "crossed" Planck constant, k is the Boltzmann constant, and E_G is the energy gap. This relation is better known in a slightly different form, namely

$$\sigma_i(0) = \sigma_i^* \exp\left(-\frac{E_G}{2kT}\right)$$

where σ_i^* denotes the electrical conductivity at temperature $T \to \infty$. Since considerably more electrons are present in the valency band than in the impurities, at sufficiently high temperatures the intrinsic conductivity will always play the main role.

In crystals-insulators deep impurities are present as a rule. If they are donors and appear in large quantities, then the valency band of the crystal can be replaced by a local level with reference to the possibility of thermal excitation of electrons. The same may happen with regard to acceptors: the acceptor states then become "traps" for conductivity electrons.

In the case where the intrinsic concentration of charge carriers is negligible and $\mathcal{N}_D$ deep donors with the ionization energy E_D are present in the crystal, the electron concentration in the conduction band is

$$n'_e = \sqrt{2\mathcal{N}_D}\left(\frac{m_e^* kT}{2\pi\hbar^2}\right)^{3/4} \exp\left(-\frac{E_D}{2kT}\right)$$

and the electrical conductivity induced by deep donors is therefore

$$\sigma_D(0) = eu_e\left(\frac{m_e^* kT}{2\pi\hbar^2}\right)^{3/4} \sqrt{2\mathcal{N}_D} \exp\left(-\frac{E_D}{2kT}\right)$$

In crystals, besides intrinsic and impurity conductivities, *ionic conductivity* (σ) may occur; it is connected with the diffusion of ions (or ion vacancies) in the crystalline lattice. An ion situated between the nodes (or a vacancy in the lattice) must overcome the potential energy barrier during its travel. The relation connecting the so-called diffusion constant $\mathcal{D}'$ with the barrier W is of the form

$$\mathcal{D}' = \nu d^2 \exp\left(-\frac{W}{kT}\right)$$

where ν is the frequency of lattice vibrations, d is the lattice constant, and k is the Boltzmann constant (see e.g., Kittel, 1966). From Einstein's relation $ukT = q\mathcal{D}'$, where u is the ion mobility and q is the ion electric charge, it follows that

$$u = \frac{q\nu d^2}{kT} \exp\left(-\frac{W}{kT}\right)$$

Thus the conductivity connected with a given type of ion can be expressed by

$$\sigma_j = \frac{\mathcal{N} q^2 \nu d^2}{kT} \exp\left(-\frac{W}{kT}\right)$$

where $\mathcal{N}$ is the number of ions (vacancies) with the charge q in a volume unit.

In magnesium-rich olivines ($\approx 90\%$ of Mg) the energy gap E_G is about 9 eV (Nitsan and Shankland, 1976a). The energy gap decreases as the temperature and pressure increase, but very slowly; it is probable that the derivatives $(\partial E_G/\partial T)_p$ and $(\partial E_G/\partial p)_T$ are of the order of 10^{-4} eV $\cdot$ K^{-1} and 10^{-2} eV $\cdot$ GPa^{-1}, respectively. Thus the intrinsic conductivity of olivines would be rather small even at high pressures and temperatures.

Inside the energy gap of an olivine crystal there exists, however, energy originating from the presence of deep impurities, namely Fe^{++} ions (band d). This energy band plays the role of the valency band. Thus the electrical conductivity of olivine should distinctly have the character of impurity conductivity.

In olivines at temperatures higher than 1400 K ionic conductivity will probably appear.

Hamilton (1965) studied the electrical conductivity of polycrystalline samples of olivine containing 10.4 and 17.5 per cent of fayalite (Fe_2SiO_4) at temperatures up to 1200 K and under uniaxial pressure reaching 4.2 GPa. The results suggest that the increase of Fe_2SiO_4 content induces the increase of conductivity and decrease of the "activation energy" ($\mathscr{E}_{ac}$).* The activation energy also decreases with the increase of pressure (the larger the content of fayalite, the faster the decrease of the energy).

The dependence of the electrical conductivity of fayalite on the temperature and pressure was studied by Akimoto and Hideyuki (1965). They found that during the phase transformation of fayalite from an olivine structure to a spinel structure the electrical conductivity distinctly changed.

The electrical conductivity of the Earth's mantle originates mainly from impurities or is of ionic origin (Nitsan and Shankland, 1976a).

Possibly in some areas of the mantle one of these conductivities plays a dominant role. It cannot be excluded that at greater depths in the mantle

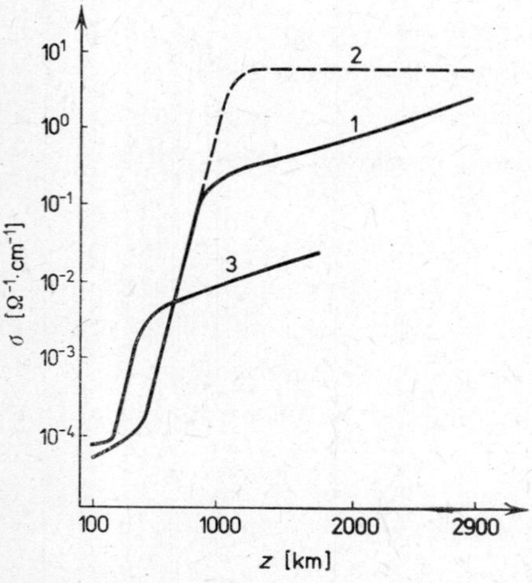

Fig. 4.4.7. Distribution of electrical conductivity in the Earth's mantle, according to McDonald (1957)—curve 1, Eckhardt *et al.* (1963)—curve 2 and Banks (1969)—curve 3.

* The measure of the energy $\mathscr{E}_{ac}$ is the slope of the straight lines which in a [$\ln \sigma$, T^{-1}] coordinate system represent the relationship $\sigma(T)$ (e.g., Shankland, 1969; Rai and Manghnani, 1978).

intrinsic conductivity may be of some significance, since the increase of pressure and temperature makes the width of the energy gap of silicates decrease, although very slowly.

The probable distribution of electrical conductivity in the Earth's mantle is shown in Fig. 4.4.7, after McDonald (1957), Eckhardt *et al.* (1963), and Banks (1969). This distribution was found from analyses of short- and long-period (secular) variations of the geomagnetic field (see e.g., Rikitake, 1966).

4.4.5 Radiative thermal conductivity in the Earth's mantle

Laboratory measurements of the radiative thermal conductivity ($\varkappa_R$) of silicate minerals at temperatures corresponding to probable temperatures in the Earth's upper mantle have given the following results: $\varkappa_R \approx 0.4 \div 2\ \mathrm{W \cdot m^{-1} \cdot K^{-1}}$ for olivine $(Mg_{0.86}, Fe_{0.14})_2SiO_4$ and $\varkappa_R \approx 2 \div 3\ \mathrm{W \cdot m^{-1} \cdot K^{-1}}$ for olivine $(Mg_{0.92}, Fe_{0.08})_2SiO_4$ at temperatures from 500 to 1900 K, and $\varkappa_R \approx 0 \div 0.8\ \mathrm{W \cdot m^{-1} \cdot K^{-1}}$ for enstatite $(Mg_{0.82}, Fe_{0.18})SiO_3$ at temperatures between 500 and 1600 K (Schatz and Simmons, 1972). On the other hand, the radiative thermal conductivity of monocrystal olivine $(Mg_{0.84}, Fe_{0.16})_2SiO_4$ calculated from the absorption spectrum increases from almost zero at room temperature to about $2\ \mathrm{W \cdot m^{-1} \cdot K^{-1}}$ ($5\ \mathrm{mcal \cdot cm^{-1} \cdot s^{-1} \cdot K^{-1}}$) at 1700 K (Shankland *et al.*, 1979).

Thus the average opacity $\langle \varepsilon \rangle$ of the upper mantle material corresponding to the above values of the conductivity $\varkappa_R$ must be at least of the order of $10\ \mathrm{cm^{-1}}$.

Variations of the opacity of the Earth's mantle according to depth can so far be estimated only qualitatively. Nevertheless, for greater depths the order of magnitude of opacity can be estimated, and this is sufficient for the rough estimation of radiative transfer. From the diagram in Fig. 4.4.7 (electrical conductivity $\sigma(z)$ in the mantle) it follows that at a depth of 1000 km the electrical conductivity is*

$$0.1\,\Omega^{-1} \cdot \mathrm{cm}^{-1} < \sigma_{1000} < 1\,\Omega^{-1} \cdot \mathrm{cm}^{-1}$$

From relations (4.4.10) it follows that at a depth of 1000 km ($\varrho \approx 4.5\ \mathrm{g \cdot cm^{-3}}$) the refractive index of temperature radiation is $n_{1000} \approx 2$. Thus according to relation (4.4.12), at a depth of 1000 km the carrier absorption

* The curve $\sigma(z)$, according to Banks (1969), implies that at a depth of 1000 km electrical conductivity is smaller by one order of magnitude.

would be

$$20 \text{ cm}^{-1} < \eta_{1000} < 200 \text{ cm}^{-1}$$

of course under the assumption that the condition $\omega\langle\tau\rangle \ll 1$ ($\langle\tau\rangle$ denotes the average relaxation time of charges) is fulfilled; in the case of $\omega\langle\tau\rangle \approx 1$ the carrier absorption would be half as large (see Clark, 1957).

Since the results of laboratory experiments clearly indicate that the opacity of the most important silicate minerals increases rapidly with temperature and pressure (see e.g., Aronson *et al.* 1970; Shankland, 1970; Shankland *et al.*, 1979), it seems reasonable to assume that the opacity of the mantle interior is of an order not smaller than 10^2 cm^{-1}. Most probably radiative thermal conductivity, connected with temperature radiation, is of greater significance (for energy transport) only in the area of the first 1000 km in the Earth's mantle.

REFERENCES

Akimoto S. and Fujisawa H. (1968), "Olivine–spinel solid solution equilibria in the system Mg_2SiO_4–Fe_2SiO_4", *J. Geophys. Res.*, **73**, 1467–1479.

Akimoto S. and Hideyuki F. (1965), "Demonstration of the electrical conductivity jump produced by the olivine–spinel transition", *J. Geophys. Res.*, **70**, 443–449.

Albarede F. (1975), "The heat flow/heat generation relationship: An interaction model of fluids with cooling intrusions", *Earth Planet. Sci. Lett.*, **27**, 73–78.

Anderson D. L. (1967a), "A seismic equation of state", *Geophys. J. R. Astron. Soc.*, **13**, 9–30.

Anderson D. L. (1967b), "Latest information from seismic observations", in: *The Earth's Mantle*, T. F. Gaskell (ed.), Academic Press, London–New York.

Anderson D. L. and Sammis C. (1970), "Partial melting in the upper mantle", *Phys. Earth Planet. Inter.*, **3**, 41–50.

Anderson O. L. (1979), "Evidence supporting the approximation $\gamma\varrho$ = const for the Grüneisen parameter of the Earth's lower mantle", *J. Geophys. Res.*, **84**, 3537–3542.

Anderson O.L. and Schreiber E. (1965), "The relation between refractive index and density of minerals related to the Earth's mantle", *J. Geophys. Res.*, **70**, 1463–1471.

Aronson J. R., Bellotti L. H., Eckroad S. W., Emslie A. G., McConnell R. K. and von Thüna P. C. (1970), "Infrared spectra and radiative thermal conductivity of minerals at high temperatures", *J. Geophys. Res.*, **75**, 3443–3456.

Aronson J. R., Emslie A. G., McConnell R. K., Eckroad S. W. and von Thüna P.C. (1967a), "The application of high-temperature mineral spectroscopy to radiative thermal conductivity upper mantle conditions", *Trans. Am. Geophys. Union*, **48**, 211.

Aronson J. R., Eckroad S. W., Emslie A. G., McConnell R. K. and von Thüna P. C. (1967b), "Radiative thermal conductivity in planetary interiors", *Nature*, **216**, 1096–1097.

Banks R. J. (1969), "Geomagnetic variations and the electrical conductivity of the upper mantle", *Geophys. J. R. Astron. Soc.*, **17**, 457–487.

Beck A. E. (1965), "Techniques of measuring heat flow on land", in: *Geophysical Monograph Series*, No. 8, Am. Geophys. Union.

Beck A. E., Darba D. M. and Schloessin H. H. (1978), "Lattice conductivities of single-crystal and polycrystalline materials at mantle pressures and temperatures", *Phys. Earth Planet. Inter.*, **17**, 35–53.

Birch F. (1969), "Density and composition of the upper mantle: First approximation as an olivine layer", in: *The Earth's Crust and Upper Mantle*, P. J. Hart (ed.), Am. Geophys. Union, Washington.

Birch F. and Clark H. (1940), "The thermal conductivity of rocks and its dependence upon temperature and composition", *Am. J. Sci.*, **238**, 529–558.

Blackman M. (1955), "The specific heat of solids", in: *Encyclopedia of Physics*, Vol. VII, Part 1: *Crystal Physics*, Springer-Verlag, Berlin–Göttingen–Heidelberg.

Boyd F. R. and England J. L. (1963), "Effect of pressure on the melting of diopside, $CaMgSi_2O_6$, and albite, $NaAlSi_3O_8$, in the range up to 50 kilobars", *J. Geophys. Res.*, **68**, 314–323.

Boyd F. R., England J. L. and Davis B. T. C. (1964), "Effects of pressure on the melting and polymorphism of enstatite, $MgSiO_3$", *J. Geophys. Res.*, **69**, 2101–2109.

Bukowinski M. S. T. (1976), "The effect of pressure on the physics and chemistry of potassium", *Geophys. Res. Lett.*, **3**, 491–494.

Bullard E. (1954), "The interior of the Earth", in: *The Earth as a Planet*, G. P. Kuiper (ed.), University of Chicago Press, Chicago.

Bullen K. E. (1963), *An Introduction to the Theory of Seismology*, 3rd ed., Cambridge University Press.

Cameron A. G. W. (1973), "Abundances of the elements in the Solar System", *Space Sci. Rev.*, **15**, 121–146.

Čermak V. and Jessop A. M. (1971). "Heat flow, heat generation and crustal temperature in the Kapuskasing area of the Canadian shield", *Tectonophysics*, **11**, 287–303.

Christensen N. I. (1979), "Compressional wave velocities in rocks at high temperatures and pressures, critical thermal gradient and crustal low-velocity zones", *J. Geophys. Res.*, **84**, 6849–6857.

Clark S. P. Jr. (1956), "Effect of radiative transfer on temperatures in the Earth", *Bull. Geol. Soc. Am.*, **67**, 1123–1124.

Clark S. P. Jr. (1957), "Radiative transfer in the Earth's mantle", *Trans. Am. Geophys. Union*, **38**, 931–938.

Clark S. P. Jr. (1963), "Variation of density in the Earth and the melting curve in the mantle", in: *The Earth Sciences*, T. W. Donnelly (ed.), University of Chicago Press, Chicago.

Crough S. T. and Thompson G. A. (1976), "Thermal model of continental lithosphere", *J. Geophys. Res.*, **81**, 4857–4862.

Davis B. T. C. and England J. L. (1964), "The melting of forsterite to 50 kilobars", *J. Geophys. Res.*, **69**, 1113–1116.

Dexter D. L. and Knox R. S. (1965), *Excitons*, J. Wiley, New York–London–Sydney.

Dudgale J. S. and MacDonald D. K. C. (1955), "Lattice thermal conductivity", *Phys. Rev.*, **98**, 1951–1952.

Dziewonski A. M., Hales A. L. and Lapwood E. R. (1975), "Parametrically simple Earth models consistent with geophysical data", *Phys. Earth Planet. Inter.*, **10**, 12–48.

Eckhardt D., Larner K. and Madden T. (1963), "Long–period magnetic fluctuations and mantle electrical conductivity estimates", *J. Geophys. Res.*, **68**, 6279–6286.

Elsasser W. M., Olson P. and Marsh B. D. (1979), "The depth mantle convection", *J. Geophys. Res.*, **84**, 147–155.

Friedlender G., Kennedy J. W. and Miller J. M. (1964), *Nuclear and Radiochemistry*, 2nd ed., J. Wiley, New York–London–Sydney.

Fukao Y. (1969), "On the heat radiative transfer and the thermal conductivity in the upper mantle", *Bull. Earthq. Res. Inst. Univ. Tokyo*, **47**, 549–560.

Fukao Y., Mizutani H. and Uyeda S. (1968), "Optical absorption spectra at high temperatures and radiative thermal conductivity of olivines", *Phys. Earth Planet. Inter.*, **1**, 57–62.

Fyfe W. S. (1964), *Geochemistry of Solids*, McGraw-Hill, New York–San Francisco–Toronto–London.

Gast P. W. (1972), "The chemical composition of the Earth, the Moon, and chondritic meteorites", in: *The Nature of the Solid Earth*, E. C. Robertson (ed.), McGraw-Hill, New York.

Gilvarry J. J. (1957), "Temperatures in the Earth's interior", *J. Atmosph. Terrest. Phys.*, **10**, 84–95.

Goettel K. A. (1976), "Models for the origin and composition of the Earth, and the hypothesis of potassium in the Earth core", *Geophys. Surv.*, **2**, 369–397.

Grim P. J. (1976), Map: Terrestrial Heat Flow, World Data Center A for Solid Earth Geophysics, Boulder, Colorado.

Gutenberg B. (1958), "Velocity of seismic waves in the Earth's mantle", *Trans. Am. Geophys. Union*, **39**, 486–489.

Gutenberg B. (1959), "The asthenosphere low-velocity layer, *Ann. Geofis.*, **XII**, 439–460.

Hamilton R. M. (1965), "Temperature variation at constant pressures of the electrical conductivity of periclase and olivine", *J. Geophys. Res.*, **70**, 5679–5692.

Hamza V. M. (1979), "Variation of continental mantle heat flow with age: Possibility of discriminating between thermal models of the lithosphere", *PAGEOPH*, **117**, 65–74.

Harlow G. E. and Shankland T. J. (1974), "Electronic structure of spinel", *Geochim. Cosmochim. Acta*, **38**, 589–599.

Hashin Z. and Shtrikman S. (1962), "A variational approach to the theory of the effective magnetic permeability of multiphase materials", *J. Appl. Phys.*, **33**, 3125–3131.

Hideyuki F., Fujii N., Mizutani H., Kanamori H. and Akimoto S. (1968), "Thermal diffusivity of Mg_2SiO_4, and Fe_2SiO_4, and NaCl at high pressures and temperatures", *J. Geophys. Res.*, **73**, 4727–4733.

Higgins G. and Kennedy G. C. (1971), "The adiabatic gradient and the melting point gradient in the core of the Earth", *J. Geophys. Res.*, **76**, 1870–1878.

Horai K. (1971), "Thermal conductivity of rock-forming minerals", *J. Geophys. Res.*, **76**, 1278–1308.

Horai K. and Baldridge S. (1972), "Thermal conductivity of nineteen igneous rocks. II. Estimation of the thermal conductivity of rock from the mineral and chemical compositions", *Phys. Earth Planet. Inter.*, **5**, 157–166.

Horai K. and Fujii N. (1971), "Thermophysical properties of lunar material returned by Apollo mission", *Proc. Lunar Geophysics Conf., October 18–20, 1971*, Houston (Lunar Science Institute), 447–475.

Horai K. and Simmons G. (1969), "Thermal conductivity of rock-forming minerals", *Earth Planet. Sci. Lett.*, **6**, 359–368.

Horai K. and Simmons G. (1970), "An empirical relationship between thermal conductivity and Debye temperature for silicates", *J. Geophys. Res.*, **75**, 978–982.

Houghton J. T. and Smith S. D. (1966), *Infra-red Physics*, 1st ed., Oxford University Press.

Hyndman R. D. Jessop A. M., Judge A. S. and Rankin D. S. (1979), "Heat flow in the Maritime Provinces of Canada", *Can. J. Earth Sci.* **16**, 1154–1165.

Jacobs J. A. (1974), *A Textbook on Geonomy*, J. Wiley, New York.

Kanamori H., Fujii N. and Mizutani H. (1968), "Thermal diffusivity measurement of rock-forming minerals from 300° to 1100°K", *J. Geophys. Res.*, **75**, 595–605.

Kaula W. M. (1968), *An Introduction to Planetary Physics*, J. Wiley, New York.

Kawada K. (1964), "Studies of the thermal state of the Earth. The 15th paper: Variation of thermal conductivity of rocks", Part 1, *Bull. Earthq. Res. Inst. Univ. Tokyo*, **42**, 631–647.

Kawada K. (1966), "Studies of the thermal state of the Earth. The 17th paper: Variation of thermal conductivity of rocks", Part 2, *Bull. Earthq. Res. Inst. Univ. Tokyo*, **44**, 1071–1091.

Keyes R. W. (1959), "High temperature thermal conductivity of insulating crystals: Relationship to the melting point", *Phys. Rev.*, **115**, 564–567.

Kieffer S. W. (1976), "Lattice thermal conductivity within the Earth and considerations of a relationship between the pressure dependence of the thermal diffusivity and the volume dependence of the Grüneisen parameter", *J. Geophys. Res.*, **81**, 3025–3030.

Kittel C. (1966), *Introduction to Solid-State Physics*, 3rd ed., J. Wiley, New York.

Klemens P. (1958), "Thermal conductivity and lattice vibrational modes", in: *Solid State Physics*, Vol. 7, F. Seitz and D. Turnbull (eds.), Academic Press, New York.

Knopoff L. and Shapiro J. N. (1969), "Comments on the interrelationship between Grüneisen's parameter and shock and isothermal equations of state", *J. Geophys. Res.*, **74**, 1439–1450.

Knopoff L. and Shapiro J. N. (1972), "The melting of metals", in: *The Nature of the Solid Earth*, E. C. Robertson (ed.), McGraw-Hill, New York.

Kraut E. A. and Kennedy G. C. (1966), "New melting law at high pressures", *Phys. Rev. Lett.*, **151**, 608–609.

Lachenbruch A. H. (1968), "Preliminary geothermal model of the Sierra Nevada", *J. Geophys. Res.*, **73**, 6977–6989.

Lachenbruch A. H. (1970), "Crustal temperature and heat production: Implications of the linear heat-flow relation", *J. Geophys. Res.*, **75**, 3291–3300.

Lachenbruch A. H., Sass J. H., Munroe R. J. and Moses T. H. (1976), "Geothermal setting and simple heat conduction models for the Long Valley caldera", *J. Geophys. Res.*, **81**, 769–784.

Langseth M. G. (1965), "Techniques of measuring heat flow through the ocean floor", in: *Geophysical Monograph Series*, No. 8, Am. Geophys. Union.

Lawson A. W. and Jamieson J. C. (1958), "Energy transfer in the Earth's mantle", *J. Geol.*, **66**, 540–551.

Lee D. W. and Kingery W. D. (1960), "Radiation energy transfer and thermal conductivity of ceramic oxides", *J. Am. Ceram. Soc.*, **43**, 594–607.

Liu H. P., Schock R. N. and Anderson D. L. (1975), "Temperature dependence of single-crystal spinel ($MgAl_2O_4$) elastic constants from 293 to 423°K measured by light-sound scattering in the Raman–Nath region", *Geophys. J. R. Astron. Soc.*, **42**, 217–250.

Loper D. E. (1978), "The gravitationally powered dynamo", *Geophys. J.*, **54**, 389–404.

Lubimova E. A. (1967), "Theory of thermal state of the Earth's mantle", in: *The Earth's Mantle*, T. F. Gaskell (ed.), Academic Press, London–New York.

MacDonald G. J. F. (1959), "Calculation on the thermal history of the Earth", *J. Geophys. Res.*, **64**, 1967–2000.

MacDonald G. J. F. (1963), "The interior of the Planets", in: *Origin of the Solar System*, R. Jastrow and A. G. W. Cameron (eds.), Academic Press, New York–London.

MacDonald G. J. F. (1964), "Dependence of the surface heat flow on the radioactivity of the Earth", *J. Geophys. Res.*, **69**, 2933–2946.

MacDonald G. J. F. and Ness N. F. (1961), "A study of the free oscillations of the Earth", *J. Geophys. Res.*, **66**, 1865–1911.

Magnitsky V. A. (1971), "Geothermal gradients and temperatures in the mantle and the problem of fusion", *J. Geophys. Res.*, **76**, 1391–1396.

Maj S. (1976), "On the linear relationships between phonon conductivity and mean sound velocity for silicate minerals", *Acta Geophys. Pol.*, **XXIV**, 157–160.

Maj S. (1978a), "A note on the geothermal gradient at a depth of 1000 km", *Acta Geophys. Pol.*, **XXVI**, 205–208.

Maj S. (1978b), "A relationship between phonon conductivity and seismic parameter ϕ for silicate minerals", *PAGEOPH*, **116**, 1225–1230.

Maj S. (1980), "Reduced heat flow–mean heat flow relationship for the continental geothermal provinces", *Acta Geophys. Pol.*, **XXVIII**, 233–240.

McDonald K. L. (1957), "Penetration of the geomagnetic secular variation through a mantle with variable conductivity", *J. Geophys. Res.*, **62**, 117–141.

McQuarrie M. (1954), "Thermal conductivity: VII. Analysis of variation of conductivity with temperature for Al_2O_3, BeO and MgO", *J. Am. Ceram. Soc.*, **37**, 91–95.

McQueen R. G., Fritz J. N. and Marsh S. P. (1964), "On the composition of the Earth's interior", *J. Geophys. Res.*, **69**, 2947–2965.

Meyer K. (1968), *Physikalisch-chemische Kristallographie*, VEB Deutsch. Verl. Grundst., Leipzig.

Nitsan U. and Shankland T. J. (1976a), "Optical properties and electronic structure of mantle silicates", *Geophys. J. R. Astron. Soc.*, **45**, 59–87.

Nitsan U. and Shankland T. J. (1976b), "Optical energy gaps in silicates", *EOS*, **57**, 160.

Pekeris C. L. (1966), "The internal constitution of the Earth", *Geophys. J. R. Astron, Soc.*, **11**, 85–132.

Pollack H. N. and Chapman D. S. (1977a), "Mantle heat flow", *Earth Planet. Sci. Lett.*, **34**, 174–184.

Pollack H. N. and Chapman D. S. (1977b), "On the regional variation of heat flow, geotherms, and lithospheric thickness", *Tectonophysics*, **38**, 279–296.

Pollack H. N. and Chapman D. S. (1977c), "The flow of heat from the Earth's interior", *Sci. Am.*, **237**, 2, 60–76.

Press F. (1970), "Earth models consistent with geophysical data", *Phys. Earth Planet. Inter.*, **3**, 3–22.

Preston F. W. (1956), "Thermal conductivity in the depths of the Earth", *Am. J. Sci.*, **254**, 754–757.

Rai C. S. and Manghnani M. H. (1978), "Electrical conductivity of ultramafic rocks to 1820 Kelvin", *Phys. Earth Planet. Inter*, **17**, 6–13.

Rao R. U. M. and Jessop A. M. (1975), "A comparison of the thermal characters of shields," *Can. J. Earth Sci.*, **12**, 347–360.

Rao R. U. M., Rao G. V. and Narain H. (1976), "Radioactive heat generation and heat flow in the Indian shield", *Earth Planet. Sci. Lett.*, **30**, 57–64.

Rikitake T. (1966), *Electromagnetism and the Earth's Interior*, Elsevier, Amsterdam–London–New York.

Ringwood A. E. (1962), "A model for the upper mantle", *J. Geophys. Res.*, **67**, 857–867.

Roy R. F., Blackwell D. D. and Birch F. (1968), "Heat generation of plutonic rocks and continental heat flow provinces". *Earth. Planet. Sci. Lett.*, **5**, 1–12.

Rybach L. (1976), "Radioactive heat production in rocks and its relation to other petrophysical parameters", *PAGEOPH*, **114**, 309–317.

Rybach L. (1979), "The relationship between seismic velocity and radioactive heat production in crustal rocks: An exponential law", *PAGEOPH*, **117**, 75–82.

Schatz J. F. and Simmons G. (1972), "Thermal conductivity of Earth materials at high temperatures", *J. Geophys. Res.*, **77**, 6966–6983.

Schreiber E. and Anderson O. L. (1966), "The determination of the critical geothermal gradient for three oxides materials", *Trans. Am. Geophys. Union*, **47**, 175.

Schreiber E. and Anderson O. L. (1967), "Pressure derivatives of the sound velocities of polycrystalline forsterite, with 6% porosity", *J. Geophys. Res.*, **72**, 762–764.

Shankland T. J. (1968), "Band gap of forsterite", *Science*, **161**, 51–53.

Shankland T. J. (1969), "Transport properties of olivines", in: *The Application of Modern Physics to the Earth and Planetary Interiors*, S. K. Runcorn (ed.), J. Wiley, London–New York–Sydney–Toronto.

Shankland T. J. (1970), "Pressure shift of infrared absorption bands in minerals and the effect on radiative heat transport", *J. Geophys. Res.*, **75**, 409–413.

Shankland T. J., Duba A. G. and Woronow A. (1974), "Pressure shifts of optical absorption bands in iron-bearing garnet, spinel, olivine, pyroxene and periclase", *J. Geophys.* Res., **79**, 3273–3282.

Shankland T. J., Nitsan U. and Duba A. G. (1979), "Optical absorption and radiative heat transport in olivine at high temperature", *J. Geophys. Res.*, **84**, 1603–1610.

Simmons G. and England A. W. (1969), "Universal equations of state for oxides and silicates", *Phys. Earth Planet. Inter.*, **2**, 69–76.

Smith F. G. (1963), *Physical geochemistry*, Addison Wesley, Palo Alto–London.

Soga N., Schreiber E. and Anderson O. L. (1966), "Estimation of bulk modulus and sound velocities of oxides at very high temperatures", *J. Geophys. Res.*, **71**, 5315–5320.

Stacey F. D. (1977), "A thermal model of the Earth", *Phys. Earth Planet. Inter.*, **15**, 341–348.

Stevels J. M. (1962), "The structure and the physical properties of glass", in: *Encyclopedia of Physics*, XIII: *Thermodynamics of Liquids and Solids*, Springer-Verlag, Berlin–Göttingen–Heidelberg.

Suffczyński M. (1965), *Electrodynamics*, PWN, Warszawa (in Polish).

Swanberg C. A., Chessman M. D., Simmons G., Smithson S. B., Grønlie G. and Heier

K. S. (1974), "Heat flow–heat generation studies in Norway", *Tectonophysics*, **2.** 31–48.

Thomsen L. (1967), "On the distributions of density and temperature in low-velocity zone", *J. Geophys. Res.*, **72**, 5649–5653.

Uffen R. J. (1952), "A method of estimating the melting point gradient in the Earth's mantle", *Trans. Am. Geophys. Union*, **33**, 893–896.

Uffen R. J. (1954), "A method of estimating the thermal conductivity of the Earth's mantle" (abstract), *Trans. Am. Geophys. Union*, **35**, 380.

Usselman T. N. (1975), "Experimental approach to the state of the core, I: The liquidus relations to the Fe-rich portion of the Fe–N–S system from 30 to 100 kb", *Am. J. Sci.*, **275**, 291–303.

Valle P. A. (1956), "Sul gradiente di temperature necessario per la formazione di 'low velocity layer'", *Ann. Geofis.*, **IX**, 371–377.

Verhoogen J. (1973), "Thermal regime of the Earth's core", *Phys. Earth Planet. Inter.*, **7**, 47–58.

Verhoogen J., Turner F. J., Weiss L. E., Wahrhaftig C. and Fyfe W. S. (1970), *The Earth: An Introduction to Physical Geology*, Holt, Rinehart and Winston, New York–Chicago–San Francisco–Atlanta–Dallas–Montreal–Toronto–London–Sydney.

Wang C. Y. (1972), "Temperature in the lower mantle", *Geophys. J.*, **27**, 29–36.

Wasserburg G. J., MacDonald G. J. F., Hoyle F. and Fowler W. A. (1964), "The relative contributions of uranium, thorium and potassium to heat production in the Earth", *Science*, **143**, 465–467.

Weber H. C. (1960), "Heat capacity", in: *McGraw-Hill Encyclopedia of Science and Technology*, Vol. 6, McGraw-Hill, New York–Chicago–San Francisco–Dallas–Toronto–London.

Williams D. W. and Kennedy G. C. (1969), "Melting curve of diopside to 50 kilobars", *J. Geophys. Res.*, **74**, 4359–4366.

Yoder H. S. and Tilley C. E. (1962), "Origin of basalt magmas: An experimental study of natural and synthetic rock system", *J. Petrology*, **3**, 342–352.

York D. and Farquhar R. M. (1971), *The Earth's Age and Geochronology*, Pergamon Press, Oxford.

Ziman J. M. (1962), *Electrons and Phonons*, Clarendon Press, Oxford.

5. The Structure of the Earth's Interior

5.1 INTRODUCTION

Studies of the Earth's interior structure, and especially of its crust and upper mantle, are necessary for the solution of several geotectonic problems, such as the origin of the continents and oceans, the direction of the evolution and motion of the Earth's crust, the understanding of orogenesis processes and the causes of earthquakes, the explanation of general laws governing the formation and distribution of mineral resources.

Among the various geophysical methods, seismic methods are the most important in studies of the Earth's structure. The history of these studies goes back to the nineteenth century; it was in 1897 that E. Wiechert, a professor of the University of Gottingen, carried out the first theoretical calculations of the Earth's density and proposed a hypothesis that the Earth is composed of two layers, not counting the crust. According to Wiechert the metallic interior of the Earth, called the *core*, is surrounded by a stone cover (*Steinmantel* in German) whose density changed little from 3.0 to 3.4 $g \cdot cm^{-3}$. According to Wiechert, the thickness of the stone cover, for which the term *mantle* was generally accepted in the nineteen forties, changed from 1200 to 1600 km.

At the beginning of the twentieth century it became obvious that any progress in the study of the Earth's structure depended on the development of seismology and seismic methods. The progress of seismology in turn depended on the development of instruments and of the theory of seismic waves. Although the foundations of instrumental seismology were already laid in the middle of the nineteenth century, significant progress was not made until 1892, when the first simple seismograph, suitable for wide application in different regions of the world, was built by J. Miln and his associates. In 1907 about one hundred seismic stations were in operation, recording practically all larger earthquakes. From that time seismology became a science, and seismograms formed the basic material used in the study of the Earth's structure.

The application of mathematical methods for the description of elastic waves radiated by an earthquake source was, together with the development of instruments, the main factor in the successes of seismology in the study of the Earth's interior achieved at the end of the nineteenth and the beginning of the twentieth century. Between 1903 and 1906 H. Bendorf in Germany and J. Miln and R. Oldham in England attempted to determine the seismic wave velocity distribution according to depth. Their estimations were of course far from precise and it was not till 1907 that E. Wiechert formulated a precise method, based on the works of the German mathematician H. Herglotz, of determining the seismic wave velocity at various depths in the Earth. This method, called the *Wiechert–Herglotz method*, is of fundamental significance for seismology and is also used, with various modifications, at the present time.

The next important stage in the study of the Earth's interior began with the work of the Yugoslav seismologist Mohorovičić, who found, while investigating the records of the Zagreb earthquake of 8 October 1909, two arrivals of P and S waves at a short epicentral distance. He concluded that at a depth of about $50 \div 60$ km there was a distinct seismic discontinuity on which seismic waves were refracted. This discontinuity was called the *Mohorovičić discontinuity* and accepted as the boundary dividing the Earths' crust from the mantle.

In 1914 the well-known German seismologist B. Gutenberg found that at a depth of 2900 km the velocity of longitudinal waves decreases suddenly from 13.2 to 8.5 $km \cdot s^{-1}$. Thus the boundary between the Earth's mantle and core was established. These discontinuities are called *discontinuities of the first order* in contrast to weaker seismic boundaries found later in the Earth's crust (the *Conrad boundary*) and also in the upper mantle (first work of B. B. Golicyn), which are called *discontinuities of the second order*.

Further works of B. Gutenberg, H. Jeffreys, I. Lehmann, K. Bullen, P. Byerly and others, published between 1920 and 1950, brought further additions and corrections to the models of the Earth's structure. In particular, these additions related to such problems as low-velocity zones in the upper mantle, the discontinuity at a depth of about 400 km, and the internal core of the Earth. These problems are still very important in the contemporary physics of the Earth's interior.

Finally, in the fifties and the sixties, methods of explosion seismology, called *deep seismic soundings*, were developed; based on modern instrumental techniques, they significantly advanced the study of the structure of the Earth's crust and upper mantle. The wide application of modern mathematical methods to the interpretation of experimental data brought about not only quantitative but also—primarily—qualitative changes in the prevailing views on the Earth's interior structure.

5.2 STRUCTURE OF THE EARTH'S CRUST

The lower boundary of the Earth's crust is determined by the Mohorovičić discontinuity, often called the *M* boundary or the *Moho boundary*. The Moho boundary is characterized by a *P* wave velocity discontinuity as large as $0.9 \div 1.0$ km · s^{-1} on the average, and occasionally it is $0.4 \div 0.5$ km · s^{-1}. The average thickness of the continental crust is about 40 km, and it is about 35 km in the areas of old platforms, while the average thickness of the oceanic crust is only about $10 \div 12$ km.

Most generally, the Earth's crust is composed of two basic layers: a sedimentary layer and a crystalline complex bounded from the bottom by the Moho discontinuity. The surface of the crystalline substratum and the Moho discontinuity are the main boundaries and the only ones that have a clear physical meaning and are generally accepted, apart from a few areas of a distinctly anomalous structure (areas of the transition crust type, island areas and mid-oceanic ridges).

The characteristic velocity of *P* waves in the crystalline substratum is on the average about 6.0 km · s^{-1}, but considerable positive and negative deviations from that value are well known. Velocities of $8.0 \div 8.2$ km · s^{-1} characterize the Moho discontinuity, where substantial deviations are also observed.

Within the crystalline complex of the crust several discontinuities are present, but their nature is little known. It is well known that the traditional division of the crust into "granitic" and "basaltic" layers is now not sufficient and has lost its basic sense. This division is still used, but mostly for an arbitrary determination of certain velocity ranges of seismic waves. The only objective parameter characterizing discontinuities in the Earth's crust is the velocity of longitudinal waves. This velocity varies between 6.0 and 7.6 km · s^{-1} within the crystalline complex.

The basic model of the crust, formed during the last few years, is the multi-layered, gradient model with the following three or four discontinuity boundaries: $5.8 \div 6.2$ km · s^{-1}, $6.3 \div 6.5$ km · s^{-1}, $6.8 \div 7.2$ km · s^{-1}, and $7.3 \div 7.7$ km · s^{-1} (e.g., Kosminskaya, 1965). The thickest layers are those with velocities from 6.5 to 7.0 km · s^{-1}. The sedimentary layers are more significant, especially in areas where they reach 2/3 of the entire crust thickness. They are characterized by average velocities of $3 \div 4$ km · s^{-1} (Fig. 5.2.1).

A characteristic feature of the Earth's crust and upper mantle, besides the already noted horizontal multi-layered structure, is their block structure. Firstly, there are continental and oceanic blocks, then blocks related to

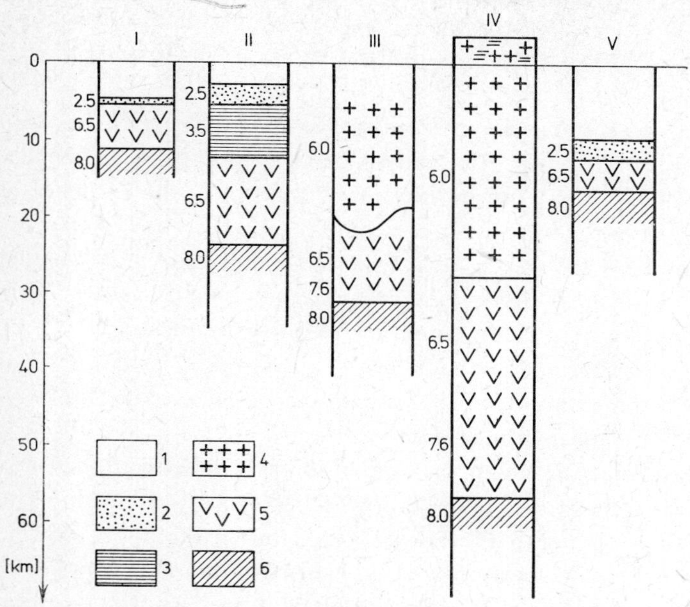

Fig. 5.2.1. Schematic models of the Earth's crust (Magnitsky, 1965): I—oceanic crust, II—depression, III—platform, IV—mountain region, V—ocean trench; 1—water, 2—clay, 3—consolidated sediments, 4—granite, 5—gabbro, 6—upper mantle; the values distinguish the velocity of *P* waves in km/s.

mountain and platform areas, and finally blocks of regional significance. Thus, studies of deep fractures are of great importance. The most effective criteria for the distinction of deep fractures are provided by deep seismic soundings.

5.2.1 Low-velocity zones in the Earth's crust

During the fifties Gutenberg (1950, 1951, 1959) formulated the hypothesis that a low-velocity zone is also present in the Earth's crust at depths of between 10 and 20 km. In his later works (e.g. 1959) Gutenberg assumed that a low-velocity zone could also exist in the lower "basaltic" layer. Studies of channel waves related to this problem have been made by Press and Gutenberg (1956), Båth (1957) and others.

Gutenberg tried to base his hypothesis on experimental studies relating to distributions of seismic velocities in rock samples under high pressures and at temperatures corresponding approximately to the conditions dominating in the "granitic" layer. In this way Gutenberg found some proofs for his hypothesis, but in his later works (1960) agreed that laboratory experiments cannot solve the problem. This is certainly due to difficulties in modelling the mineral composition of the "granitic" layer and also the physical conditions dominating in it.

In spite of intensive development of the study of the crust by deep seismic soundings during the fifties and sixties, the concept of Gutenberg still remained a hypothesis, and no new proof confirming the existence of low-velocity zones in the crust were presented.

Gutenberg's idea was taken up again by Landisman and Mueller (Mueller and Landisman, 1966; Landisman and Mueller, 1966), who presented the results of a reinterpretation of the published data of the study of the crust by seismic methods. They proved that a certain type of strong reflections are connected with the lower boundary of "the low-velocity zone in the upper lithosphere".

In general, it should be noted that the problem of investigating zones, characterized by velocity inversion in the Earth's crust is much more complex than that relating to the low-velocity zone in the upper mantle. This results from a much greater local differentiation of the structure of the crust in comparison with the structure of the upper mantle, and also from great difficulties in the interpretation of the travel times of seismic waves in the crust.

5.2.2 The nature of the layers and boundaries in the Earth's crust

The nature of the layers and discontinuity boundaries in the Earth's crust is usually studied by comparing the velocities of seismic waves travelling in the crust with the elastic wave velocities measured in rock samples in laboratory experiments under high pressures and temperatures (Birch, 1958, 1960, 1961; Volarovich *et al.*, 1967; and others).

From laboratory experiments it follows that seismic wave velocities in the crust correspond to wave velocities in such rocks as granite, basalt and gabbro under the pressure of a few kbar. Seismic velocities under the Moho discontinuity correspond approximately to the velocities in such rocks as peridotite, dunite and eclogite (Birch, 1963). The observed changes of velocities in the crust may also be induced by structural changes and/or by the influence of pressure and temperature (Birch, 1963).

More recent comparisons of seismic wave velocities with elastic wave velocities in rock samples do not contradict in general the classic division of the Earth's crust into the "granitic" and "basaltic" layers. The main difficulties, however, appear when a detailed interpretation of the nature of consecutive discontinuities, seismically distinguishable, is attempted.

An interpretation of the layer with P wave velocities of $7.3 \div 7.7 \text{ km} \cdot \text{s}^{-1}$, sometimes appearing in the lower parts of the crust, is especially difficult. These velocities are clearly too high in comparison with those in basalt or gabbro, which are probably the main components of the lower crust. Seismic velocities higher than in granite samples and lower than in basalt are also often observed.

The velocity of P waves depends primarily on the density and mean atomic mass. The comparison of mean atomic masses with the relation $v_P(\varrho)$ — P wave velocity as a function of density — under higher pressures (Birch, 1960, 1961; Knopoff, 1970; Magnitsky, 1965) shows that there is little probability of granite becoming a component of the lower parts of the crust, since its density is too low. Rocks with densities at least as high as that of diorite would be acceptable. From the values of P wave velocities below the Moho boundary it can be concluded that eclogites and peridotites, especially garnet peridotites, may form the composition of the upper mantle. Dunites, on the other hand, are characterized by too low P wave velocities and thus there is little probability of their entering into the composition of the upper mantle.

Such a petrographic interpretation of the crust layers is certainly not unique since the great majority of rocks are characterized by a mean atomic mass between 21 and 22 $\text{g} \cdot \text{mole}^{-1}$, regardless of their composition. Thus the denisty ϱ is a basic parameter determining the velocity v. It happens occasionally, however, that rocks have significantly different compositions while their densities and velocities are identical. Thus a seismic interpretation is not unique, and only laboratory experiments relating to the determination of the stability of various minerals under high pressures and temperatures corresponding to those in the crust and upper mantle could clarify the situation to some extent.

A separate problem for investigation, both from a petrological and from a seismic point of view, is the explanation of "weak" boundaries, which cannot be determined precisely as yet, but whose presence is indicated by seismic measurements. They are usually boundaries with small discontinuities in velocity gradients, which cannot be distinguished kinematically but are more or less distinctly reflected in a dynamic pattern of the wave field (Figs. 5.2.2 *a*, *b*, *c*). It seems that this problem could be solved by great laboratory syntheses establishing the ranges of phase changes and the stability of particular minerals. If the crust layers are of homogeneous chemical composition, then at great depths a stepwise change of velocity would not be possible (Kosminskaya, 1965). It is only possible when the change of chemical composition is sufficiently sharp. But such

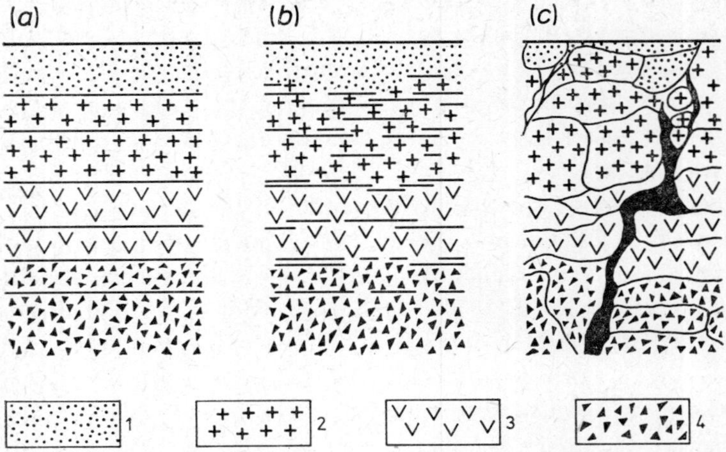

Fig. 5.2.2. Models of the Earth's crust (Kosminskaya, 1965): 1—sedimentary layer, 2—upper crustal layer ("granitic" layer), 3—lower crustal layer ("basaltic" layer), 4—upper mantle.

a change of chemical composition is little probable, especially in the upper depths intervals (10÷15 km) of the crystalline crust. Thus a plausible explanation would be that the boundaries result from considerable changes of temperature and internal stress in different zones of the Earth. In this sense, the seismic boundaries form a type of front lines of metamorphic processes, which extend over large areas, forming quasi-layers, approximately horizontal, in the heterogeneous complex of the consolidate crust (Riznichenko and Kosminskaya, 1963).

According to this theory, deep seismic boundaries should be considered as a secondary feature of the medium, imposed on the details of the crust structure. At the same time, the present distribution of seismic boundaries and the corresponding layers indicates traces of the conditions of the initial configuration, which are only partially obliterated. Thus, in spite of the exclusively physical nature of these boundaries, their investigation is one of the most important tasks in the solution of problems relating to the evolution of the crust and upper mantle of the Earth.

Between the surface of the consolidated substratum and the Moho discontinuity there are various seismic boundaries whose characteristics, both dynamic and kinematic, are not generally established. If the surface of the consolidated substratum is characterized by P wave velocities between 6.0 and 6.2 $km \cdot s^{-1}$ and the typical values for the Moho discontinuity are between 8.0 and 8.2 $km \cdot s^{-1}$, then the possible seismic boundaries in the consolidated complex may have any value of the velocity between

6.3 and 7.8 km · s^{-1}. It is difficult, therefore, to determine precisely what value of *P* wave velocity corresponds to the Conrad boundary or surface of the "basaltic layer", i.e., the boundary between the granitic-gneissic and granulitic-alkalinous complex of the crust model of Belousov (1966). If, in spite of these difficulties, the term "Conrad boundary" is still used in some works, then from a point of view of velocity distribution it has an arbitrary and approximate meaning. This boundary is usually determined from wave criteria and corresponds mainly to the surface of a fast increase of *P* wave velocities with depth.

The nature of seismic waves connected with boundaries in the consolidated complex of the crust is very often difficult to evaluate. Mostly, however, these waves have the character of reflected-refracted waves, as a result of usually weak velocity stratification of this part of the crust. Low- and high-velocity zones present in the Earth's crust additionally complicate the whole pattern of the discrete structure of the crust.

The nature of deep seismic boundaries in the consolidated complex is specific and entirely different from stratigraphic and lithologic boundaries appearing in sedimentary series. This dissimilarity results from the fact that deep boundaries are most probably boundaries of the phase type. This especially concerns the Moho boundary. The interpretation of strong waves reflected from the Moho discontinuity (Fig. 5.2.3) in the area of near-critical and over-critical angles suggests that in most cases it is a zone several kilometres thick, in which one, two or more discontinuities can be distinguished. It is often difficult to indicate which of these boundaries forms the base of the Earth's crust. This problem becomes more and more complex in the light of recent results of experimental petrology, relating to phase transitions under high pressures and temperatures. The discontinuous and unsystematic character of the records of waves reflected from Moho, the strong attenuation of certain waves and the appearance of others, reflects the true nature of the lower boundary of the crust, which is a complex transition zone to the upper mantle. In general, several types of transition zones between the crust and the upper mantle can be distinguished (Guterch *et al.*, 1980).

The results of determining the continuous distribution of velocity as a function of depth lead to similar conclusions with regard to the structure and thickness of the transition zone between the crust and the upper mantle. In this case the Moho boundary can be defined as the zone of the largest gradients of velocity, or as a depth at which the velocity increases suddenly from about 7.5 km · s^{-1} to 8.1÷8.5 km · s^{-1} (Giese *et al.*, 1967; Giese and Stein, 1971). An interesting model of the transition zone between the crust and the upper mantle was proposed by Meissner (1973). It is a model with several velocity inversion layers, confirmed by a detailed interpretation of the dynamics and kinetics of waves reflected from the Moho discontinuity.

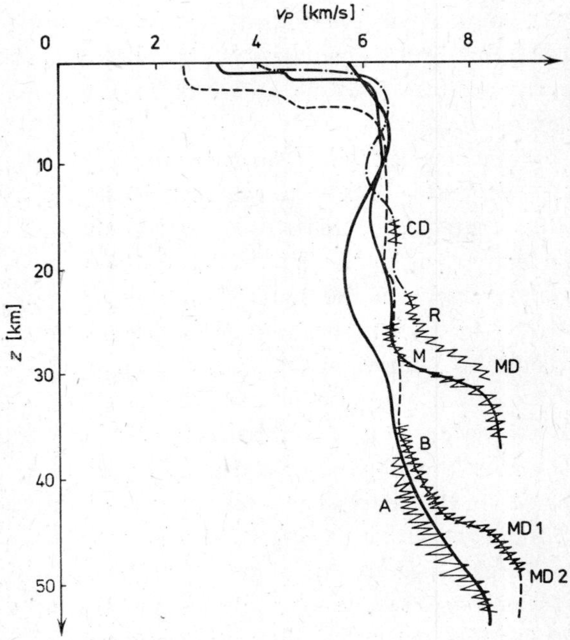

Fug. 5.2.3. Example of velocity–depth curves for different areas (Meissner, 1973): B—Buchara, M—Molasse, A—Alps, R—Renish Massif, the zigzag line—indication of lamella type layering, CD—Conrad discontinuity, MD—Moho discontinuity, MD_1 and MD_2—discontinuities in the transition zone between crust and upper mantle.

From studies of deep vertical reflections it follows that in most cases the nature of the waves reflected from deep crustal discontinuities is in conflict with the classical model of the Earth's crust (Meissner, 1973).

A detailed interpretation of several groups of waves reflected in the transition zone between the crust and the upper mantle leads to intersections of the crust with either the discontinuous or multiple Moho boundary or with several other discontinuities occurring near it within a depth interval of several kilometers (Guterch, 1977; and others). Such a pattern of the crust base may suggest that the results are fragmentary and not unique. In reality, however, this pattern reflects very well the complex nature of the transition zone between the crust and the upper mantle. In the light of recent results of experimental petrology, the transition zone should be interpreted as a complex phase transformation zone, generally of the basalt–eclogite type. Ito and Kennedy (1970, 1971) demonstrated that a transformation of this type is not only possible, which was proved by earlier works (Yoder and Tilley, 1962; Ringwood and Green, 1966), but that it occurs under pressures and temperatures close to those dominating in the lower crust and the upper mantle. Experiments have shown that the basalt–eclogite transformation clearly occurs in two stages: from basalt to garnetic granulite and from garnetic granulite to eclogite (Fig. 5.2.4). Two transformation stages should form two seismic discontinuities

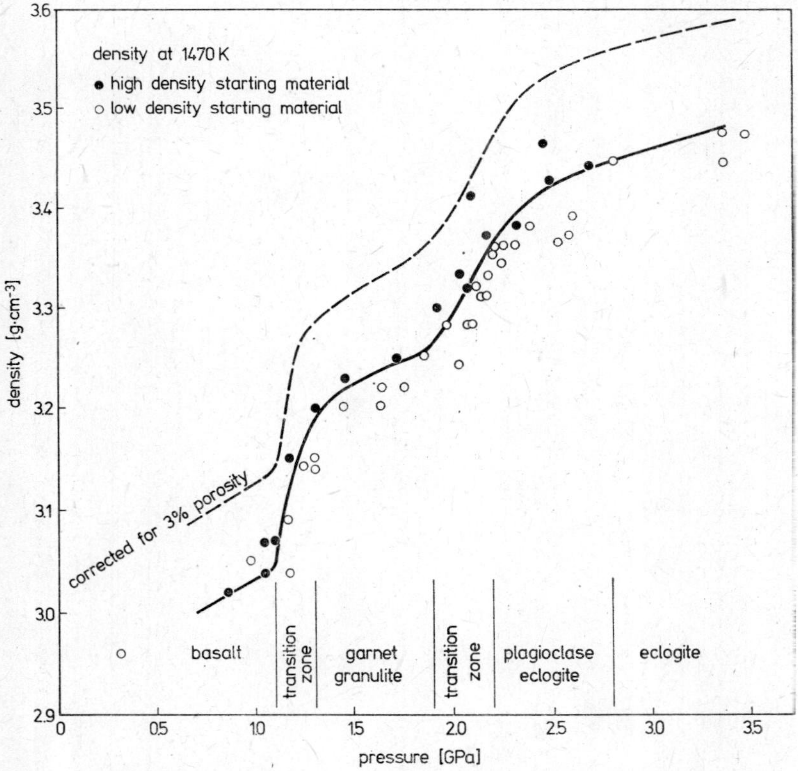

Fig. 5.2.4. Model of the basalt–eclogite transformations (Ito and Kennedy, 1970).

are either sharp or gentle, or one may be gentle and the other sharp. The nature of a discontinuity depends on the mineralogical composition of the "basalt" undergoing phase transformation.

In general, the petrological interpretation of seismic measurements of the transition zone between the crust and the mantle, based on a phase transformation of the basalt–eclogite type, is still very difficult and not

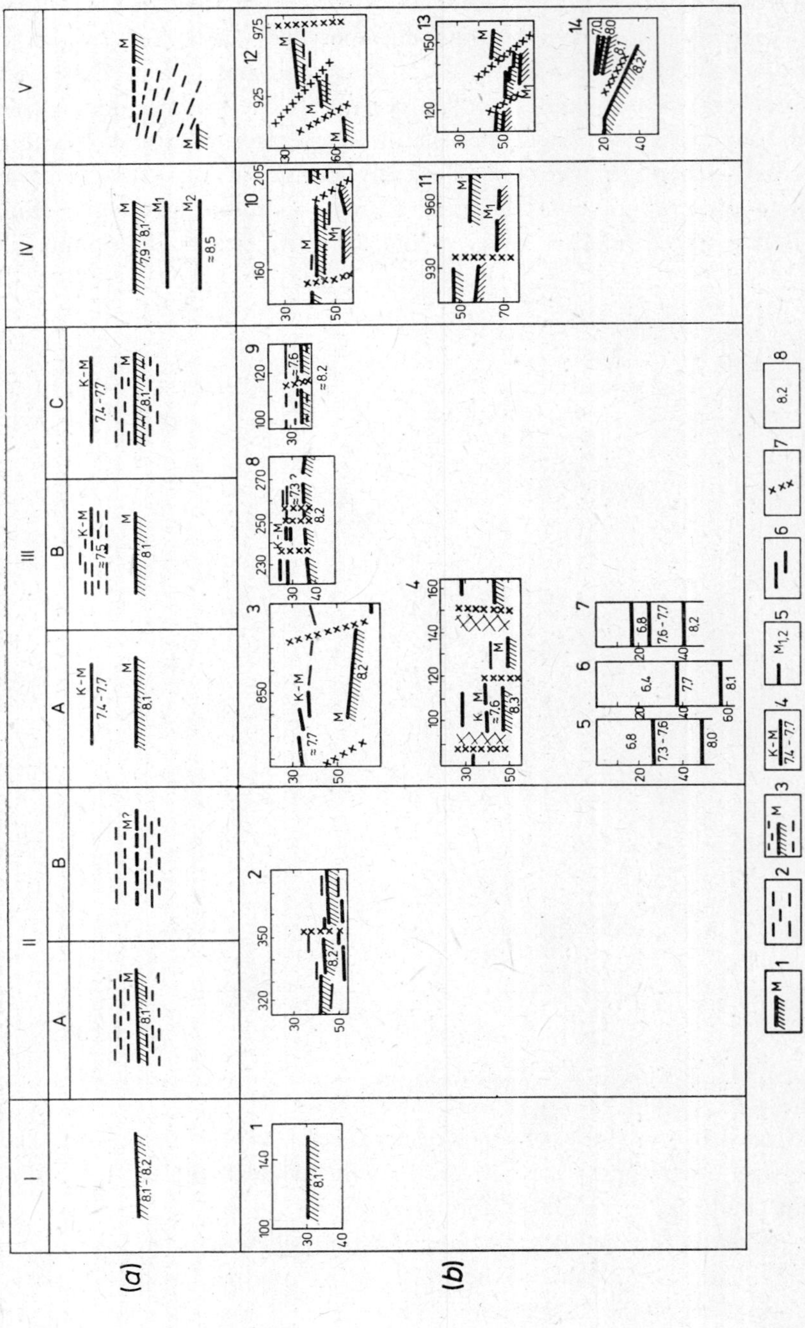
(a)
(b)
I
II
III
IV
V
A
B
C
K–M
M
M₁
M₂
8.1–8.2
7.4–7.7
7.9–8.1
≈8.5
≈7.5
1
2
3
4
5
6
7
8
9
10
11
12
13
14

unique (Guterch *et al.*, 1975). The processes of eclogite formation in nature are very complex (Smulikowski, 1964, 1967). Studies carried out by Smulikowski have shown that eclogites may originate from supracrustal rocks, including sedimentary rocks. To some extent this apparent discrepancy is explained by laboratory experiments, which show that elastic wave velocities in crustal eclogites are considerably lower than those in eclogites from the Earth's upper mantle.

5.2.3 Types of transition zone between the crust and the upper mantle

The transition zone between the crust and the upper mantle under the continents has a highly complex structure and cannot be described by a uniform model. It is only in a few cases that this zone is formed by the sharp Moho boundary. Usually the transition zone is of a diffusive character, and is from a few to a dozen kilometres thick. An attempt to determine the types of the crust–upper mantle transition zone, the examples being taken mainly from the European continent is shown in Fig. 5.2.5 (Guterch *et al.*, 1980). It can be noted that there is a distinct correlation between the type of transition zone, the tectonics and the history of the geological structures in a given area. Such a correlation is of great significance for the explanation of the physical processes occurring at the boundary between the crust and the Earth's upper mantle.

Five principal types of the crust–upper mantle transitions are distinguished.

Type I corresponds to the classical Moho boundary pattern—a sharp seismic boundary with a change of *P* wave velocity to 8.1 $km \cdot s^{-1}$. It is a seismic boundary characteristic for structures with a reduced thickness of the Earth's crust in the Mediterranean area.

Fig. 5.2.5. Principal types of the crust–upper mantle transitions in Central and Eastern Europe: (*a*) models of transition, (*b*) typical examples; A, B, C—types of Earth's crust; 1—sharp transition (Moho discontinuity), 2—diffusive transition, 3—diffusive transition with one distinct boundary, 4—discontinuity in the Earth's crust with high velocity of *P* waves, 5—double Moho discontinuity, 6—other weak seismic discontinuities, 7—deep fractures, 8—values distinguish *P* wave velocities in km/s; numbers of crustal section: 1—Panonia Block (Hungary Depression), 2—Ukrainian shield, 3—Eastern Carpathians, 4—Marginal zone of the Precambrian Platform of Eastern Europe in Poland, 5—East-African rift, 6—Baikal rift, 7—Rhine Graben, 8 and 9—Paleozoic Platform of Central and Western Poland, 10—Dniepr–Donetz trench, 11—South-Western marginal zone of the East European Platform and Carpathians, 13—Crimea orogen, 14—Corsica region.

The structure of a transition zone of type II is a certain complication of type I. In some cases a distinct 8.1 km $\cdot$ s^{-1} boundary is accompanied by a series of unstable reflection boundaries, while in other cases only series of reflection boundaries appear and no unique determination of *P* wave velocity is possible; hence no precise determination of the Moho boundary is possible either. This transition type is characteristic for platforms, especially for old platforms. In this case the materials of the mantle and those of the lower crust are undoubtedly mixed, and the share of the mantle material increases with depth.

In a crust–upper mantle transition zone of type III a layer with velocity between 7.4 and 7.7 km $\cdot$ s^{-1} appears, bounded from the top and from the bottom by distinct seismic boundaries; the lower of these is the Moho boundary with a typical velocity of 8.1 km $\cdot$ s^{-1}. The sharpness of these boundaries defines three types of transition: (1) both boundaries are sharp, (2) the lower boundary is sharp while the upper is "diffusive", (3) the lower boundary is diffusive and the upper is sharp. The appearance of a layer with velocity between 7.4 and 7.7 km $\cdot$ s^{-1} implies that the floor of the Earth's crust is undergoing transformation, probably leading to the formation of a new Moho boundary. This type of the crust–upper mantle transition zone is characteristic for structures tectonically active, including young orogenes as well as rift systems and some areas of young platforms.

Type IV of the crust–upper mantle transition is characterized by several sharp "Moho boundaries" with various seismic wave velocities (from 8.0 to 8.5 km $\cdot$ s^{-1}). The petrographic interpretation of these boundaries is clearly not unique. This type of transition most probably represents a pattern characteristic for the stage appearing after the processes characterizing transition type III are ended.

Type V of the crust–upper mantle transitions is characteristic for areas with horizontal motions of crustal blocks of either regional or continental extension. This type of transition is known in limited areas of contacts of different lithospheric blocks, e.g., the Carpathians and the East European Platform, the region of Corsica, and others.

5.3 THE UPPER MANTLE OF THE EARTH

A typical velocity cross-section of the upper mantle, with selected curves of velocity distribution according to different authors, is shown in Fig. 5.3.1 (Hales, 1971). A comparison of the curves of Gutenberg and Jeffreys, reflecting their long-lasting discussion, is of particular interest. The models of Gutenberg always contained a low-velocity layer at a depth of about

150 km, while in the models of Jeffreys the velocity of *P* waves systematically increased with depth. On the basis of these models, three contemporary models of velocity distribution in the upper mantle, prepared by Hales (1971), are presented: for shield areas (ER 1), for platforms (ER 2) and for tectonically active regions (NTS 1). In all these models a low-

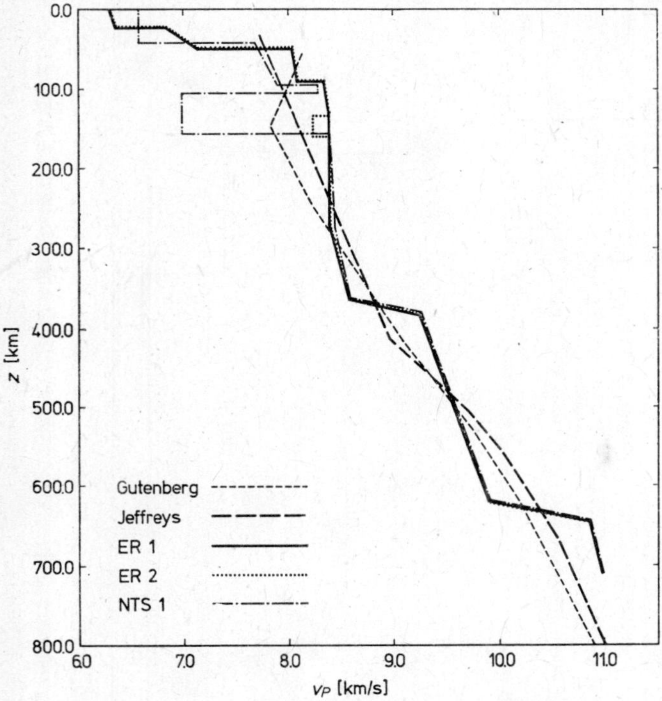

Fig. 5.3.1. Velocity distribution according to different authors (Hales, 1971).

velocity zone exists, which is especially distinct in tectonically active areas. Furthermore, two very sharp seismic boundaries are observed. The first boundary occurs at depths of between 350 and 400 km with a velocity change from about 8.7 to 9.3 $km \cdot s^{-1}$, and the second discontinuity is situated at depths of between 650 and 700 km with a corresponding velocity change from about 9.9 to 10.9 $km \cdot s^{-1}$.

Numerous refraction studies of the crust and the upper mantle carried out by explosion seismology methods have failed to show directly a decrease in *P* wave velocity directly beneath the Moho boundary, while a number of data indicate the existence of one or more seismic discontinuities in the lower lithosphere (Green and Hales, 1968; Lewis and Meyer, 1968; Ryaboy, 1966; and others). Special studies concerning the Earth's structure carried out by Lukk and Nersesov (1965) on the basis of an analysis of data from earthquakes have also shown that a distinct discontinuity is present in the lower lithosphere at a depth of about 85 km.

Many authors who studied the travel times of body waves from earthquakes noted the appearance of velocities of $8.5 \div 8.8\ \mathrm{km \cdot s^{-1}}$ indicating the existence of a discontinuity in the upper mantle at a depth of about $100 \div 150$ km.

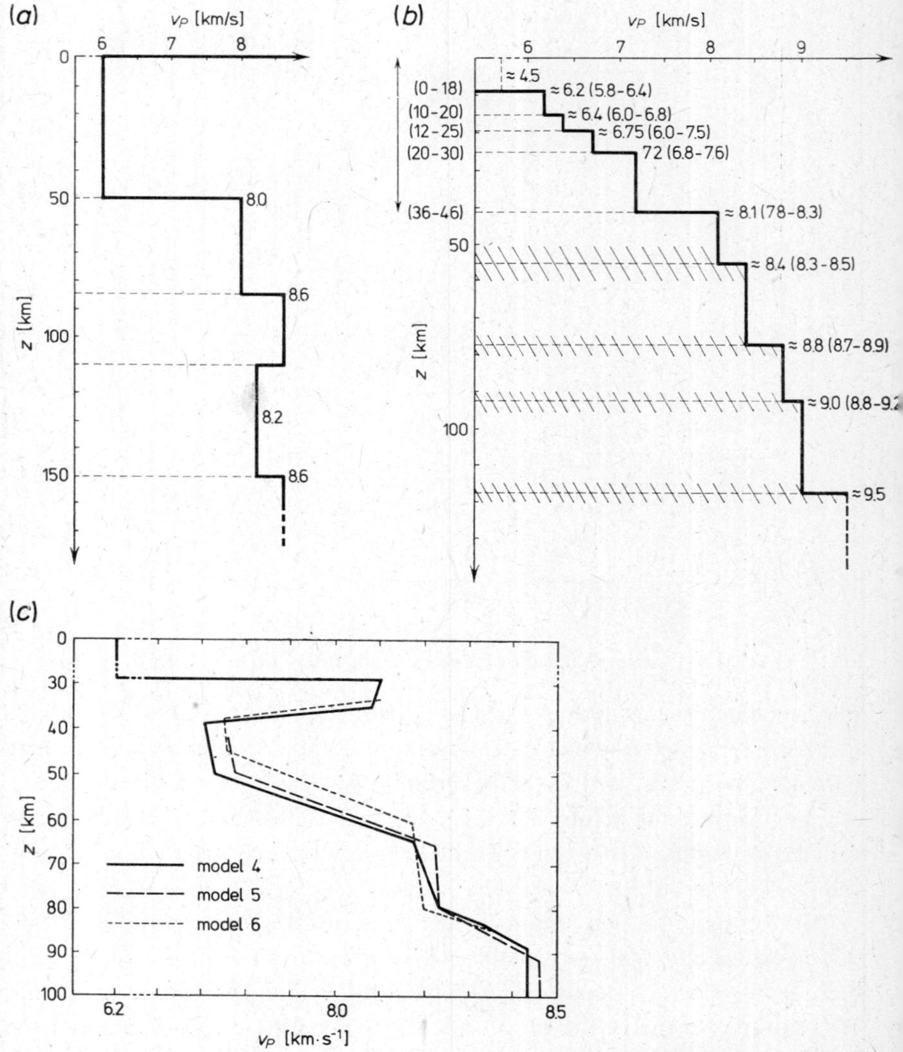

Fig. 5.3.2 (*a*) Model of the lower lithosphere (Lukk and Nersesov, 1965), (*b*) model of the lower lithosphere (Ryaboy, 1966), (*c*) model of lower lithosphere (Hirn *et al.*, 1973).

High velocities, and thus probably also high densities, were also found in the lower oceanic lithosphere. A discontinuity with a velocity of about 8.77 km · s^{-1} was found by Hales *et al.* (1970) at a depth of 57 km in the Mexican Gulf region. The presence of a layer with a velocity of 8.5÷8.8 km · s^{-1} was also confirmed by Zverev (1970) at a depth of 20 km beneath the Pacific Ocean between Sachalin and the Hokkaido Islands.

Considering some experimental data relating to sudden changes of seismic wave velocities from 8.0 to 8.3÷8.5 km · s^{-1} at a depth of some 80÷120 km, Hales (1969) concluded that the corresponding seismic discontinuities probably result from transitions of spinel into garnetic peridotite.

Recent studies of the lower lithosphere in France and in the Soviet Union carried out by explosion seismology methods have also indicated low-velocity zones in the lithosphere (Hirn *et al.*, 1973). A model of the lower lithosphere in France is shown in Fig. 5.3.2*a*.

Examples of models of the lower lithosphere, with the physical properties considered above, are shown in Figs. 5.3.2*a*, *b*, c. Additionally, a model of the upper mantle in south-eastern Europe based exclusively on the characteristics of seismic wave amplitudes is also presented (Vaněk, 1968) (Fig. 5.3.3).

From the limited number of studies carried out so far it follows that the lower lithosphere structure may be represented by two typical models. In the first model one or two boundaries with velocities of about 8.3÷8.5 km · s^{-1} are present in the lower lithosphere at depths of between 80 and 120 km. In this model the low-velocity zone in the upper mantle is rather indistinct. In the second model several seismic discontinuities, most probably of the nature of transition zones, appear in the lower lithosphere. In this model low-velocity layers may also exist locally, but they must be connected with the asthenosphere situated at considerably greater depths in the Earth's upper mantle. The first model of the lower lithosphere corresponds to tectonically fairly stable continental and oceanic areas, while the second model is connected with tectonically active areas or with regions of young platforms.

From detailed analyses of numerous models of the Earth constructed recently, from new data resulting primarily from seismic and laboratory studies, carried out by methods of high pressure and temperature physics, it follows that the Earth's mantle can be approximately divided into the following three large zones, with which the principal geodynamic processes are connected:

1. The lithosphere, a zone with relatively high density and rigidity, characterized by considerable lateral inhomogeneity.

2. The asthenosphere together with the transition zone in the mantle, relatively homogeneous, where phase transformations are the most characteristic processes.

3. The lower mantle.

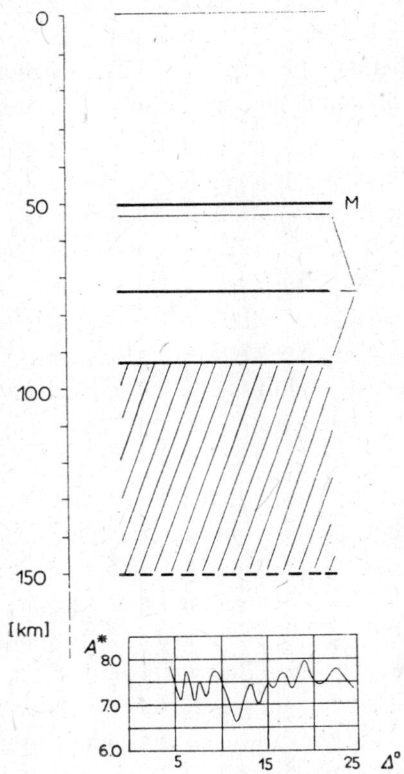

Fig. 5.3.3. The upper mantle model derived from seismic wave amplitudes (Vaněk, 1968): M—Moho discontinuity, thick and thin lines—strong and weak discontinuities in the lower lithosphere, dashed area—low-velocity zone, A^*—amplitude, $\Delta°$—distance.

As can be seen the first two zones correspond to the Earth's crust and upper mantle (Fig. 5.3.4).

Of major importance in contemporary geodynamic and neotectonic considerations is unquestionably the lithosphere. Its structure, as already indicated, is highly differentiated. Consequently, for the proper determination of its properties in a given region the most correct global definition must be taken into consideration. A definition of the lithosphere and its

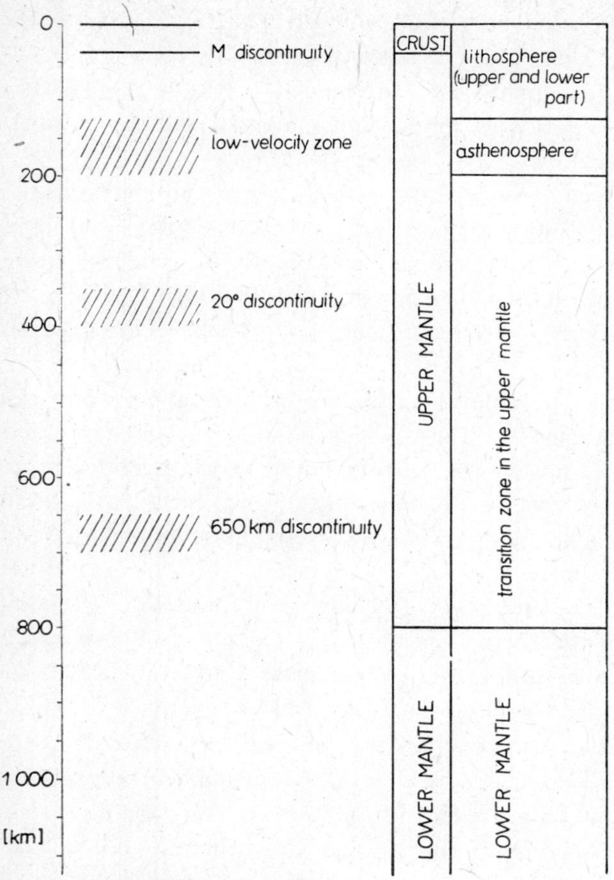

Fig. 5.3.4. Scheme of tectonosphere.

principal parameters were given by Press (1970) and Kanamori and Press (1970), who analysed statistically several thousand models of the Earth. They found that at a depth of about 70 km beneath the oceans a distinct decrease of the elastic properties of the medium is observed. This physical property defines the lower boundary of the lithosphere. A similar decrease of elastic properties, though not so distinct, is also observed beneath shield and old platform areas, at depths twice as great, i.e., of about 140 km. These values, that is 70 and 140 km, define the typical thickness of the lithosphere on the scale of the whole globe and form a framework for the classification of lithospheric plates distinguished on tectonospheric cross-sections. The numerous deviations from these values which can be observed either characterize individual large tectonic regions of the Earth or define anomalous zones, playing a leading role in geodynamics.

Lithospheric plates are characterized by a distinctly higher density than that of the asthenosphere. The gravitational instability of the "heavy" lithosphere covering the "light" asthenosphere is of extreme importance for complex convection processes in the upper mantle, and thus also for geodynamic processes.

The second zone, composed of the asthenosphere and the transition zone of pyrolite or peridotite composition, extends at depths from about 150 to 850 km. Its physical state changes with depth. The asthenosphere, extending down to a depth of about 250 km, is usually identified with the seismic channel or with the low-velocity zone, and probably is the source of basaltic magma.

In the entire second zone, defined in the above way, phase transformation processes are very characteristic. These processes induce vertical movements with which horizontal movements, corresponding to rift hypotheses, are always connected. Phase processes of especial intensity occur at depths of between 350 and 400 km and probably correspond to transformations of the olivine–spinel type.

Thus a modern interpretation of geodynamic processes, including vertical movements of the Earth's crust, obviously requires a consideration of the state of the entire lithosphere, whose present structure affects the mechanism of forces transforming the upper tectonosphere.

Any attempt to classify the complex structure of the lithosphere requires the most precise data available, which can only be supplied by explosion seismology. At present only limited information, relating mainly to the depth of the Moho discontinuity and its internal structure, is available. The physical state of the Moho discontinuity zone is strongly connected with the state of the lower lithosphere. Thus, when the structure of the transition zone between the crust and the upper mantle is relatively well known, as well as the state of progress of the probable physical processes occurring in it (e.g., internal erosion of the crust or phase transformations), an approximate estimation of the thickness and physical properties of the entire lithosphere can be made.

The main conclusion which can be drawn from these considerations is the statement that a correct formulation of the tectonic development theory requires first of all the consideration of the physical properties of individual lithospheric plates undergoing geodynamic processes in a given area.

The results of intensive studies, published in recent years, indicate that a spherically homogeneous model of the structure of the tectonosphere composed of the crust and the upper mantle must be replaced by a model that is spherically heterogeneous, even with reference to a specified type of tectonic structure. The dimensions of lateral heterogeneities in the upper mantle, which are now subject to intensive studies, vary from several thousands to hundreds and tens of kilometres. Thus we have regional and continental as well as local heterogeneities connected with given geological structures. Distinct relationships between surface structures and deep structures in the upper mantle have been confirmed. Studies of heterogeneities in the structure of the crust and the upper mantle become indispensable for the clarification of the laws of tectonosphere evolution and of the general rules of the formation and distribution of mineral resources in geological structures.

5.4 THE LOWER MANTLE AND THE EARTH'S CORE

The area extending at depths of between 1000 and 2900 km in the Earth is called the *lower mantle*. Since *P* and *S* waves are not restricted in their travel in the lower mantle, its matter must be in a solid state. Discontinuities and heterogeneities are probably present there. Johnson (1969) found velocity gradient discontinuities at depths of 1230, 1540, 1910 and 2370 km. Little is known so far about the nature of these discontinuities and their extension on a global or regional scale.

The existence of the Earth's core was first ascertained by Wiechert in 1897. Using seismic observations, Oldham in 1906 proved its reality, and in 1913 Gutenberg determined the depth of the core to be 2900 km. Later Jeffreys corrected this value to 2898 ± 4 km (Bullen, 1963). The disappearance of *S* waves at the lower mantle–core boundary indicates that the Earth's core is in a liquid state.

In 1936 Lehman presented a hypothesis, based on seismic observations, that there exists an inner core, in which the velocity of *P* waves clearly increases in comparison with the outer core. This hypothesis was fully confirmed by the later works of Jeffreys and Gutenberg. The inner core boundary is situated at a depth of about 5120 km. Further studies, including data relating to free Globe oscillations, have shown that the Earth's inner core is in a solid state.

During the last twenty years or so, the structure of the core–mantle boundary has been considered in a number of seismic studies. The interpretation of *PcP* waves, and especially of the coefficient of their reflection, provides substantial information related to the structure of this boundary. In general theories concerning the physical state of the Earth, the commonly accepted view is that the core–mantle boundary is a boundary of the first type, where the velocity of body waves and the density change stepwise. Numerous models of the Earth's interior assume a stepwise change in density at the core boundary by a factor of about 1.7. One of Bullen models of type A gives the following values for the body wave velocities above and beneath the mantle–core boundary:

$$v_P = 13.6\ \text{km}\cdot\text{s}^{-1}, \quad v_S = 8.1\ \text{km}\cdot\text{s}^{-1}, \quad \varrho = 5.66\ \text{g}\cdot\text{cm}^{-3}$$

and

$$v_P' = 8.1\ \text{km}\cdot\text{s}^{-1}, \quad v_S' = 0.0\ \text{km}\cdot\text{s}^{-1}, \quad \varrho' = 9.7\ \text{g}\cdot\text{cm}^{-3}$$

Many authors believe, however, that the density ratio ϱ'/ϱ is equal to one, though the confirmation of this hypothesis remains difficult. Studies of the reflection coefficient of *PcP* waves could help to solve this problem. Båth (1954) calculated the coefficient of reflection of longitudinal waves from the core as a function of the density ratio and the angle of incidence for fixed values of the velocities

$$v_P = 13.7\ \text{km}\cdot\text{s}^{-1}, \quad v_S = 7.5\ \text{km}\cdot\text{s}^{-1}$$

$$v_P' = 8.0\ \text{km}\cdot\text{s}^{-1}, \quad v_S' = 0\ \text{km}\cdot\text{s}^{-1}$$

The *PcP* wave changes the phase at epicentral distances of 30° and 82° when there is no velocity change at the mantle–core boundary. When the density ratio is 1.7, according to Bullen's model A′, a phase reversal occurs at an epicentral distance of 90°. From an analysis of observational data Båth concluded that the density at the core–mantle boundary changes stepwise. But more recent studies (Buchbinder, 1968; Kogan, 1968; Ibrahim, 1971), based mainly on analyses of records of nuclear explosions, have contradicted the conclusions of Båth. It has been found that the density in the lower mantle is approximately equal to that in the outer core. The problem is still under discussion and remains open.

The velocity distribution of seismic waves in the lower mantle and in the Earth's core is still considered in many papers. The proposed changes with reference to the established models relate first of all to the lowest parts of the mantle and to the core. Some authors indicate that a negative gradient of *P* wave velocity occurs in the lower part of zone *D′* with a thickness of about 200 km (Ergin, 1967; Taggart and Engdahl, 1968; Bolt, 1970). The possibility of the existence of velocity negative gradients in this zone also for *S* waves has been confirmed by studies of diffracted waves. Cleary (1969) confirmed the negative velocity gradient in zone *D′* and determined the velocity value as 6.8 km · s⁻¹. Other similar results give values between 6.8 and 7.0 km · s^{-1}. In Jeffrey's model it is 7.3 km · s^{-1}. The velocities of *S* waves in the lower mantle are also determined from analyses of the Earth's free oscillations. Recent seismological observations based, among other things, on the travel times of *PcP* reflected waves (Taggart and Engdahl, 1968), on the *ScS–S* time differences (Jordan and Lynn, 1974) and on the amplitude ratios of *PcP* and *ScS* waves, as well as similar studies concerning the penetration into the core of *PKP*, *PKIKP* and *SKS* waves (Hales and Roberts, 1971; Haddon, 1972), suggest that large changes in the velocity gradient can be expected in the transition zone.

A precise determination of the core radius is closely connected with the problem of the distribution of seismic wave velocities in the lower mantle. In the model of Jeffreys–Bullen the radius extends to 3473 km, while Taggart and Engdahl (1968) obtained the value $R = 3477 \pm 2$ km from analyses of travel times of *PcP* waves. From the travel time differences between *ScS* and *S* waves Hales and Roberts (1970) found the values $R = 3490 \pm 4.7$ km and 3486 ± 4.6 km for two models of *S* wave velocity distribution in the lower mantle.

From analyses of the Earth's free oscillations it also follows that the core radius is probably greater. Jordan and Anderson (1974) found the value $R = 3485$ km from free oscillation data. This seems the best contemporary estimation of the Earth's core radius.

The sharpness of the mantle–core boundary was questioned by Dorman *et al.* (1965), who studied the differences between the observed and the theoretical periods of spheroidal free oscillations. Trying to explain the difference between the observed and the theoretical coefficient of reflection of *PcP* waves, Phinney and Alexander (1966) concluded that the transition zone between the mantle and the core is of a multi-layer nature, with thickness from a few tens to 160 km. To explain the observed properties of the reflection coefficient of *PcP* waves, Berzon and Pasechnik (1972) also assumed that the mantle–core transition zone forms a multi-layered system which can be modelled by a system of thin layers with clearly differentiated velocities of longitudinal waves. In such a case the reflection coefficients can exceed several times the reflection coefficient from the single boundary as a result of interference in a certain band of the wave frequencies. Thus also the amplitudes of waves reflected from a system of thin-layered boundaries could considerably exceed the amplitudes of waves reflected from the single core–mantle boundary. On the basis of studies of the amplitudes of *PcP* waves Buchbinder (1973) proposed a model of the mantle–core boundary in the form of a liquid layer, a few kilometres thick, with density close to that in the mantle. The velocity of longitudinal waves in this layer should be identical with the velocity in the core. On the basis of studies of spectral ratios of various wave types (type *P*/*PcP*, ScS_{n+1}/ScS_n and others) Ibrahim (1973) proposed several models of the mantle–core boundary. In these models he assumed the presence of (from two to four) layers with lower velocities of *P* and *S* waves and with higher densities in comparison with those accepted in a standard model. The outer core is here characterized by the non-zero modulus μ and by a maximum shear velocity $v_S = 1.4 \text{ km} \cdot \text{s}^{-1}$.

In general, the structure of the transition zone between the lower mantle and the core is still not clear. Despite considerable progress in studies of this zone made in recent years including the interpretation of various seismic waves and the Earth's free oscillations, there is still a lack of unequivocal data concerning the discrete structure of the lower mantle–core transition zone. The known solutions admit models with continuously changing velocities of seismic waves as well as models with large gradients of velocity. The lower part of the lower mantle, of a thickness estimated to be about 200 km, is probably highly heterogeneous. Seismic waves are dispersed there. From their interpretation it follows that a distinct lateral heterogeneity is situated in the lower part of the lower mantle. The horizontal changes of velocity in comparison with the average velocity are estimated to be about 1 per cent. Haddon and Cleary (1974) have shown that the computed amplitudes and travel times of diffracted waves scattered on the randomly distributed fluctuations of elastic moduli and density may explain the observed seismological data. The outer zones of the Earth's core, in the direct vicinity of the lower part of the mantle, also display dispersive properties, testifying to the existence of lateral heterogeneities also in this region.

Precise analyses of travel times of *P* and *S* seismic waves, of their amplitudes and of the form of their records are of primary importance for the solution of these problems. The results of such studies form a basis for the correct determination of the chemical and physical properties of the widely defined transition zone between the lower mantle and the core.

REFERENCES

Båth M. (1954), "The density ratio at the boundary of the Earth's core", *Tellus*, **6**, 403–414.

Båth M. (1957), "A continental channel wave guided by the intermediate layer in the crust", *PAGEOPH*, **38**, 19–31.

Belousov V. V. (1966), *The Earth's Crust and Upper Mantle of the Continent*, Nauka, Moskva (in Russian).

Berzon J. S. and Pasechnik J. P. (1972), "Dynamic characteristic of *PcP* waves for thin-layered model of transition zone between mantle and Earth's core", *Fizika Zemli*, **6**, 21–33 (in Russian).

Birch F. (1958), "Interpretation of the seismic structure of the crust in the light of experimental studies of the wave velocities in rocks", in: *Contributions in Geophysics. In Honour of B. Gutenberg*, H. Benioff *et al.* (eds.), Pergamon Press, London.

Birch F. (1960), "The velocity of compressional waves in rocks to 10 Kilobars", Part 1, *J. Geophys. Res.*, **65**, 1083–1102.

Birch F. (1961), "The velocity of compressional waves in rocks to 10 Kilobars", Part 2, *J. Geophys. Res.*, **66**, 2199–2224.

Birch F. (1963), "Some geophysical applications of high-pressure research", in: *Solids under Pressure*, W. Paul and D. M. Warschauer (eds.), McGraw-Hill, New York.

Bolt B. A. (1970), "*PdP* and *PKKP* waves and difracted *PcP* waves", *Geophys. J. R. Astron. Soc.*, **20**, 367–382.

Buchbinder G. G. R. (1968), "Properties of the core–mantle boundary and observations of *PcP*", *J. Geophys. Res.*, **73**, 5901–5923.

Buchbinder G. G. R. (1973), "Problems related to *PcP* waves and the core–mantle boundary illustrated by two nuclear events", *Bull. Seismol. Soc. Am.*, **63**, 2047–2070.

Bullen K. E. (1963), *An Introduction to the Theory of Seismology*, Cambridge Univ. Press, Cambridge.

Carder D. S., Gordon D. W. and Jordan J. N. (1966), "Analysis of surface-foci travel times", *Bull. Seismol. Soc. Am.*, **56**, 815–840.

Cleary J. (1969), "The *S* velocity at the core–mantle boundary, from observations of diffracted *LS* waves, *Bull Seismol. Soc. Am.*, **59**, 1399–1405.

Dorman J., Ewing J. and Alsopl E. (1965), "Oscillations of the Earth, new core–mantle boundary model based on low-order free vibrations", *Proc. Nat. Acad. Sci. USA*, **54**, 364–368.

Ergin K. (1967), "Seismic evidence for a new layered structure of the Earth's core", *J. Geophys. Res.*, **72**, 3669–3687.

Giese P., Prodehl C. and Behnke C. (1967), "Ergebnisse refraktionsseismischer Messungen 1965 zwischen dem Französischen Zentralmassiv und den Westalpen", *Z. f. Geophysik*, **33**, 215–261.

Giese P. and Stein A. (1971), "Versuch einer einheitlichen Auswertung tiefenseismischer Messungen aus dem Bereich zwischen der Nordsee und den Alpen", *Z. f. Geophysik*, **37**, 237–272.

Green R. W. E. and Hales A. L. (1968), "The travel times of *P* waves to 30° in the central United States and upper mantle structure", *Bull. Seismol. Soc. Am.*, **58**, 267–289.

Gutenberg B. (1950), "Structure of the Earth's crust in the continents", *Science*, **111**, 29–30.

Gutenberg B. (1951), "Revised travel times in Southern California", *Bull. Seismol. Soc. Am.*, **41**, 143–163.

Gutenberg B. (1959), *Physics of the Earth's Interior*, Academic Press, New York–London.

Gutenberg B. (1960), "Low-velocity layers in the earth, ocean and atmosphere", *Science*, **131**, 959–965.

Guterch A. (1977), "Structure and physical properties of the Earth's crust in Poland in the light of new data of DSS", *Publ. Inst. Geophys. Pol. Acad. Sci., A-4*, **115**, 347–357.

Guterch A., Materzok R., Pajchel J. and Perchuć E., (1975), "Seismic structure of the Earth's crust along the VII international profile in light of investigation by deep seismic sounding method", *Przegląd Geologiczny*, **4**, 153–163 (in Polish).

Guterch A., Sollogub V. B. and Chekunov A. B. (1980), "Properties of transition zone between crust and upper mantle", in: *Structure of the Earth's Crust in Central and Eastern Europe, Geophys. Monogr.*, V. B. Sollogub, A. Guterch and Prosen (eds.), Naukova Dumka, Kiev, 158–160.

Haddon R. A. W. (1972), "Corrugations on the mantle–core boundary on transition layers between inner and outer core", *Trans. Am. Geophys. Union*, **53**, 600.

Haddon R. A. W. and Cleary J. R. (1974), "Evidence for scattering of seismic *PKP* waves the near mantle–core boundary", *Phys. Earth Planet. Inter.*, **8**, 211–234.

Hales A. L. (1969), "A seismic discontinuity in the litosphere", *Earth Planet. Sci. Lett.*, **7**, 44–46.

Hales A. L. (1971), "The travel times of *P* seismic waves with special reference to upper mantle distribution", A paper presented at the final U.M.C. Symposium in Moscow, August, 1971.

Hales A. L., Hesley C. E. and Nation J. B. (1970), "*P* travel times for an Oceanic Path", *J. Geophys. Res.*, **75**, 7362–7381.

Hales A. L. and Roberts J. L. (1970), "Shear velocities in the mantle and the radius of the core", *Bull. Seismol. Soc. Am.*, **60**, 1427–1436.

Hales A. L. and Roberts J. L. (1971), "The velocities in the outer core", *Bull. Seismol. Soc. Am.*, **61**, 1051–1059.

Hirn A., Steinmetz L., King R. and Fuchs K. (1973), "Long-range profiles in Western Europe", *Z. Geophys.*, **39**, 363.

Ibrahim A. K. (1971), "The amplitude ratio *PcP/P* and the core–mantle boundary", *PAGEOPH*, **91**, 114–133.

Ibrahim A. K. (1973), "Evidences for a low-velocity core–mantle transition zone", *Phys. Earth Planet. Inter.*, **7**, 187–198.

Ito K. and Kennedy G. C. (1970), "The fine structure of the basalt–eclogite transition", *Mineral. Soc. Am. Spec. Pap.*, **3**.

Ito K. and Kennedy G. C. (1971), "An experimental study of the basalt–garnet granulite–eclogite transition", in: *The Structure and Physical Properties of the Earth's Crust, Geophys. Monogr.*, J. G. Heacock (ed.), Am. Geophys. Union, Washington, **14**, 303–314.

Johnson L. R. (1969), "Array measurements of *P* velocities in the lower mantle", *Bull. Seismol. Soc. Am.*, **59**, 973–1008.

Jordan T. H. and Anderson D. L. (1974), "Earth structure from free oscillations and travel times", *Geophys. J. R. Astron. Soc.*, **36**, 411–459.

Jordan T. H. and Lynn W. S. (1974), "A velocity anomaly in the lower mantle", *J. Geophys. Res.*, **79**, 2679–2685.

Kanamori H. and Press F. (1970), "How thick is the litosphere", *Nature*, **226**, 330–332.

Knopoff L. (1970), "Problems of continental rift structures", in: *Graben Problems*, J. H. Illies and St. Mueller (eds.), E. Schweizerbart'sche Verlagsbuchhandlung, Stuttgart.

Kogan S. D. (1968), "Arrival times and amplitudes of *PcP* waves from surface sources", *Fizika Zemli*, **3**, 3–10 (in Russian).

Kosminskaya J. P. (1965), "Results of researches on the international geophysical projects", *Geofiz. Bull.*, **14**, 168–187 (in Russian).

Landisman M. and Mueller St. (1966), "Seismic studies of the Earth's crust in continents", II, *Geophys. J. R. Astron. Soc.*, **10**, 539–548.

Lewis B. T. R. and Meyer R. P. (1968), "A seismic investigation on the upper mantle to the west of Lake Superior", *Bull. Seismol. Soc. Am.*, **58**, 565–596.

Lukk A. A. and Nersesov I. L. (1965), "Structure of the upper mantle from observation of intermediate earthquakes", *Dokl. AN SSSR*, **162**, 3, 559–562 (in Russian).

Magnitsky V. A. (1965), *Structure and Physics of the Earth's*, Nedra, Moskva (in Russian).

Meissner R. (1973), "The 'Moho' as a transition zone", *Geophys. Surv.*, **1**, 195–216.

Mueller St. and Landisman M. (1966), "Seismic studies of the Earth's crust in continents", I, *Geophys. J. R. Astron. Soc.*, **10**, 525–538.

Phinney R. A. and Alexander S. S. (1966), "*P* wave diffraction theory and the structure of the core–mantle boundary", *J. Geophys. Res.*, **71**, 5959–5975.

Press F. (1970), "Regionalized Earth models", *J. Geophys. Res.*, **75**, 6515–6581.

Press F. and Gutenberg B. (1956), "Channel *P* waves in the Earth's crust", *Trans. Am. Geophys. Union*, **37**, 6, 754–756.

Ringwood A. E. and Green D. H. (1966), "An experimental investigation of the gabbro–eclogite transformation and some geophysical implications", *Tectonophysics*, **3**, 383–427.

Riznichenko J. V. and Kosminskaya J. P. (1963), "Nature of stratification of the Earth's crust and upper mantle", *Dokl. AN SSSR*, **153**, 2, 323–325 (in Russian).

Ryaboy V. Z. (1966), "Kinematic and dynamic characteristics of deep waves associated with boundaries in the crust and upper mantle (from a DSS study on the profile Kopet Dagh–Aral Sea)", *Izv. AN SSSR, Ser. Fiz. Zemli*, **3**, 74–82 (in Russian).

Smulikowski K. (1964), "Le problem des écologites", *Geologia Sudetica*, **1**, 13–77.

Smulikowski K. (1967), "Ecologites of the Śnieżnik Mts. in the Sudetes", *Geologia Sudetica*, **111**, 7–156.

Taggart J. N. and Engdahl J. (1968), "Estimation of *PcP* travel times and the depth to the core", *Bull. Seismol. Soc. Am.*, **58**, 1293–1304.

Vaněk J. (1968), "Amplitude curves of seismic body waves and the structure of the upper mantle in Europe", *Tectonophysics*, **5**, 235–243.

Volarovich M. P., Kurskeyev A. K., Tomashevskaya J. S., Tuzova I. L. and Urazayev B. M. (1967), "Relation between velocities of longitudinal waves and density of rocks under high comprehensive pressure", *Izv. AN SSSR, Ser. Fiz. Zemli*, **5**, 14–20 (in Russian).

Yoder H. S. and Tilley C. E. (1962), "Origin of basalt magmas, an experimental study of natural and synthetic rock system", *J. Petrology*, **3**, 342–532.

Zverev S. M. (1970), "Fine structure of the upper mantle for some regions of Pacific Ocean from DSS data", in: "Results of researches on the international geophysical projects", *Verkhnaya Mantya*, **7**, 74–82 (in Russian).

6. Phase Transformations: Elements of the Theory and Geophysical Applications

6.1 POSSIBILITY OF PHASE TRANSFORMATIONS IN THE EARTH'S INTERIOR

The layered structure of the Earth is the result of depth-dependent changes in pressure and temperature. Their influence on mineral composition is of vital importance for the Earth's structure. In several media, and especially in the minerals forming the Earth's mantle, an increase in pressure causes changes in crystalline lattices, from less densely to more densely packed ones. Such a change is usually called a *polymorphic phase transformation* (alotropic transformation), and it is a phase transformation of the first order, thermodynamically included in the same class as melting, evaporation or sublimation.

The polymorphic phase transformation connected with an increase in pressure results in an increase of the medium density (usually $\Delta\varrho/\varrho \leqslant 10\%$), in a change of the mechanical parameters (moduli of elasticity), and in the absorption or generation of a certain amount of heat $c_{1,2}$, which is the specific heat of a transformation from a less dense phase 1 to a more dense phase 2. For transformations generating heat (exothermic transitions) it follows that $c_{1,2} > 0$. Changes in the mechanical properties of the medium can be observed as changes in seismic wave velocity; the thermal effect may result in local changes of the geothermal gradient.

Polymorphic phase transformations can be considered either on a microscopic scale, taking into account quantum-mechanical models of interactions, or on the basis of phenomenological thermodynamics. The first method is successfully used to study simple crystallic structures such as those of the ZnS type, haloids of alkaline metals, and crystals of solidificated inert gases. In particular, the theory of interactions between ions forming crystalline lattices was developed by Born and Huang (1954). The theory was used by many authors (e.g., Lombardi and Jansen, 1964; Voronov and Goncharova, 1966; Anderson and Liebermann, 1970) for describing phase transformations occurring in haloids of alkaline metals. They crystallize in regular systems and a surface-centred cubic structure of the NaCl type is related to the low-pressure phase (for a fixed temperature), while a denser body-centred cubic structure of the CsCl type is relevant to the high-pressure phase. Quantum-mechanical treatment of a crystalline lattice permits the establishment of phase equilibrium conditions; thus, in particular the transition pressure can be determined. The basic kinetic phenomena here are the activation of a new phase nucleus and the growth of new phase grains. These processes depend on several factors; in particular, they depend upon the type of material, the admixtures, the mechanical properties of the medium (monocrystal or polycrystalline medium, fractured medium, medium with a high density of dislocations), the geometrical conditions and the thermodynamical parameters p and T.

It is obvious that the phase state of minerals forming the Earth's mantle depends on the actual thermodynamical conditions and also on the history of the pressure and temperature changes which affected these minerals during consecutive geological epochs.

The Earth's crust and upper mantle form the most heterogeneous area of physical, chemical and mineralogical changes of the Earth's matter. At the same time it is that area of the Earth where the most intensive displacement of matter in a vertical direction takes place: the formation of mid-oceanic ridges, island arcs and oceanic trenches, uplift of mountains, and volcanism. Any movement of this type changes the thermodynamical conditions of a given mass element of terrestrial matter. The subduction (or uplift) of that element into (or from) the Earth's interior is always connected with an increase (or decrease) of pressure and temperature, and structural changes are certainly present. Phase transformations of the first order, where the density of the medium under transformation undergoes discontinuous change, should be mentioned first of all. To some extent transformations of the second order, when the density is continuous but various material constants are abruptly changed, especially the coefficients of elasticity, could also be of importance. Changes of the nature of chemical bonds, leading to mineral composition changes, form a separate group of transitions, e.g., change of spinel into the postspinel phase—a transformation concerning the layer of phase equilibrium at a depth of about 680 km. Since this reaction is accompanied by the change of specific volume, it can be considered from the point of view of its mechanical consequences as a phase transition of the first order.

In deep layers of the Earth and planets of the Earth's group, as well as in the interiors of large planets, an important role is possibly played by phase transformations of the molecular phase → metallic phase type. The molecular liquid hydrogen → metallic liquid hydrogen transformation is probably very important for the internal structure of Jupiter and Saturn (see Section 2.5). For planets of the Earth's group models in which iron and magnesium silicates are transformed into a metallic state have been considered.

The most important evidence relevant to the possible existence of phase tranformation boundaries in the Earth's interior comes from seismology. The discontinuities observed in seismic wave velocity at various depths are explained by discontinuities in the material parameters. From the relations describing the velocities v_P and v_S of elastic compressional wave P and shear wave S

$$v_P = \sqrt{\frac{K+\frac{4}{3}\mu}{\varrho}}, \quad v_S = \sqrt{\frac{K}{\varrho}} \tag{6.1.1}$$

it immediately follows that discontinuities in the velocity of these waves are connected with discontinuities either in density ϱ or in the bulk modulus K or in shear modulus μ. Abrupt changes in ϱ, K and μ, corresponding to discontinuity boundaries of seismic waves, may result from changes either in chemical composition or in mineral composition. A seismic discontinuity may also be connected with a phase transition of the second order, where ϱ is continuous but K and μ change abruptly. For a particular case of the lower mantle–outer core boundary an abrupt change in the velocity of P waves is observed, while the shear waves S vanish completely. This effect is explained by the existence of a phase boundary of the first order between the solid (lower mantle) and fluid (outer core) phases. Since in fluid $\mu = 0$, S waves do not exist.

Since the Earth is a heterogeneous body in the sense of chemical and mineral composition, the phase transformation of any mineral should always be considered with regard to the influence of the medium in which the mineral is placed. In practice it is usually assumed that the only factors which affect a particular mineral are pressure and temperature. This approach permits us to make one of the following two assumptions: either the medium is composed entirely of one mineral in which phase transformation takes place, or the mineral is built in a scattered form into the matrix which surrounds it and does not interact chemically. The mineral composition of the matrix might be such that under a different pressure and temperature another of its components might undergo phase transformation, while the built-in mineral would be a neutral component of the surrounding matrix. In these consecutive transformations the matrix does not have to be counted as a component in the phase rule of Gibbs. The behaviour of minerals in the presence of H_2O, CO_2 and other substances which are very volatile under high pressure is an important exception.

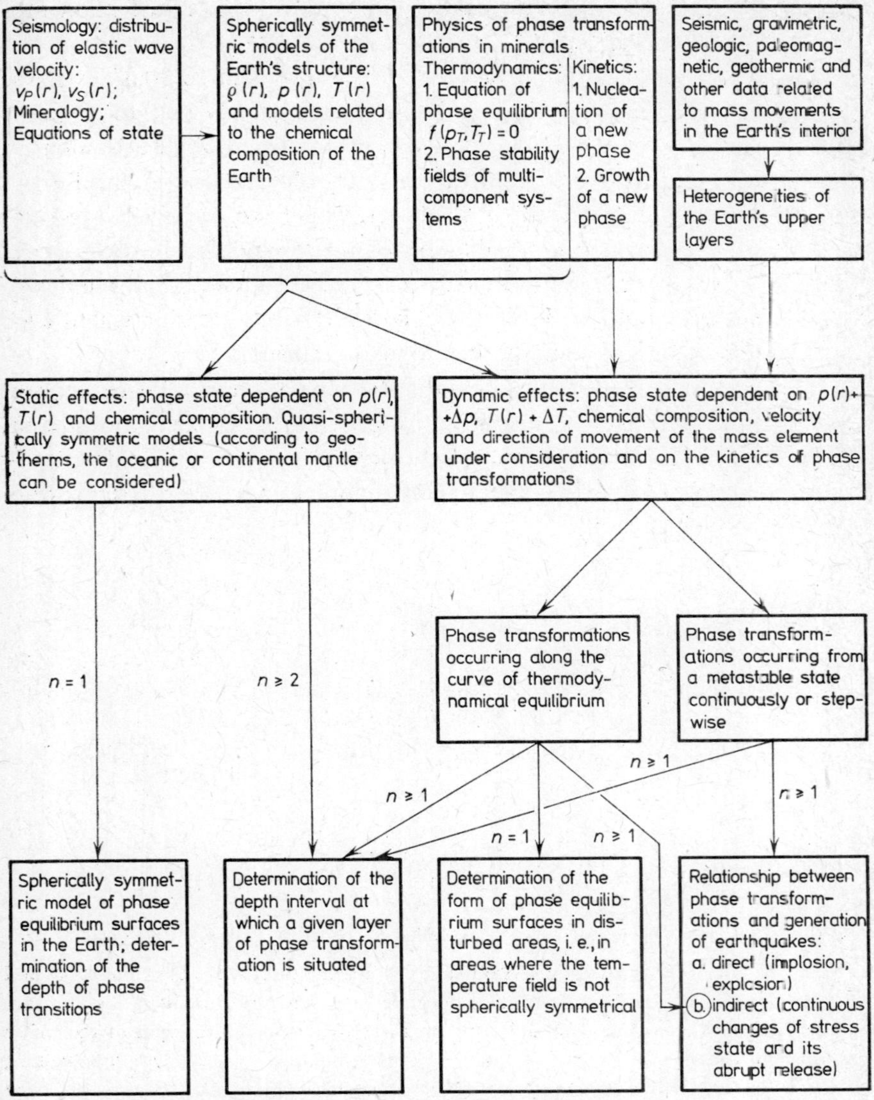

Fig. 6.1.1. Diagram showing methods of phase transformation studies and their applications to phase transformation problems in the Earth's interior; ϱ, p, T, v_P and v_S are the spherically symmetric density, pressure, temperature, P wave velocity and S wave velocity in the Earth's interior; p_T and T_T are the pressure and temperature of phase equilibrium; Δp and ΔT are the corrections of pressure and temperature taking into account their deviation from the spherically symmetric model; n is the number of components coexisting in the phase equilibrium.

The presence of volatiles can substantially influence the liquefaction processes of some rocks.

From a geophysical point of view, it is convenient to consider phase transformations separately in relation to static effects and to dynamic phenomena connected with mass displacements in the Earth's interior. The present state of the studies of phase transformations in the Earth's interior is shown schematically in Fig. 6.1.1, where two principal directions of research are marked: determination of phase equilibrium areas under the conditions of "the static model of the entire Earth", and studies of phase transformations in those areas of the Earth's interior where such transformations might be influenced by mass transportation.

Under static conditions the area of phase separation in a one-component system ($n = 1$), in which pressure and temperature gradients are present, forms a surface. The surfaces of phase equilibrium in the Earth's interior correspond to concentric spheres with different radii. For systems with

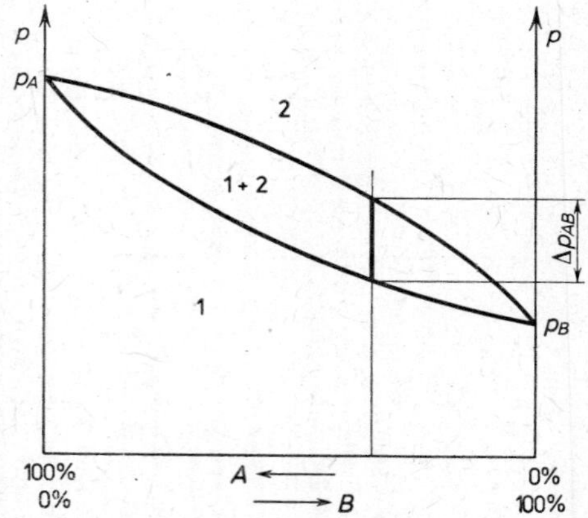

Fig. 6.1.2. Schematic isothermic phase diagram for a mixture of the components A and B. The symbols 1, 2 and 1+2 denote, respectively, the areas of stability of low-pressure phase 1, high-pressure phase 2 and their mixture; p_A and p_B correspond to the phase equilibrium pressure of a system containing only component A or only component B; $\Delta p_{AB} < |p_A - p_B|$ is the pressure interval in which both phases can coexist at a fixed concentration.

$n \geqslant 2$ components the surface of phase separation becomes a layer with a definite thickness. The schematic phase diagram of a two-component system shown in Fig. 6.1.2 makes it possible (the pressure distribution in the Earth's interior being known) to determine the equilibrium levels r_A or r_B of phases 1 and 2 of component A or B, or to determine the depth interval of the coexistence of phases 1 and 2 of the components A and

B. It should be noted that the applicability of the phase diagram shown in Fig. 6.1.2 to phase transformations in the Earth's interior is not entirely correct, since the area of phase stability corresponds to a certain fixed temperature, and thus the diagram is applicable only to phase transformations caused by changes of pressure. The extreme points p_A and p_B of the phase diagram, corresponding to the equilibrium pressure of the phase 1 → phase 2 transition of the components *A* and *B*, change their position on the *p* axis with the change of temperature according to the Clausius–Clapeyron equation for a given component. Under the conditions which exist in the Earth's interior transformations caused solely by pressure or solely by temperature cannot be considered, as these factors can not be separated. Under laboratory conditions experiments of the type *p*–variable and $T \approx$ const or $p \approx$ const and *T*–variable can readily be performed. An ideal laboratory experiment would be one in which the behaviour of minerals could be studied under the same pressure *p* and temperature $T(p)$ as exist in the Earth's interior. The function $T(p)$ can readily be established by using model estimations of the pressure $p(r)$ and temperature $T(r)$ distributions in the Earth's interior. At the present state of laboratory technique, however, such experiments cannot be performed, as the pressure would have to be changed up to about 350 GPa and the temperature up to about 4400 K, i.e., to the values of *p* and *T* existing in the Earth's centre, and furthermore the proper relation $T(p)$ would have to be maintained at the same time. The required relation $T(p)$ can be obtained only for pressure corresponding to that in the Earth's crust and upper mantle. In experiments with shock waves which provide the highest pressure, corresponding even to that in the Earth's core, the required relation $T(p)$ cannot be maintained. The results of phase transformation studies obtained from shock wave experiments should be extrapolated very cautiously to terrestrial conditions, under which pressure and temperature act in an entirely different time scale.

6.2 THERMODYNAMICS OF PHASE TRANSFORMATIONS

The formation of various minerals at different depths in the Earth's interior and phase transformations of the type solid → solid appropriate to these minerals should be studied experimentally, under pressure and temperature conditions similar to those dominating at given depths in the Earth, as well as theoretically on the basis of the thermodynamics and kinetics of phase transformations.

The basic equation relating to one another the number f of thermodynamical degrees of freedom of the system, the number n of components of the system, and the number r of phases coexisting in equilibrium is the phase rule of Gibbs (see e.g., Landau and Lifshitz, 1958)

$$f = n+2-r \tag{6.2.1}$$

From this equation it follows that in a one-component system ($n = 1$) appearing in two phase states ($r = 2$) only one degree of freedom is possible ($f = 1$). In a one-component system existing at the same time in two polymorphic variants and in a fluid phase the number of degrees of freedom is zero and a triple point occurs.

A general thermodynamic classification of phase transformations was given by Ehrenfest in the twenties and is described in the works of Swalin (1962), Münster (1969) or Rao and Rao (1978), and others. The Ehrenfest classification is connected with the behaviour of the Gibbs function G, which determines the thermodynamic state of a body where the parameters of state are pressure p and temperature T. From the definition of the Gibbs function and according to the first principle of thermodynamics it follows that

$$G(p, T) = F+pV = U-TS+pV, \quad \mathrm{d}G = V\mathrm{d}p-S\mathrm{d}T \tag{6.2.2}$$

where F is the free energy, U is the internal energy, S is the entropy, and V is the specific volume. A phase transformation is called a transformation of the nth order when the function of Gibbs together with its derivatives up to the $(n-1)$st order remains continuous during the transformation, while the nth derivative becomes discontinuous. Using this definition and noting that in thermodynamic equilibrium the pressure and temperature along the border of both phases are equal, the relations between the pressure and the transformation temperature for transitions of a given order can be obtained.

1. Phase transformation of the first order. The function of Gibbs remains continuous, and in equilibrium its changes in the states of both phases must be equal

$$\mathrm{d}G_1 = \mathrm{d}G_2 \tag{6.2.3}$$

or

$$V_1\mathrm{d}p-S_1\mathrm{d}T = V_2\mathrm{d}p-S_2\mathrm{d}T \tag{6.2.4}$$

The derivatives of the Gibbs function change stepwise, and they are the entropy and the volume

$$\left(\frac{\partial G_i}{\partial T}\right)_p = -S_i, \quad \left(\frac{\partial G_i}{\partial p}\right)_T = V_i, \quad \text{where } i = 1, 2 \tag{6.2.5}$$

From eqn. (6.2.4) the equation of Clausius–Clapeyron is obtained

$$\frac{dp}{dT} = \frac{S_2 - S_1}{V_2 - V_1} = \frac{c_{1,2}}{T} \frac{1}{V_1 - V_2} = \frac{c_{1,2}}{T} \frac{\varrho_1 \varrho_2}{\varrho_2 - \varrho_1} \tag{6.2.6}$$

where

$$c_{1,2} = -(S_2 - S_1)T \tag{6.2.7}$$

is the transformation heat ($c_{1,2} > 0$ for transformations occurring with heat radiation or exothermic transformations) and ϱ_1 and ϱ_2 are the densities of the two phases.

2. Phase transformation of the second order. The function of Gibbs and its first derivatives remain continuous, and thus

$$d\left(\frac{\partial G_1}{\partial T}\right)_p = d\left(\frac{\partial G_2}{\partial T}\right)_p \quad \text{or} \quad dS_1 = dS_2 \tag{6.2.8}$$

$$d\left(\frac{\partial G_1}{\partial p}\right)_T = d\left(\frac{\partial G_2}{\partial p}\right)_T \quad \text{or} \quad dV_1 = dV_2 \tag{6.2.9}$$

From eqn. (6.2.8), expressing the continuity of entropy during a phase transformation of the second order, it follows that the latent heat of this transition is zero, while from eqn. (6.2.9) it follows that there is no stepwise change in the specific volume.

Condition (6.2.8) is equivalent to the equation

$$\left(\frac{\partial S_1}{\partial T}\right)_p dT + \left(\frac{\partial S_1}{\partial p}\right)_T dp = \left(\frac{\partial S_2}{\partial T}\right)_p dT + \left(\frac{\partial S_2}{\partial p}\right)_T dp \tag{6.2.10}$$

Using the thermodynamic relation

$$\left(\frac{\partial S}{\partial T}\right)_p = \frac{c_p}{T} \tag{6.2.11}$$

where c_p is the specific heat under constant pressure, and

$$\left(\frac{\partial S}{\partial p}\right)_T = -\alpha V, \quad \alpha = \frac{1}{V}\left(\frac{\partial V}{\partial T}\right)_p \tag{6.2.12}$$

where α is the coefficient of thermal expansion, eqn. (6.2.10) can be transformed into

$$\frac{dT}{dp} = VT \frac{\alpha_2 - \alpha_1}{c_{p2} - c_{p1}} \tag{6.2.13}$$

Similarly, from eqn. (6.2.9), expressing the continuity of density during a phase transformation of the second order, we obtain

$$\left(\frac{\partial V_1}{\partial T}\right)_p \mathrm{d}T+\left(\frac{\partial V_1}{\partial p}\right)_T \mathrm{d}p = \left(\frac{\partial V_2}{\partial T}\right)_p \mathrm{d}T+\left(\frac{\partial V_2}{\partial p}\right)_T \mathrm{d}p \tag{6.2.14}$$

Using definition (6.2.12) of the coefficient α and the definition

$$\varkappa = -\frac{1}{V}\left(\frac{\partial V}{\partial p}\right)_T \tag{6.2.15}$$

of the coefficient of volume compressibility $\varkappa$, eqn. (6.2.14) can be written in the form

$$\frac{\mathrm{d}T}{\mathrm{d}p} = \frac{\varkappa_2-\varkappa_1}{\alpha_2-\alpha_1} \tag{6.2.16}$$

Equations (6.2.13) and (6.2.16) are the Ehrenfest equations describing the relation between transformation pressure and temperature for a phase transformation of the second order, expressed by changes of the material coefficients.

Since phase transformations of the first order are certainly the most important transformations occurring in the Earth's interior, we should consider them in more detail.

The Clausius–Clapeyron equation (6.2.6) in the close vicinity of a given point (p_0, T_0) can be expresssed by an approximate linear relation

$$p-p_0 = \frac{c_{1,2}}{V_1-V_2}\,\frac{T-T_0}{T_0} \tag{6.2.17}$$

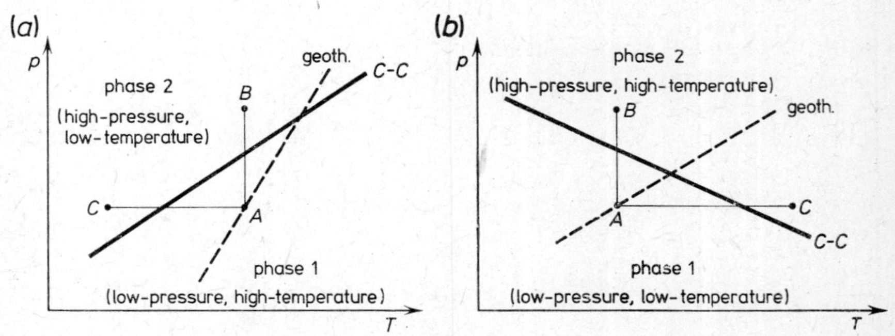

Fig. 6.2.1. Diagrams of areas of stability for the low- and high-pressure phase 1 and 2. The continuous Clausius–Clapeyron line *C–C* is the line of phase equilibrium, and the dashed geotherm line shows the relation $p(T)$ in the Earth's interior. *AB* denotes pressure transformations, *AC* temperature ones, while the transition along the geotherm is of a mixed character. Fig. (*a*) shows the most typical case when $V_2 > V_1$ and $c_{1,2} > 0$ or the case when $V_2 < V_1$ and $c_{1,2} < 0$. Fig. (*b*) corresponds to the case when $V_2 > V_1$ and $c_{1,2} < 0$ or $V_2 < V_1$ and $c_{1,2} > 0$.

shown in Fig. 6.2.1. This straight line divides the plane (p, T) into the areas of stability of two different phases. According to the slope of this phase equilibrium line the phases are defined as being low-pressure and at the same time high-temperature, and high-pressure and low-temperature (Fig. 6.2.1*a*), or as low-pressure and low-temperature, and high-pressure

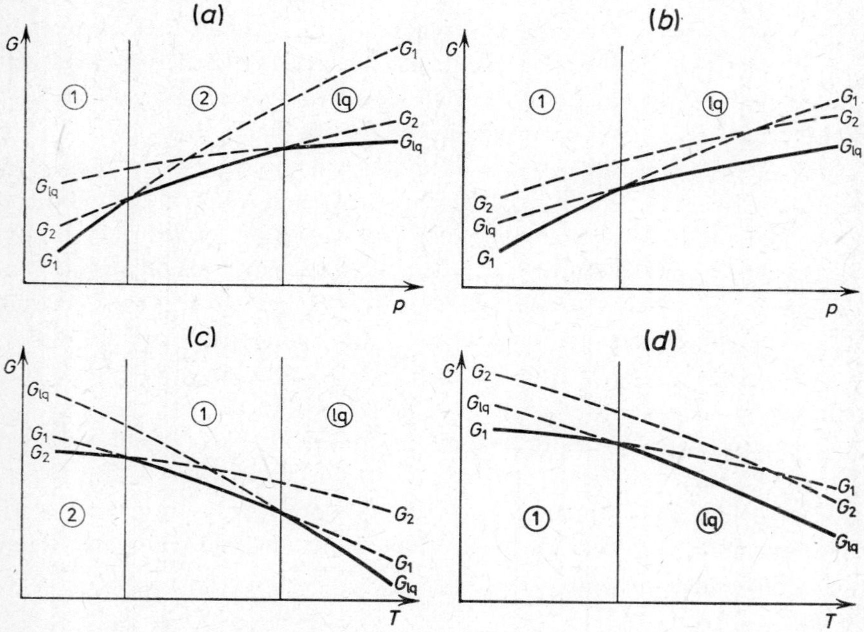

Fig. 6.2.2. Diagrams for determining stability intervals for different phases according to the function of Gibbs. G_1, G_2, and G_{lq} denote the function of Gibbs of phase 1, 2, and liquid respectively. The continuous line corresponds to a stable state, while the dashed lines correspond to metastable states. Intersections of the lines determine the points of phase equilibrium. The cases (*a*) and (*b*) denote pressure transformations (for $T = \text{const}$), and (*c*) and (*d*) denote temperature transformations (for $p = \text{const}$). The diagrams (*a*) and (*c*) represent such a medium which at a given temperature (*a*), or a given pressure (*c*), is stable in two polymorphic forms. The diagrams (*b*) and (*d*) represent a medium in which at a given temperature (*b*), or a given pressure (*d*), only one stable polymorphic form can exist.

and high-temperature (Fig. 6.2.1*b*). A reversible transition from the state of one of the phases, point A, into the state of another phase can be performed either along the isothermic path AB or along the isobaric one AC. In this case the transformations are called either pressure or temperature transitions. A transition from a low-pressure to a high-pressure phase under the conditions in the Earth's interior is determined by the intersection of the geotherm $p = p(T(z))$ with the curve of Clausius–Clapeyron.

The pattern described above can become much more complicated if the material parameters of both phases are strongly temperature and pressure dependent, since then the applicability of eqn. (6.2.17) becomes meaningless. In particular, when the transformation heat $c_{1,2}$ changes its sign, i.e., when the endothermic transformation becomes an exothermic one, the Clausius–Clapeyron curve has a minimum or maximum (see Fig. 6.5.2). A complex pattern of isotherms in some areas of the Earth combined with this type of phase equilibrium curve could lead to essential deformations of the phase equilibrium surfaces in the Earth's mantle (see Section 6.5.2).

To the state of thermodynamic equilibrium corresponds the occurrence of such a phase for which (under given pressure and temperature conditions) the Gibbs function has a minimum. This condition is illustrated in Fig. 6.2.2, where for the known Gibbs functions the intervals of pressure and temperature stability, which include two different solid phases and one liquid phase, can be determined. Figs. 6.2.2*a* and *b* correspond to pressure transformations, while Figs. 6.2.2*c* and *d* correspond to temperature transitions.

The determination of the areas of mineral stability is of essential importance for understanding the structure of various layers in the Earth. Experimental studies are conducted out simultaneously with theoretical work, and the latter provides many interesting results, especially for some simple structures of ion crystals.

6.3 KINETICS OF PHASE TRANSFORMATIONS

The kinetics of phase transformations depends on two phenomena—the formation of new phase nuclei (nucleation) and their growth. The direct consequence of these two effects is the phenomenon of hysteresis: at a constant temperature the transformation to a high-pressure phase occurs above the pressure p_T of thermodynamical equilibrium, and the reverse transformation occurs below this pressure. Similarly, under constant pressure the transformation to a low-temperature phase occurs below the temperature T_T of thermodynamical equilibrium, and the reverse transformation takes place above this temperature. The dependence of phase transformation velocity on the temperature and the corresponding width of the hysteresis loop is of significant importance for minerals in the Earth's mantle. Polymorphic species of minerals corresponding to overcooled high-pressure states are often observed in rocks accessible at the Earth's surface. Such minerals originated in the mantle and were transported to the Earth's crust.

The construction of a general theory of nucleation, applicable to the growth of new phase grains of any shape, does not seem to be possible. The case most often considered is a spherically symmetrical growth of the grains (e.g., Cahn, 1956; Turnbull, 1956; Delmon, 1969). This model was used by Hanneman (1968) who studied the influence of high pressure upon the phase transformation process. Cahn (1956) derived the rate laws for transformations in which a new phase nucleates on the grain boundary surface, grain edges or grain corners, and in which the new phase grows with a constant radial velocity. Sung and Burns (1976a, b) and Sung (1979) used the model of spherical grain growth to study the kinetics of the olivine–spinel reaction. Tobin (1974) developed the theory of phase transformations for the case where the nucleation is homogeneous and the growth is two-dimensional. An extensive description of the nucleation and growth of new phase grains was given by Rao and Rao (1978). In particular, the calculation results of Nabarro relating to the nucleation of spheroidal grains with any degree of oblatness, as obtained in 1940, are given there in detail.

The nucleation of a new phase can occur with the same probability at any random point of the medium, or it can occur at certain selected points. The process is called homogeneous nucleation when the formation of new phase nuclei is related to random fluctuations of the atom density. The density fluctuation can be so great that the packing of atoms corresponds to the state of a different phase from the phase state of the medium surrounding the fluctuation. Such fluctuation is called the "embryo" of a new phase. The growth or decline of this embryo is governed by statistical laws, and thus it can either decline or increase its volume. If the number n of atoms in the embryo is higher than a certain critical number n_c, such an embryo is called the "nucleus" of a new phase. Under unchanged thermodynamic conditions the nucleus does not decline but undergoes further growth.

The fluctuations which occur during the process of inhomogeneous nucleation behave similarly, and the only difference in relation to homogeneous nucleation is the fact that imperfections of the crystalline structure (impurities, dislocation lines, crystalline grain boundaries) form areas energetically favourable for fluctuation occurrence. Such imperfections of the structure, forming privileged points for the onset of nucleation, are called "potential sites" or "potential grains" of the new phase. A nucleus of slightly greater size than the critical one is called the "activated potential site" of a new phase. Real phase transformations are always phenomena in which both types of nucleation occur simultaneously, but it may also happen that one of them will clearly dominate.

Let us consider the nucleation for the case where the shape of the nuclei is characterized only by one parameter r of length dimension (a sphere or a symmetrical polyhedron). The nucleation of a new phase depends on the change ΔG of the Gibbs function connected with the formation of a new phase grain

$$\Delta G = ar^3(-G_v+\mathscr{E})+br^2G_s \tag{6.3.1}$$

where $-G_v$ is the decrease of the Gibbs function in relation to a volume unit, G_s is the increase of the Gibbs function in relation to an increase of the boundary surface of the new phase grain corresponding to the surface unit, $\mathscr{E}$ is the energy related to mechanical strains induced in the surrounding medium by the nucleus of the new phase, whose density is different

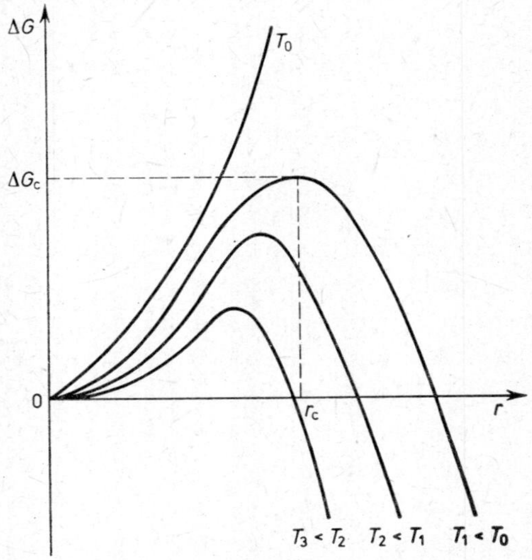

Fig. 6.3.1. The change ΔG of Gibbs function versus the grain size r of the new phase, given by formula (6.3.1). For $r < r_c$ the embryo of the new phase, for $r > r_c$ the nucleus of the new phase are observed. The temperature T is a parameter. T_T is the equilibrium temperature. For $T \to T_T$ the critical radius $r_c \to \infty$. According to Rao and Rao (1978).

from that of the surrounding medium, and the coefficients a and b are related to grain shape, e.g., for a sphere $a = \frac{4}{3}\pi$ and $b = 4\pi$, and for a cube $a = 1$ and $b = 6$. The function $\Delta G(r)$, given by formula (6.3.1) and shown in Fig. 6.3.1, has the maximum ΔG_c for the new phase embryo with a critical size r_c

$$\Delta G_c = \frac{4}{27}\frac{b^3}{a^2}\frac{G_s^3}{(-G_v+\mathscr{E})^2}, \qquad r_c = \frac{2b}{3a}\frac{G_s}{G_v-\mathscr{E}} \tag{6.3.2}$$

Introducing the number n of atoms forming the embryo, instead of its characteristic size r, formulas (6.3.1) and (6.3.2) can be written in an equivalent form. Then, instead of ar^3 the factor nv_a will appear, where v_a is the volume corresponding to a single atom forming the embryo, and instead of br^2 the factor $gn^{2/3}$ will appear, where g is the shape factor. It is convenient in this case to recalculate the quantities G_v, G_s, and $\mathscr{E}$ for a single atom. From formula (6.3.2) for r_c we can calculate the number n_c of atoms in an embryo with the critical size.

The critical quantities ΔG_c and r_c depend on how far the actual values of pressure and temperature exceed the values corresponding to a state of thermodynamical equilibrium. Let us consider the transformation from a low-pressure to a high-pressure phase (the reverse transformation can be described in a similar way) with the phase equilibrium curve shown in Fig. 6.3.2. Let us define the overpressure Δp as the difference between the actual pressure and the pressure of phase equilibrium and the over-

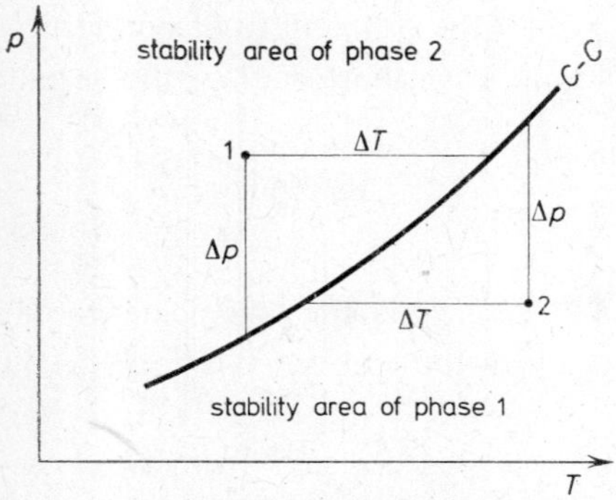

Fig. 6.3.2. Diagram defining the overpressure Δp and overcooling ΔT of phase 1 in the stability area of phase 2 and the underpressure Δp and overheating ΔT of phase 2 in the stability area of phase 1. C–C is the Clausius–Clapeyron equilibrium line.

cooling ΔT as the difference between the actual temperature and the temperature of phase equilibrium. The state (T, p) can be considered as a deviation from the equilibrium line C–C either by Δp or by ΔT. Since approximately $\Delta p \propto \Delta T$, both approaches are equivalent and the overpressure or overcooling can replace each other in various formulas. When the transformation can be considered as a process occurring under isothermic conditions, it is convenient to study the effect of the overpressure Δp upon the nucleation and the growth of a new phase (Hanneman, 1968; Leliwa-Kopystyński *et al.*, 1974, 1975).

Studies of the transformation processes in relation to ΔT and T (see e.g., Rao and Rao, 1978) are of a more general character, because they take into account the deviation from an equilibrium state and the temperature dependent exponential terms simultaneously. According to approximate estimations, it is possible to assume that the surface component of the Gibbs function G_s is constant and that the mechanical energy $\mathscr{E}$ can be neglected in comparison with the volume component G_v. The volume Gibbs function G_v is therefore approximately the only variable quantity in eqns. (6.3.2). Since $G_v \propto \Delta p \propto \Delta T$, the critical radius of a new phase grain $r_0 \propto (\Delta p)^{-1} \propto (\Delta T)^{-1}$ and the corresponding critical value of Gibbs function $G_c \propto (\Delta p)^{-2} \propto (\Delta T)^{-2}$.

The nucleation rate $I = \mathrm{d}\gamma/\mathrm{d}t$ expresses the time changes of the nucleus density $\gamma(t)$ for the new phase. The density can be described by different equations dependent also on the mechanism of nucleus origination of the new phase homogeneous or heterogeneous. The influence of temperature on the nucleation ratio is expressed first of all by two factors $\exp(-\Delta G_c/kT)$ and $\exp(-E_a/kT)$. The number of embryos with a critical size is proportional to the first factor, while the probability of connection with the embryo of the surrounding atoms is proportional to the second factor. Thus the nucleation rate is

$$I \propto \exp\left(-\frac{\Delta G_c + E_a}{kT}\right) \tag{6.3.3}$$

where ΔG_c is defined by relation (6.3.2) and E_a is the activation energy needed for the migration of an atom across the embryo interface of the new phase.

The growth of a single grain of the new phase results from two competitive processes: the migration of atoms of the primary phase to the nucleus of the new phase and the escape of atoms from the nucleus to the surrounding area. Both these effects, under certain simplifying assumptions, can be described by the approximate formula

$$u \propto \exp\left(-\frac{G_a^{12}}{kT}\right)\left[1 - \exp\left(-\frac{\Delta G}{kT}\right)\right] \tag{6.3.4}$$

where u denotes the linear rate of growth, G_a^{12} and ΔG are defined in Fig. 6.3.3: G_a^{12} and G_a^{21} are the free energies of activation enabling the atoms to cross the interface from phase 1 to phase 2 and from phase 2 to phase 1 and ΔG is the change in free energy of the atom upon transfer. Rao and Rao (1978) proposed a slightly different formula for the linear rate of growth,

$$u \propto \exp\left(-\frac{E_a}{kT}\right)\left\{1-\exp\left[\frac{c_{1,2}}{k}\left(\frac{1}{T_T}-\frac{1}{T}\right)\right]\right\} \tag{6.3.5}$$

where $c_{1,2}$ is the heat transformation and T_T is the equilibrium temperature. From this formula it immediately follows that phase transformation is not possible at the equilibrium temperature $T = T_T$.

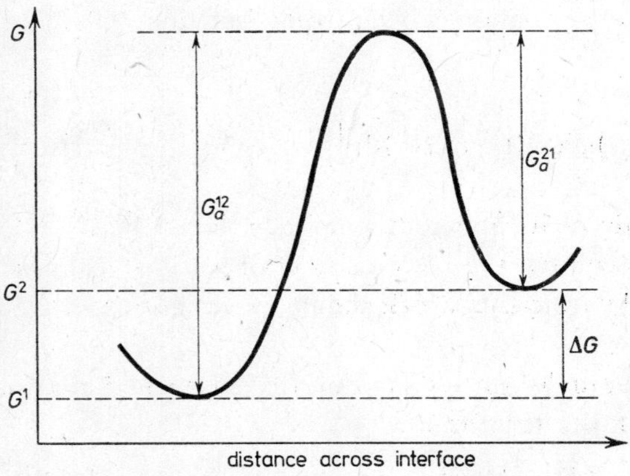

Fig. 6.3.3. Diffusion barriers across the interface for the growth of nuclei. According to Rao and Rao (1978).

There are several methods of calculating the transformation progress function $\nu(t)$, e.g., the function determining the fraction of transformable body mass which has undergone transformation from the moment $t = 0$, at which thermodynamic conditions p_T and T_T corresponding to phase equilibrium occurred, to a moment t.

All these methods are based on the work carried out in the forties by Avrami, whose main result is expressed by the equation

$$\nu = 1-\exp(-v_{ex}) \tag{6.3.6}$$

where the "extended volume" v_{ex} is defined to be the total volume of all domains calculated with no regard to impingement by the volume contributed by the phantom domains which would nucleate in regions already transformed. For the calculation of v_{ex} the knowledge of time dependence of a single domain volume of the new phase and of the nucleation rate is essential since

$$v_{\text{ex}} = \int_0^t v_{t-\tau} I(\tau)\,\mathrm{d}\tau \tag{6.3.7}$$

where $v_{t-\tau}$ is the domain volume at a moment t of the new phase which originated from the activated embryo and began its growth at the moment τ. Under the assumption that the growth rate is time independent, this volume is given by the formula (see e.g., Turnbull, 1956)

$$v_{t-\tau} = gu^3(t-\tau)^3 \tag{6.3.8}$$

where g is the shape factor and u is the linear rate of growth. Using relations (6.3.7) and (6.3.8), the transformation advancement function (6.3.6) can be expressed in the form

$$\nu(t) = 1-\exp\left[-\int_0^t g(\tau)u^3(\tau)(t-\tau)^3 I(\tau)\,\mathrm{d}\tau\right] \tag{6.3.9}$$

Equation (6.3.9), in spite of its apparent simplicity, is not suitable for direct applications. This requires the introduction of various empirical assumptions and numerical coefficients which should be verified experimentally.

In the simplest case, i.e., under the assumption that g, u and I are time independent, eqn. (6.3.9) takes the form

$$\nu(t) = 1-\exp(-Kt^4), \quad K = \tfrac{1}{4}gu^3 I \tag{6.3.10}$$

which gives a proper description of some polymorphic transformations.

Equation (6.3.9) has been applied for to description the pressure hysteresis loop in quasi-isothermic phase transformations in haloids of alkaline metals, expecially in rubidium chloride RbCl (Leliwa-Kopystyński *et al.*, 1974). The used method is sufficiently general and, after proper modifications of the formulas, can be applied numerically to any polymorphic transformations. It has been assumed that the domain volume of the new phase is given by

$$v_{t-\tau} = l\int_\tau^t \alpha(\vartheta)(\vartheta-\tau)^{l-1}\mathrm{d}\vartheta \tag{6.3.11}$$

and that between the nucleation rate $\mathrm{d}\gamma/\mathrm{d}\tau$, the density γ of appearance of the new phase nuclei in a transforming medium and the density Γ (Γ = const) of potential sites of the new phase the following relation is valid:

$$I(\tau) \equiv \frac{\mathrm{d}\gamma}{\mathrm{d}\tau} = \beta(\tau)[\Gamma-\gamma(\tau)] \tag{6.3.12}$$

or

$$\frac{d\gamma}{d\tau} = \Gamma\beta(\tau)\exp\left[-\int_0^\tau \beta(\vartheta)d\vartheta\right]$$

where α and β are the coefficients of grain growth of the new phase and of potential site activation, respectively.

The power exponent l ($l = 1$ or 2 or 3) in eqn. (6.3.11) corresponds to three different mechanisms of grain growth of the new phase. For $l = 1$ the growth is one-dimensional, e.g., in two directions from the interface of two crystalline grains. The interface in this case is a potential site which was activated and became a nucleation surface of the new phase. For $l = 2$ the grain growth is of two-dimensional character, e.g., cylindrical-symmetrically along a dislocation line or a grain edge, becoming potential sites of the new phase. The case $l = 3$ corresponds to a three-dimensional growth around point potential sites, which can be formed, for example, by the atoms of admixtures or vacancies of the crystal lattice. The case $l = 3$ corresponds also to homogeneous nucleation in which the new phase grows spherical-symmetrically around randomly appearing density fluctuations.

It has been assumed that the coefficients α and β are power-dependent on the overpressure Δp as follows

$$\alpha \propto (\Delta p)^q, \quad \beta \propto (\Delta p)^r, \quad q, r = \text{const} \geqslant 0 \tag{6.3.13}$$

In experiments in which the sample is placed inside a chamber with volume linearly variable in time, the overpressure Δp is also a function of time (though in general the function is nonlinear, especially near the pressure p_T of phase equilibrium). The dependence of the growth coefficient α of a new phase domain on the overpressure Δp can be explained by relation (6.3.5). In the case where $(c_{1,2}/k)(1/T_T - 1/T) \ll 1$ and simultaneously $\Delta T = T - T_T \ll T_T$ we have approximately

$$u \propto \Delta T \quad \text{or} \quad u \propto \Delta p \tag{6.3.14}$$

the second relation resulting directly from the linearized equation of Clausius–Clapeyron, according to which $\Delta p \propto \Delta T$. From the comparison of formula (6.3.8) and (6.3.11) it follows that the coefficient α corresponds to the rate of the volume increase u^3, thus to preserve relation (6.3.8) it should be assumed that $\alpha \propto (\Delta p)^3$ or that relation (6.3.13) is valid with the exponent $q = 3$. In the particular case where it is assumed that $\alpha = \text{const}$ and that the domain growth is three-dimensional around point potential sites, relation (6.3.11) describing the volume increase of this domain is reduced to the generally accepted formula (6.3.8).

The assumption that the coefficient β of activation of potential sites is a power function of Δp (6.3.13) cannot be justified theoretically. Numerical attempts made in relation to the exponent r show that $\beta \approx$ const can be assumed. This conclusion is in agreement with eqn. (6.3.3), according to which an approximate proportional relation of the type $I \propto (\Delta T)^r$ cannot exist.

Formulas (6.3.11) and (6.3.13) substituted into eqn. (6.3.7) and then in (6.3.6) lead to the following equation for the function of transformation advancement:

$$\nu(t) = 1-\exp\left\{-l\Gamma \int_0^t \left[\int_\tau^t \alpha(\vartheta)(\vartheta-\tau)^{l-1} d\vartheta\right] \times \right.$$

$$\left. \times \beta(\tau)\exp\left[-\int_0^\tau \beta(\vartheta)d\vartheta\right] d\tau\right\} \tag{6.3.15}$$

This formula has been fitted numerically to the experimentally observed pattern of the high-pressure branch of the hysteresis loop in the phase transformation occurring in RbCl, for which the equilibrium pressure at room temperature between the crystal structure of NaCl type and that of CsCl type is $p_T = 0.52$ GPa.

An interesting approach to the problem of olivine → spinel phase transformations in relation to the Earth's mantle was proposed by Sung and Burns (1976a). They used the following equation for the function of transformation advancement:

$$\nu(t) = 1-\exp(-Kt^l) \tag{6.3.16}$$

which differs from relation (6.3.10) only in the value of the exponent l, not necessarily $l = 4$, and in the form of the coefficient K. Sung and Burns calculated the relations $K = K(p, T)$ for four cases of the nucleation of spinel in olivine: homogeneous nucleation with incoherent interfaces, homogeneous nucleation with coherent interfaces, nucleation on grain surfaces and nucleation on dislocations.* Then they determined the areas of domination of specific nucleation mechanisms on the plane p, T (or on the equivalent plane z, T, where z is the depth in the mantle). Although the values of the adopted numerical coefficients were of low accuracy (which is usual in computations using material constants of minerals from the Earth's interior), the calculations led to the important conclusion that the dominant mechanism for the olivine → spinel transition is heterogeneous nucleation described by eqn. (6.3.16) with the exponent $1 \lesssim l \lesssim 2$.

* The last two cases are heterogeneous nucleations.

6.4 PHASE TRANSFORMATIONS IN STATIC MODELS OF THE EARTH WITH SPHERICAL SYMMETRY

In generally accepted models of the Earth it is assumed that the Earth's properties depend primarily on the distance r from the centre. Thus, neglecting local heterogeneities of the Earth's mantle and the strong differentiation of the crust as well as convection movements, it can be assumed that the Earth is spherically symmetric. The Earth's oblateness is not really of importance for models used to determine phase equilibrium surfaces in the mantle. Thus, the positions of equilibrium surfaces are usually studied as functions of the depth z or distance r from the centre of the spherical Earth ($z+r = R = 6371$ km). The function of z is of more convenient use for the external layers (the crust and the mantle), while the function of r is better for the internal area (the core).

All surfaces of phase transformation selected so far are connected with discontinuities of seismic wave velocities. It can be said in general that the surfaces of seismic discontinuities become more and more frequent with the increase of r, i.e., when approaching the Earth's surface. Some of the surfaces of seismic discontinuities are considered in terms of composition changes, while some others are interpreted as phase transformation boundaries. It should be noted that the interpretation is not unequivocal and different authors propose different models for the mineral and chemical composition of the Earth, especially for the external layers.

According to seismic data (lack of S waves) the matter of the outer core is in a liquid state. Thus the deepest surfaces of phase discontinuity in the Earth's interior are most probably equilibrium surfaces of the liquid–solid type between the inner and the outer core ($z = 5160$ km or $r = 1211$ km) and between the outer core and the lower mantle ($z = 2890$ km or $r = 3481$ km). Numerous models of the Earth take into account a possible density discontinuity between the inner and outer core. If it is assumed that composition of the entire core depends continuously on r, then an abrupt density change on that boundary can be caused only by the change of phase state. The boundary between the outer core and the lower mantle is considered not only as a boundary of phase state changes but also as a discontinuity of chemical composition: the most important component of the core is most probably Fe, while the mantle is composed mainly of oxygen compounds of Mg, Fe and Si. Thus from a thermodynamical point of view, the boundary between the outer core and the lower mantle is not a phase transformation surface, in spite of the fact that it separates the liquid from the solid state.

One of the numerous models of the temperature distribution in the Earth's interior is shown in Fig. 6.4.1. The temperature distribution $T(r)$ is shown together with the melting temperature distribution $T_m(r)$ computed by Stacey (1977). The calculations of T were made by using the state equation of Mie–Grüneisen, while the values of T_m were found from the Lindemann equation of melting. The discontinuity in the melting tempera-

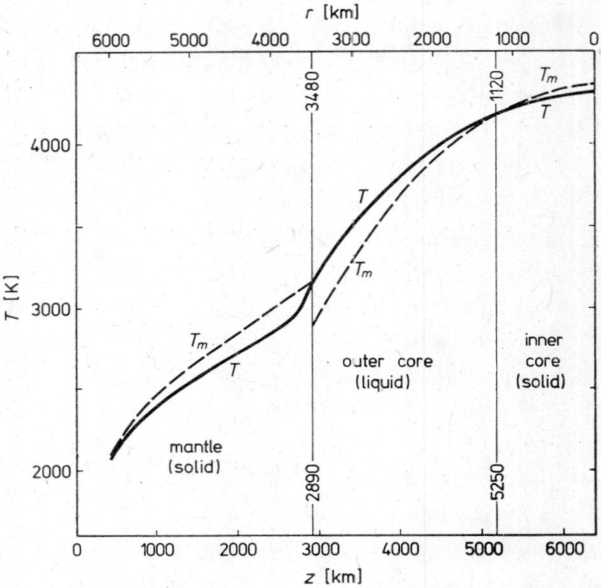

Fig. 6.4.1. Distribution of temperature in the Earth's interior. The continuous line denotes the real temperature T, and the dashed line denotes the melting temperature T_m (numerical data taken from Stacey, 1977).

ture T_m at the boundary between the outer core and the lower mantle results directly from the Lindemann equation, according to which the melting tempertature is expressed by the density (T in kelvin and ϱ in $kg \cdot m^{-3}$)

$$T_m = T_m(\varrho) = 1265 \exp(11.366 - 1102\,\varrho^{-3/5} - \tfrac{2}{3} \ln \varrho) \qquad (6.4.1)$$

and from the fact that at this boundary the discontinuity is caused not only by a phase transformation but also by a discontinuity in chemical composition. The preservation of the continuity of T_m (formula (6.4.1)) at the boundary between the outer and the inner core results from the very probable assumption that this boundary corresponds only to a phase transition.

A discontinuity of seismic wave velocities in the lower mantle (1000 km $< z <$ 2890 km) was not found. This means that there is no discontinuity of ϱ, K and μ in tle lower mantle, and thus the area can be considered as a homogeneous one with regard to phase composition. This is also the area where probably oxygen compounds of magnesium and iron appear in the most densely packed structures.

Numerous seismic discontinuities have been found in the upper mantle (33 km $< z <$ 1000 km*), where, therefore, several phase boundaries of the solid–solid type may be deduced. The correspondence between the respective phase transitions and given seismic discontinuities is based on various data relating to the chemical and mineral composition and on laboratory experiments concerning the stability fields of polymorphic forms of minerals under various conditions of pressure and temperature. Usually, when the phase stability domains found in laboratory experiments are applied to the matter in the Earth's mantle, they have to be extrapolated to higher temperatures. In general, it can be stated that the results are not entirely univocal and the physical nature of several seismic boundaries raises serious doubts. A detailed description of hypothetical phase transitions in the Earth's mantle is given in Chapter 7.

6.5 PHASE TRANSFORMATIONS IN AREAS UNDERGOING GLOBAL DEFORMATIONS

6.5.1 *Introduction*

A convenient description, and one close to reality, of the Earth on a global scale is the spherically symmetric model of the distribution of phase equilibrium surfaces (or layers) in the Earth's interior. This idealized pattern is distorted, however, by the existing heterogeneities in the real Earth. The most important heterogeneities on a global scale are the differences in geological structure and thermodynamic conditions observed in the crust and the upper mantle between the continental and oceanic areas. The continental and oceanic geotherms given by Stacey (1977) are shown in Fig. 6.5.1, together with the curves of phase equilibrium of two hypothetical components A and B. It is assumed that the components A and B can form isomorphic crystals, and that they can occur in the state of low-pressure phase 1, in the state of high-pressure phase 2, or in the state of a mixture

* The depth of the Mohorovičić discontinuity is arbitrarily taken to be 33 km in spherically symmetric models of the Earth.

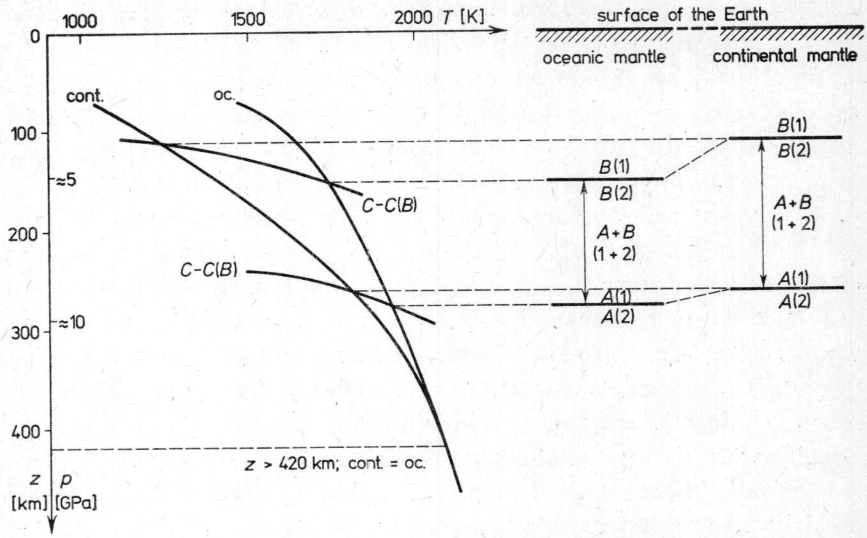

Fig. 6.5.1. The left side: the oceanic (oc.) and continental (cont.) geotherms according to Stacey (1977), together with the curves of Clausius–Clapeyron (*C–C*) for two hypothetical components *A* and *B* which can form isomorphic cristals. The right side: the depth of the phase equilibrium levels $A(1) \to A(2)$ and $B(1) \to B(2)$, and the estimate of the maximum possible thickness of a layer in which the components *A* and *B* can coexist in equilibrium as a mixture of phase 1 and phase 2 (partial transformed medium).

of phases 1+2 (see Fig. 6.1.2). In Fig. 6.5.1 the following proportion between the depth and pressure is assumed:

$$p = \frac{1}{29} z \quad \text{(1 GPa corresponds to 29 km)}$$

This relation is quite suitable for the upper mantle. As can be seen in Fig. 6.5.1, the levels of phase equilibrium for a one-component medium, e.g., only *A* or only *B*, are different for subcontinental and for suboceanic areas. As far as the layer of coexistence of phases 1 and 2 for the two-component system $A_m + B_{1-m}$ (*m* being the mole fraction) is concerned, we can only say that, in general, it is situated somewhere between the levels of phase equilibrium of pure components, just as the pressure interval Δp_{AB} in Fig. 6.1.2 is between p_B and p_A, but at the same time it is known that $\Delta p_{AB} < p_A - p_B$. The areas of coexistence of the components *A* and *B* in the state of phases 1 and 2 are shown in Fig. 6.5.1 symbolically by arrows whose length cannot exceed the interval enclosed between the phase equilibrium levels of pure components. For the exact determination of the position of the coexistence layer of phases 1 and 2, we should know the mole fraction *m* (it is possible that $m = m(z)$) as well as the changes of phase diagram (Fig. 6.1.2) as functions of temperature.

Though global tectonics and dynamic processes in the Earth's interior are not discussed in this volume, we will now make use of some concepts of global plate tectonics to analyse the possible phase transformations in the Earth's mantle. This concerns, in particular, the radial component of mass displacements in the contact regions of lithospheric plates and in the regions in which the plates are being formed along oceanic ridges.

In the upper mantle areas situated between suboceanic and subcontinental regions or in areas of midoceanic ridges, the pattern of phase equilibrium surfaces or layers is distorted and, in general, not necessarily monotonic or even continuous. The non-monotonic pattern may result from a distorted temperature field: undulations of layer shape may be caused by a strong deviation of the local temperature field from that of the surrounding area. A discontinuity in the pattern of these layers may be caused by the slightly different composition of the matter dividing the suboceanic from the subcontinental areas, e.g., the matter of a down-going lithospheric slab.

Further on we will consider phase transformations in the area dividing the continental from the oceanic mantle. The models presented below and applied to a down-going lithospheric slab can be used for any convection mass movement in the Earth's mantle, where the area in question is displaced in the pressure and temperature field by the vertical component of movement.

6.5.2 The shape of phase transformation surfaces

The results presented below are based on the works of Leliwa-Kopystyński and Bakun-Czubarow (1978, 1980). The following assumptions were made to find an approximate shape of phase equilibrium surfaces in the vicinity of down-going lithospheric slab:

1. The curves of phase equilibrium are nowhere identical with the geotherm.

2. A phase transformation occurs only when a given mass element passes the corresponding phase equilibrium level (metastable states do not occur).

3. Only phase transformations in one-component systems are considered, e.g., in olivine with the formula Mg_2SiO_4 or Fe_2SiO_4 but not with the composition $(Mg_xFe_{1-x})_2\ SiO_4$.

4. The equations of phase equilibrium of the minerals considered,

i.e., the equations of Clausius–Clapeyron, are those given by Leliwa-Kopystyński and Bakun-Czubarow (1978).

5. Oceanic and continental geotherms are those given by Stacey (1977).

6. The temperature distribution along a down-going slab and in its vicinity is that given by Minear and Toksöz (1970a, b).

Assumptions 1 and 2 are of considerable importance as they permit us to limit our considerations to deformations of the surfaces of phase separation (surfaces of transformation) without simultaneous considerations of structure changes in the phase separation layer. The omission of these assumptions would result in the existence, even in a one-component medium, of the structure of the phase transformation layer considered in Sections 6.5.3 and 6.5.4.

We consider the shape of phase equilibrium surfaces for the transformations with the following equilibrium equations (p is in GPa and T in K):

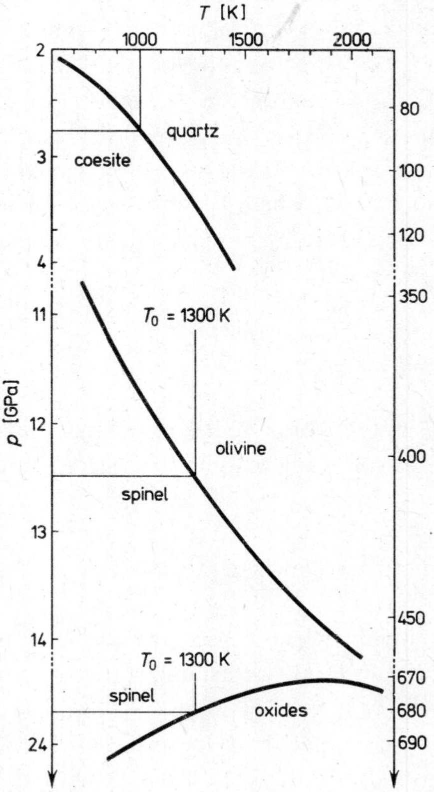

Fig. 6.5.2. Phase equilibrium curves: SiO_2, quartz → coesite: Mg_2SiO_4, olivine → spinel; Mg_2SiO_4 (spinel) → 2 MgO (periclase) + SiO_2 (stishovite).

(a) SiO_2: quartz → coesite*

$$p = 1.42+1.27\times10^{-3}T-0.69\times10^{-6}T^2+0.78\times10^{-9}T^3 \quad (6.5.1)$$

for 600 K < T < 1600 K

(b) Mg_2SiO_4: olivine → spinel**

$$p = 7.82+4.57\times10^{-3}T-0.70\times10^{-6}T^2+0.007\times10^{-9}T^3 \quad (6.5.2)$$

for 700 K < T < 2000 K

(c) $Mg_2SiO_4 \rightarrow 2MgO+SiO_2$: spinel → periclase+stishovite

$$p = 25.95-2.81\times10^{-3}T+0.89\times10^{-6}T^2-0.059\times10^{-9}T^3 \quad (6.5.3)$$

for 700 K < T < 2000 K

The curves of phase equilibrium given above and shown in Fig. 6.5.2 were found from the equation of Clausius–Clapeyron

$$\frac{dT}{dp} = \frac{\Delta S}{V_1-V_2} \quad (6.5.4)$$

in which it was assumed, before its integration, that the transformation entropy ΔS changes linearly with temperature

$$\Delta S = \Delta S_0\left[1+\frac{\Delta c_p}{T_0\Delta S_0}(T-T_0)\right] \quad (6.5.5)$$

and the specific volumes V_1 and V_2 of the two phases are linearly dependent on temperature and pressure

$$V_i = V_{i,0}\left[1-\frac{1}{K_i}(p-p_0)+\alpha_i(T-T_0)\right], \quad i = 1 \text{ or } i = 2 \quad (6.5.6)$$

The appropriate material coefficients, namely the difference in the specific heat of the two phases under a fixed pressure Δc_p, the bulk moduli K_i, the coefficients of thermal expansion α_i, the transformation entropy ΔS_0, and the specific volumes $V_{i,0}$ under reference conditions p_0 and T_0, were selected from various publications. It should be noted that the pressure p_0 between 3 and 30 GPa and the temperature T_0 higher than 1000 K are interesting from the point of view of the Earth's mantle thermodynamic conditions, which should be considered as the reference conditions. The material parameters corresponding to these conditions are usually known

* Recently Weaver *et al.* (1979), in a review paper concerning SiO_2, give the following linear equation of equilibrium for this transformation: $p = 1.57+1.3\times10^{-3}T$.

** Sung and Burns (1976a) for the olivine → spinel transition in the Earth's mantle used the equation $p = 7.2+3.3\times10^{-3}T$.

only approximately and the determination of the Clausius–Clapeyron numerical relations such as eqns. (6.5.1)–(6.5.3) is rather difficult.

The phase equilibrium surfaces for three selected transformations in the vicinity of a down-going lithospheric slab are shown in Fig. 6.5.3 (of

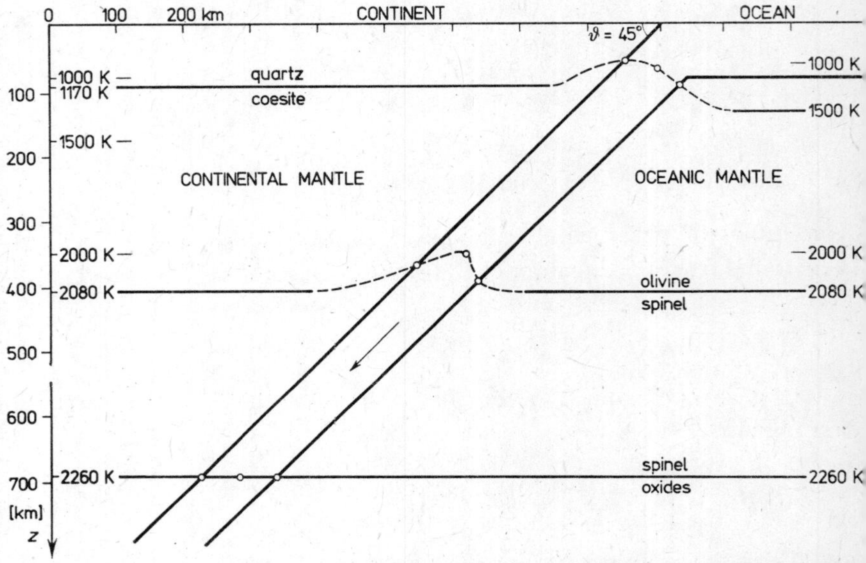

Fig. 6.5.3. Phase equilibrium surfaces in the area of a down-going lithospheric slab: SiO_2, quartz → coesite; Mg_2SiO_4, olivine → spinel; Mg_2SiO_4, spinel → 2 MgO, periclase + SiO_2, stishovite.

TABLE 6.5.1

Depth of three hypotethical surfaces of phase separation in the Earth's mantle. An interval of ±3 km for the spinel → oxides transformation shows that a change of ±0.1 GPa in transformation pressure causes only small undulation of the phase separation surface in different areas of the mantle

Transformation	Oceanic	Descending slab			Continental
		Lower surface	Middle surface	Upper surface	
Quartz → coesite	5.5 GPa 159 km	3.2 GPa 92 km	2.3 GPa 57 km	1.9 GPa 57 km	3.2 GPa 92 km
Olivine → spinel	14.3 GPa 414 km	13.5 GPa 390 km	12.0 GPa 350 km	12.7 GPa 368 km	14.3 GPa 414 km
Spinel → oxides	(23.5 ± 0.1) GPa,		(680 ± 3) km		

course, under the assumption that the appropriate minerals are present in the down-going slab and in the surrounding static mantle). The construction of the phase equilibrium surfaces was based on geotherm diagrams similar to those shown on the left side of Fig. 6.5.1 and supplemented by temperature distributions on the upper, middle and lower surfaces of the down-going slab (according to Schubert *et al.*, 1975). The depths of the phase separation surfaces shown in Fig. 6.5.3 are given in Table 6.5.1.

6.5.3 *Structure of the phase transformation layer: transformations occurring in a state of thermodynamic equilibrium*

According to the considerations presented in the previous section, phase transformation surfaces in the Earth's interior, especially in the areas where movement of matter takes place, are not spherically symmetric. In reality even the term "transformation surface" is not proper, as most probably the "surfaces" are of finite thickness and thus form layers. The finite thickness of such a transformation layer may result from one or several of the following causes:

1. A multi-component system is undergoing transformation, e.g., silicate $(Mg_xFe_{1-x})_2SiO_4$, $0 < x < 1$ (Figs. 6.1.2 and 6.5.1).
2. The transformation causes such a distortion of thermodynamic conditions that in the area where the mineral undergoing transformation is displaced there is some interval of depth in which the geotherm distorted by the thermal effect of the phase transformation is identical with the respective curve of Clausius–Clapeyron.
3. A transformation from a metastable state is taking place.

A description of the joint effect of these factors on the structure of a transformation layer would require the use of very complex multi-parameter models. Thus it is advisable to introduce a mathematical description in which the effects are considered independently. Phase transformations in orthosilicates $(Mg_{0.9}Fe_{0.1})_2SiO_4$ were investigated by Sung and Burns (1976a, 1976b). They presented diagrams of phase stability under the thermodynamic conditions of the mantle and of a down-going lithospheric slab. From these diagrams the thickness of transformation layers in different regions can be estimated.

We will now consider a transformation in a system in which only one component undergoes phase transition, and this takes place under conditions of thermodynamic equilibrium. The results are partially based on a paper by Leliwa-Kopystyński *et al.* (1979). We assume that the component undergoing transformation attains the transformation conditions

$p_{T,0}$ and $T_{T,0}$ at a depth of $z = z_0$ (or $\zeta = 0$, where $z = z_0 + \zeta \sin\vartheta$, and the axis ζ is directed according to the movement direction, see Fig. 6.5.4), and then moves, gradually undergoing transformation is such a manner that the thermal effect of the transformation and its influence on local pressure changes permanently maintain the conditions of thermodynamic equilibrium. Fig. 6.5.4 shows: (*a*)—the shape of a phase equilibrium surface which forms a layer in the areas of displacements, and for comparison (*b*)—the shape of the transformation layer for a multi-component system.

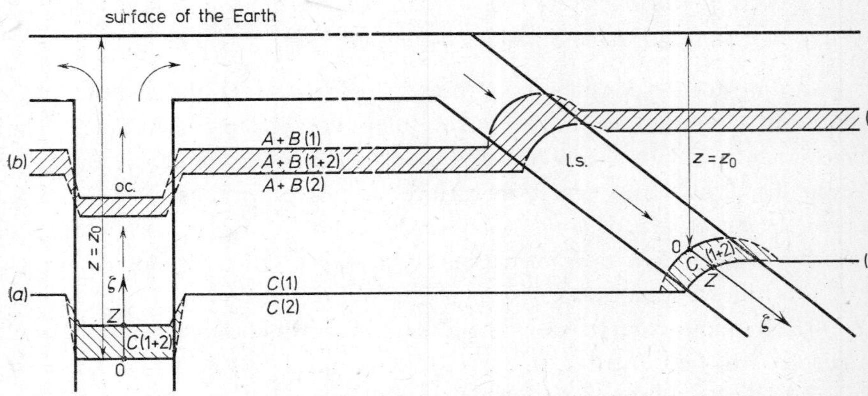

Fig. 6.5.4. (*a*) pattern of the phase separation surface for a one-component system; in moving areas the surface becomes a layer; (*b*) pattern of the phase coexistence layer for a multi-component system. The symbols: *A*, *B* and *C* are the minerals, 1 is the state of the low-pressure phase, 2 is the state of the high-pressure phase, l.s. is the down-going lithospheric slab, oc. is the midoceanic ridge, z_0 or $\zeta = 0$ is the level at which the element of transformable matter under consideration reaches the phase equilibrium state, ζ is the coordinate measured according to the direction of movement, and Z is the local thickness of the transformation layer.

We seek the total thickness Z of the transformation layer and its structure, i.e., the function of reaction advancement $\nu(\zeta)$ determining what fraction of a given element of mass has undergone transformation during the displacement of the mass from level 0 to level ζ. We make the following assumptions:

1. The layer considered is sufficiently thin for assuming that the gravity acceleration is $g(\zeta) = \text{const} = g(z_0)$.

2. The density $\varrho(\zeta)$ depends on the chemical and mineral composition but does not depend on pressure (incompressible medium).

3. The pressure in the layer is determined by the hydrostatic formula

$$dp(\zeta) = g\varrho(\zeta)\sin\vartheta\, d\zeta \tag{6.5.7}$$

where ϑ is the angle between the direction of layer movement and the horizontal plane.

4. The temperature in the layer changes according to the formula

$$T(\zeta) = T_{T,0} + \zeta\beta\sin\vartheta + T_v(\zeta) \tag{6.5.8}$$

where β is the geothermic gradient along the increasing ζ, and $T_v(\zeta)$ is the temperature correction resulting from the occurrence of phase transformation. We assume that β = const which means that changes of the geothermic gradient in the transformation layer can be neglected.

5. The linearized equation of Clausius–Clapeyron is fulfilled. The equation makes it possible to express the phase equlibrium pressure $p_T(\zeta)$ at a depth ζ by the temperature $T(\zeta)$ at the same depth

$$p_T(\zeta) = p_{T,0}\left(1 + \frac{c_{1,2}\varrho_1}{p_{T,0}r}\,\frac{T(\zeta)-T_{T,0}}{T_{T,0}}\right), \qquad r = 1 - \frac{\varrho_1}{\varrho_2} \tag{6.5.9}$$

where $c_{1,2}$ = const is the heat of transformation from phase 1 to phase 2 ($c_{1,2} > 0$ for exothermic transitions; $c_{2,1} = -c_{1,2}$), and ϱ_1 and ϱ_2 are the densities of the respective phases.

From the assumption that the transformation occurs under conditions of thermodynamic equilibrium it follows that the hydrostatic and transformation pressures are identical:

$$\mathrm{d}p(\zeta) = \mathrm{d}p_T(\zeta) \quad \text{for } 0 \leqslant \zeta \leqslant Z \tag{6.5.10}$$

Condition (6.5.10) is shown in Fig. 6.5.5. It defines the required depth interval Z and the corresponding interval of pressure under which the phase transformation occurs. To the left from point 1 in Fig. 6.5.5 or to the right from point 2 the hydrostatic pressure p is lower or higher, respectively, from the phase equilibrium pressure p_T, and thus the respective areas correspond only to the state of one phase. It should be noted that the value Z is not determined univocally by condition (6.5.10). Additional considerations are needed for its determination; in particular, it may happen that $Z = 0$. It will be shown later that such a situation will arise if a transformation is of isothermic character, i.e., if $T_v(\zeta) \equiv 0$—thus in a static area. Hence follows the important conclusion that the phase boundary in a one-component system is always a sharp geometrical boundary and not a finite thickness area in which two phases can coexist. Introducing formulas (6.5.7) and (6.5.9) to eqn. (6.5.10) the following relation is obtained:

$$g\varrho(\zeta)\sin\vartheta = \frac{c_{1,2}\varrho_1}{rT_{T,0}}\left(\beta\sin\vartheta + \frac{\mathrm{d}T_v}{\mathrm{d}\zeta}\right) \tag{6.5.11}$$

When the density ϱ and temperature T are expressed by the function ν, relation (6.5.11) becomes the equation of the transformation advancement function $\nu(\zeta)$.

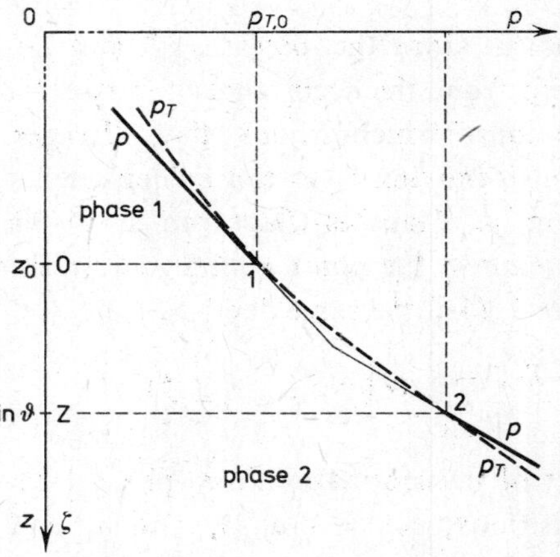

Fig. 6.5.5. Schematic diagram showing the depth interval and corresponding pressure interval in which the hydrostatic pressure p is identical with the phase equilibrium pressure p_T. To the left from point 1 and to the right from point 2 the hydrostatic pressure p is represented by straight lines with the slopes determined by $g\varrho \sin\vartheta$, where ϱ is given by formula (6.5.14) with $\nu = 0$ and $\nu = 1$, respectively. In the area between points 1 and 2 condition (6.5.10) and consequently eqn. (6.5.11) is fulfilled.

To find the relation $\varrho = \varrho(\nu)$ we note that the total volume m/ϱ of an element of mass m of a down-going slab is

$$\frac{m}{\varrho} = \frac{m'}{\varrho'} + \frac{m_1}{\varrho_1} + \frac{m_2}{\varrho_2}, \quad \text{where } m = m' + m_1 + m_2 = \text{const} \tag{6.5.12}$$

The constant m' is the mass of the matrix contained in an element of mass, ϱ' is the density of the matrix, and m_1 and m_2 are the partial masses of phases 1 and 2, respectively. These masses are variable during the transformation but $m_1 + m_2 = m_{1,0} = \text{const}$, where $m_{1,0}$ is the element of mass available for transformation (transformable) at the level $\zeta = 0$. The function of transformation advancement $\nu(\zeta)$ and the parameter μ, which is the fraction of the transformable mass at the beginning of the transformation, are defined as follows

$$\nu(\zeta) = \frac{m_2(\zeta)}{m_1(\zeta) + m_2(\zeta)} = \frac{m_2(\zeta)}{m_{1,0}}, \qquad \mu = \frac{m_{1,0}}{m} = \text{const} \tag{6.5.13}$$

From relations (6.5.12) and (6.5.13) the equation for the density $\varrho = \varrho(\zeta)$

is obtained

$$\frac{1}{\varrho} = \frac{1-\mu}{\varrho'} + \frac{\mu(1-\nu)}{\varrho_1} + \frac{\mu\nu}{\varrho_2} = \frac{\mu}{\varrho_1 L}(1-rL\nu) \tag{6.5.14}$$

where

$$L = \frac{\mu\varrho'}{(1-\mu)\varrho_1+\mu\varrho'} \tag{6.5.15}$$

and $0 \leqslant Lr < r$, if $\varrho_2 > \varrho_1$, i.e., if the transformation leads to a denser phase.

To find the changes of temperature $\mathrm{d}T_\nu/\mathrm{d}\zeta$ caused by the thermal effect of phase transformation, the energy balance should be considered

$$bc_{1,2}\mathrm{d}m_2 = [c'm'+c_1 m_{1,0}(1-\nu)+c_2 m_{1,0}\nu]\mathrm{d}T \tag{6.5.16}$$

where c', c_1 and c_2 denote the specific heat of the matrix and phase 1 and 2, respectively. The coefficient $b \in [0, 1]$ has been introduced to characterize the deviation of the transformation process from an adiabatic one. If the transformation is adiabatic, then $b = 1$, and for the isothermic case $b = 0$. The latter is possible only if the transformation is sufficiently slow for assuming that the heat released during the transformation is dissipated rapidly enough to leave the temperature unchanged. The quasi-adiabatic process ($b \approx 1$), on the other hand, is possible when the transformation is relatively rapid and the heat generation effect is larger than the energy losses caused by thermal conductivity. Using definition (6.5.13) we can write eqn. (6.5.16) in the form

$$bc_{1,2}\mathrm{d}\nu = \left[c'\left(\frac{1}{\mu}-1\right)+c_1(1-\nu)+c_2\nu\right]\mathrm{d}T$$

and therefore

$$\frac{\mathrm{d}T_\nu}{\mathrm{d}\zeta} = bM\frac{c_{1,2}}{c_1}\frac{1}{1+CM\nu}\frac{\mathrm{d}\nu}{\mathrm{d}\zeta} \tag{6.5.17}$$

The parameters M and C are of a similar form to that of L and r

$$M = \frac{\mu c_1}{(1-\mu)c'+\mu c_1}, \qquad C = \frac{c_2-c_1}{c_1} \tag{6.5.18}$$

and we always have $0 \leqslant M < 1$ and $C > -1$.

Substituting relations (6.5.14) and (6.5.17) in formula (6.5.11), we obtain the required equation for the function $\nu(\zeta)$

$$A\frac{1}{1-rL\nu} = 1+B\frac{1}{1+CM\nu}\frac{\mathrm{d}\nu}{\mathrm{d}\zeta} \tag{6.5.19}$$

where the constants are

$$A = \frac{rgT_{T,0}}{c_{1,2}}\frac{L}{\mu}, \qquad B = bM\frac{c_{1,2}}{c_1\beta\sin\vartheta} \tag{6.5.20}$$

The solution of eqn. (6.5.19) when $\nu(0) = 0$ gives the relation between the transformation advancement function ν and the coordinate ζ

$$\zeta = \frac{B}{(A-1)CM-rL}\left[\frac{rL+CM}{CM}\ln(1+CM\nu)-A\ln\left(1+\frac{rL}{A-1}\nu\right)\right] \tag{6.5.21}$$

and also the thickness Z of the transformation layer, which can be found immediately from formula (6.5.21) for $\nu = 1$.

For a full description of the transformation layer structure the equations expressing the distribution of temperature T and of transformation pressure p_T (according to the assumption that $p_T \equiv p$ in a transformation area) should also be given.

The integration of eqn. (6.5.17) and substitution of the result in relation (6.5.8) gives

$$T = T_{T,0}+\zeta\beta\sin\vartheta+b\frac{c_{1,2}}{c_1 C}\ln(1+CM\nu) \tag{6.5.22}$$

Equations (6.5.8), (6.5.9), and (6.5.22) give

$$P_T = p_{T,0}\left\{1+\frac{c_{1,2}\varrho_1}{T_{T,0}p_{T,0}r}\left[\beta\zeta\sin\vartheta+b\frac{c_{1,2}}{c_1 C}\ln(1+CM\nu)\right]\right\} \tag{6.5.23}$$

The last terms in eqns. (6.5.22) and (6.5.23) represent the corrections connected with phase transformations.

Thus, even for a such highly simplified model as that described above, the results depend on several parameters which are not well known.

As an example we shall consider a special case assuming that

$$\frac{rL}{A-1} \ll 1 \quad \text{and} \quad |CM| \ll 1 \tag{6.5.24}$$

Introducing conditions (6.5.24) into eqns. (6.5.21) and (6.5.22), and using the first term of the expansion $\ln(1+x) \approx x$, the formulas for the transformation function ν and the temperature correction T_ν are obtained

$$\zeta = \frac{B}{A-1}\nu, \qquad T_\nu = b\frac{c_{1,2}}{c_1}M\nu \tag{6.5.25}$$

A justification of the approximations defined by formulas (6.5.24) should be given. Since the densities ϱ', ϱ_1 and ϱ_2 are close to each other, we have $L \approx \mu$, according to definition (6.5.15). The value of the parameter $r = 1-$

$-\varrho_1/\varrho_2$ for minerals is not greater than 20%, according to the data listed in Table 7.1.1 (the only exception is the coesite → stishovite transformation). Thus it can be assumed that $Lr < 0.2\,\mu$. To estimate the lowest value of parameter A, the values $g = 10\ \mathrm{m \cdot s^{-2}}$, $T_{T,0} = 1600$ K and $r = 0.05$, can be assumed. Now we obtain

$$A = \frac{rgT_{T,0}}{c_{1,2}}\,\frac{L}{\mu} > \frac{0.8 \times 10^6}{c_{1,2}}$$

where $c_{1,2}$ is expressed in $\mathrm{J \cdot kg^{-1}}$. If $|c_{1,2}| \leqslant 10^5\ \mathrm{J \cdot kg^{-1}}$, which is usually true for minerals, the parameter $A \geqslant 8$ and we can assume that $A-1 \approx A$. The second condition in (6.5.24), i.e., that $CM \ll 1$, is fulfilled when the specific heat of the matrix and of both phases c', c_1 and c_2 are close to each other. In this case $M \approx \mu$ and $|C| \ll 1$.

Using these estimations and introducing the definition of parameters A and B into formulas (6.5.25), we obtain the relations for $\nu(\zeta)$ and T_ν in the final form

$$\zeta = b\mu \frac{c_{1,2}^2}{c_1 g T_{T,0} r \sin\vartheta}\,\nu = b\mu Z_{\max}\nu \qquad (6.5.26)$$

$$T_\nu = b\mu \frac{c_{1,2}}{c_1}\,\nu = b\mu T_{\nu\,\max}\nu \qquad (6.5.27)$$

valid if the following conditions are fulfilled:

$$\begin{aligned} &\varrho' \approx \varrho_1 \quad \text{and } 1 - \frac{\varrho_1}{\varrho_2} > 0.05 \\ &c' \approx c_1 \quad \text{and } \frac{|c_2 - c_1|}{c_1} < 0.2 \\ &|c_{1,2}| < 10^5\,\mathrm{J \cdot kg^{-1}} \end{aligned} \qquad (6.5.28)$$

The coefficients placed in front of ν in formulas (6.5.26) and (6.5.27) describe the total thickness of the transformation layer and the total increase of temperature during transformation.

If the medium is composed only of a transforming material ($\mu = 1$) and the transformation is fully adiabatic ($b = 1$), then the maximum thickness of the transformation layer and the maximum increase of temperature are

$$Z = Z_{\max} = \frac{c_{1,2}^2}{c_1 g T_{T,0} r \sin\vartheta}, \qquad T_\nu = T_{\nu\,\max} = \frac{c_{1,2}}{c_1} \qquad (6.5.29)$$

For a transformation of Mg_2SiO_4 of the olivine → spinel type in the area of a down-going slab we assume that $c_{1,2} = 10^5\ \mathrm{J \cdot kg^{-1}}$, $c_1 = 10^3\ \mathrm{J \cdot kg^{-1} \cdot K^{-1}}$, $r = 0.1$, and that the temperature $T_{T,0}$ changes from about 1000 K in the central part of the slab to about 1700 K on the lower (oceanic) surface; on the upper (continental) surface $T_{T,0} \approx 1400$ K. Introducing these values into formulas (6.5.29) we obtain the following estimates:

$$Z_{\max}\sin\vartheta = \begin{cases} 7100 \text{ m on the continental side} \\ 10000 \text{ m in the middle of the slab} \\ 5900 \text{ m on the oceanic side} \end{cases}$$

$T_{\max} = 100$ K in the entire transformation layer.

The data related to the heat of the spinel → oxides transformation are controversial but certainly $|c_{1,2}| < 10^4 \text{ J}\cdot\text{kg}^{-1}$, which is at least one order of magnitude smaller than that for the olivine → spinel transformation. As a result, the thickness of the transformation layer and the temperature increase for the spinel → oxides transformation become very small and may be neglected—of course only in a model in which the transformation is considered in the state of equilibrium. The possible existence of a spinel → oxides transformation layer can be affected by its multi-component character ($(Mg, Fe)_2SiO_4$ composition) or by the existence of metastable states.

6.5.4 Structure of the phase transformation layer: the transformation from a metastable state

As it was considered in Section 6.3, two factors control phase transformations: the activation of grains of a new phase and the growth of the grains. The kinetics of these processes depends both on the properties of the transforming material and on the thermodynamical conditions under which the material is placed. In the previous chapter kinetic effects were neglected as a result of the assumption that the transformation occurs exactly in the state of equilibrium. Thus the results presented in Section 6.5.3, and especially formula (6.5.21) and its approximated version (6.5.29), give the lowest possible estimate of the thickness of the transformation layer. In reality the transformation occurs with some delay and is more intensive when the given element of the moving area has crossed the phase equilibrium surface and has been for some time, under thermodynamic conditions corresponding to the stable state of a new phase. An attempt to give a method of calculating the transformation probability distribution was made by Leliwa-Kopystyński (1979). The probability distribution was computed as a function of the time interval from the moment of crossing the equilibrium surface by the given area element and of the size of the transforming domain. The determination of the transformation probability distribution as a function of time is equivalent to its determination as a function of the distance ζ from the phase equilibrium surface, since the velocity of displacement of the element under condideration can be regarded as constant. (According to the equation of continuity, the relative change of the displacement velocity $\Delta v/v$ of an element of mass undergoing transformation is equal to $\Delta v/v = 1-\varrho_1/\varrho_2$, which is usually only a few per cent.)

The pattern of phase transformation in the area of a down-going lithospheric slab and of an ascending mid-oceanic ridge is shown in Fig. 6.5.6. The transformation goes stepwise in such a manner that to every

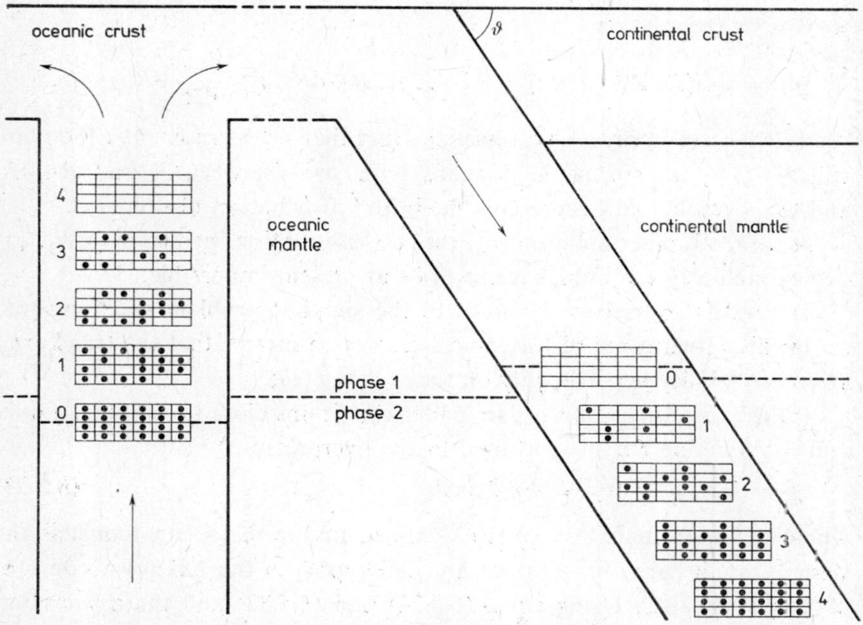

Fig. 6.5.6. Schematic diagram of phase transformations in a down-going lithospheric slab or mid-oceanic ridge area occurring in a stepwise manner. For the sake of clarity it was assumed that all transforming domains are of the same size. The consecutive positions of the area being considered are marked by the numbers 0, 1, 2, 3 and 4. Domains of the low-pressure phase are marked by open rectangles and those of the high-pressure phase are shown by rectangles with a dot.

transformation act corresponds a change of the phase state of randomly selected domain. In general, it can be stated that the probability dP of transformation of a domain with a size* between R and $R+\mathrm{d}R$ and at a time between t and $t+\mathrm{d}t$ depends on the number of domains dn with this size,

* For the sake of simplicity we assume that all domains are of the same shape and are described by one parameter R (e.g., with the dimension of length or volume).

on the time interval $\mathrm{d}t$, and on a certain function f characterizing transformation kinetics

$$\mathrm{d}P = \mathrm{d}P[(R, R+\mathrm{d}R), (t, t+dt)] = f\mathrm{d}n\mathrm{d}t \tag{6.5.30}$$

The function f depends on the transformation temperature $T_T(t)$ (to which corresponds the phase equilibrium pressure $p_T(t)$), on the pressure difference $p(t)-p_T(t)$ above the equilibrium pressure, and on the whole set Σ of different material parameters (in general Σ depends on R)

$$f = f\{T_T(t), [p(t)-p_T(t)], \Sigma\} \tag{6.5.31}$$

The time-dependent distribution of domains $\mathrm{d}n$ as a function of their size is given by the relation

$$\mathrm{d}n = \mathrm{d}n[(R, R+\mathrm{d}R), t] = \{1-P(R, R+\mathrm{d}R), t]\} \frac{\mathrm{d}n_0}{\mathrm{d}R} \mathrm{d}R \tag{6.5.32}$$

where the term in braces expresses the fact that at the moment t there are still $(1-p)\%$ transformable domains with sizes between R and $R+\mathrm{d}R$, and the factor $\mathrm{d}n_0/\mathrm{d}R$ represents the initial distribution of domains.

A general determination of the transformation probability $P(R, t)$ depending on two variables seems to be at present impossible.

We restrict ourselves therefore to the simplest problem assuming that

(a) all domains are of the same size, which means that the transformation probability depends only on time, $P = P(t)$;

(b) the transformation occurs under conditions close to isothermic ones and the function f is proportional to the overpressure*

$$f = A[p(t)-p_T(t)] = A\varrho g\zeta \sin\vartheta \tag{6.5.33}$$

where the coordinate $\zeta = vt$ is as defined in Fig. 6.5.4, v = const is the velocity of the area in question and A = const is the unknown constant of proportionality. Using eqns. (6.5.32) and (6.5.33) and the assumption that the domains which undergo transformation are identical, we can write eqn. (6.5.30) in the form

$$\frac{\mathrm{d}P}{\mathrm{d}t} = A\varrho g v \sin\vartheta (1-P)t \tag{6.5.34}$$

and hence a classic formula for the probability is obtained

$$P = 1-\exp\left(-\frac{A\varrho g \sin\vartheta}{2v}\zeta^2\right) \tag{6.5.35}$$

* It was found by Leliwa-Kopystyński *et al.* (1974) that phase transformations in haloids of alkaline metals are well described by the model of a transformation occurring under isothermic conditions, in which the coefficients of activation and growth of a new phase depend on overpressure proportionally to $(p-p_T)^{\mathrm{const}}$.

It should be noted that no assumption was made with regard to the size of the domains except that they are identical. If the domains are considered as elementary grains of a new phase, stopping their growth when they reach heterogeneities in the medium, then formula (6.5.35) expresses simply the function of reaction advancement, i.e., $P(\zeta) \equiv \nu(\zeta)$, which is the fraction of transformable mass which at a depth ζ is already transformed.

From formula (6.5.35) it follows that an increase in velocity v leads to slower changes of the function $P(\zeta)$, and consequently to an increase in the thickness of the transformation layer. For $v \to 0$ the transformation layer is reduced to the phase separation surface. The value of the parameter A should be sought both from seismological evidence, i.e., from data relating to seismic discontinuities, and from laboratory experiments on phase transformations of minerals under isothermic conditions.

6.6 PHASE TRANSFORMATIONS AND GENERATION OF EARTHQUAKES

Phase transformations of the first order are related to specific volume changes of the order of a few per cent (see Table 7.1.1). Two qualitatively different mechanisms related to energy release during volume changes induced by phase transformations can be proposed:

1. Abrupt release of mechanical energy caused by a stepwise transformation in a given domain:
   (a) implosion when $\varrho_2 > \varrho_1$, e.g., during a transformation process in a descending lithospheric plate,
   (b) explosion when $\varrho_2 < \varrho_1$, e.g., during a transformation process in an ascending area under a midoceanic ridge.
2. Continuous local changes of stress state caused by transformations of a continuous character.

The process of energy radiation and the energy efficiency of rapid phase transformation were studied by Randall (1964a, b; 1966), who showed that even a small change of specific volume makes the energy of implosion or explosion several times as great as the energy released during the process of stress release in the same volume. A change of volume by 1 per cent at a depth of 60 km causes an energy release of 3×10^{15} J $\cdot$ km^{-3}, and at a depth of 600 km it is as much as 8×10^{15} J $\cdot$ km^{-3}, while the seismic energy connected with a fault focus mechanism is several hundred times smaller.

Attempts were made to explain the mechanism of some deep earthquakes by the implosion phenomenon. Evison (1967) studied the polarities of first motion in seismic waves and found that they often represent compression in the focus area and thus may be originated by rapid phase transformations of the implosion type. The results, however, are not univocal, as the origin of a given earthquake can be explained by different source mechanisms.

Seismic data obtained in recent years indicate an indirect rather than direct influence of phase transformations on the generation of earthquakes. Two different series of phenomena can be proposed here, in general identical for transitions of the low-pressure phase → high-pressure phase and the opposite transition (Leliwa-Kopystyński, 1975). Thus when a moving area passes through the phase equilibrium surface one of the two following phenomena occurs:

1. Continuous phase transition takes place → the stresses between the region that undergoes transformation and its surroundings increase → relative shear movement between the transforming and nontransforming layers occurs → earthquake source.

2. Transition of the considered area into a metastable state → sudden phase transformation leading to an implosion or explosion → earthquake source.

Using observational data from two deep earthquakes Vaišnys and Pilbeam (1976) have proposed the following mechanisms of their occurrence:

3. The new phase is nucleated several tens of kilometres below the equilibrium depth, creating a stress disturbance which triggers off the nucleation and reaction of the rapid rate ($\approx 10^2$ s). The transformation causes a decrease in yield strength and effective viscosity allowing fast shear deformations to occur.

The mechanism of type 3 combines in a sense mechanism 1, accepting fast shear deformations as a direct cause of earthquakes, with mechanism 2, assuming the occurrence of a sudden phase transformation in a selected area.

REFERENCES

Anderson O. L. and Liebermann R. C. (1970), "Equations for the elastic constants and their pressure derivatives for three cubic lattices and some geophysical applications", *Phys. Earth. Planet. Inter.*, **3**, 61–85.

Born M. and Huang K. (1954), *Dynamical Theory of Crystal Lattices*, Clarendon Press, Oxford.

Cahn J. W. (1956), "The kinetics of grain boundary nucleated reactions", *Acta Metall.*, **4**, 449–459.

Delmon B. (1969), *Introduction à la Cinétique Hétérogène*, Technip, Paris.

Evison F. F. (1967), "On the occurence of volume change at the earthquake source" *Bull. Seismol. Soc.*, **57**, 9–25.

Hanneman R. E. (1968), "Effect of high pressure on phase transformation rates", *General Electric Technical Information Series*, **68–C–132**, 1–6, Schenectady, New York.

Landau L. D. and Lifshitz E. M. (1958), *Statistical Physics*, Pergamon Press, London (transl. from the Russian).

Leliwa-Kopystyński J. (1975), "High pressure polymorphic phase transitions and its geophysical aspect", *IFTR Reports*, Warsaw, 1–110 (in Polish).

Leliwa-Kopystyński J. (1979), "Probability of the earthquake initiated by the phase transformation from the metastable state", in: *High-Pressure Science and Technology*, B. Vodar and Ph. Marteau (eds.), Vol. 2, Pergamon Press, Oxford–New York.

Leliwa-Kopystyński J. and Bakun-Czubarow N. (1978), "Influence of material parameters on solid–solid phase transitions: its implications for the minerals of the Earth's mantle", *Bull. Acad. Pol. Sci.*, **XXVI**, No. 3–4, 127–138.

Leliwa-Kopystyński J. and Bakun-Czubarow N. (1980), "The effect of material parameters on the shape of the phase separation surfaces within the Earth's mantle", *Phys. Earth. Planet. Inter.*, **22**, 244–254.

Leliwa-Kopystyński J., Lacam A. and Peyronneau J. (1979), "Polymorphic transformations in the Earth's mantle, structure of the phase boundary", in: *High-Pressure Science and Technology*, K. D. Timmerhaus and M. S. Barber (eds.), Vol. 2, Plenum, New York.

Leliwa-Kopystyński J., Peyronneau J. and Lacam A. (1974), "Étude cinétique de l'effet d'hystérésis dans les transitions polymorphiques: influence de la surpression", *J. Phys.*, **35**, 609–614.

Leliwa-Kopystyński J., Peyronneau J. and Lacam A. (1975), "Transformations solide/solide sous hautes pressions; modèles cinétiques de croissance", *J. Phys.*, **36**, 1273–1278.

Lombardi E. and Jansen L. (1964), "Stability of crystals of rare gas atoms and alkali-halides in terms of three body interactions", *Phys. Rev.*, **136**, 1011–1023.

Minear J. W. and Toksöz M. N. (1970a), "Thermal regime of a downgoing slab", *Tectonophysics*, **10**, 367–390.

Minear J. W. and Toksöz M. N. (1970b), "Thermal regime of downgoing slab and new global tectonics", *J. Geophys. Res.*, **75**, 1397–1419.

Münster A. (1969), *Statistical Thermodynamics*, Vol. I, Springer-Verlag, Berlin, and Academic Press, New York.

Randall M. J. (1964a), "On the mechanism of earthquakes", *Bull. Seismol. Soc. Am.*, **54**, 1283–1289.

Randall M. J. (1964b), "Seismic energy generated by a sudden volume change", *Bull. Seismol. Soc. Am.*, **54**, 1291–1298.

Randall M. J. (1966), "Seismic radiation from a sudden phase transition", *J. Geophys. Res.*, **71**, 5297–5302.

Rao C. N. R. and Rao K. J. (1978), *Phase Transition in Solids*, McGraw-Hill, Chatham.

Schubert G., Yuen D. A. and Turcotte D. L. (1975), "Role of phase transitions in a dynamic mantle", *Geophys. J. R. Astron. Soc.*, **42**, 705–735.

Stacey F. D. (1977), "A thermal model of the Earth", *Phys. Earth Planet. Inter.*, **15**, 341–348.

Sung C. M. (1979), "Kinetics of the olivine → spinel transition under high pressure and temperature: experimental results and geophysical implications", in: *High-Pressure*

Science and Technology, K. D. Timmerhaus and M. S. Barber (eds.), Vol. 2, Plenum, New York.

Sung C. M. and Burns R. G. (1976a), "Kinetics of the olivine → spinel transition: implications to deep-focus earthquake genesis", *Earth Planet. Sci. Lett.*, **32**, 165–170.

Sung C. M. and Burns R. G. (1976b), "Kinetics of high-pressure phase transformations: implications to the evolution of the olivine → spinel transition in the downgoing lithosphere and its consequences on the dynamics of the mantle", *Tectonophysics*, **31**, 1–32.

Swalin R. A. (1962), *Thermodynamics of Solids*, J. Wiley and Sons, New York–London.

Tobin M. C. (1974), "Theory of phase transitions kinetics with growth site impingement. Homogeneous nucleation", *J. Polymer Sci., Polymer Physics Edition*, **12**, 399–406.

Turnbull D. (1956), "Phase changes", in: *Solid State Physics*, F. Seitz and D. Turnbull (eds.), Vol. 3, Academic Press, New York.

Vaišnys J. R. and Pilbeam C. C. (1976), "Deep-earthquake initiation by phase transformations", *J. Geophys. Res.*, **81**, 985–988.

Voronov F. F. and Goncharova V. A. (1966), "Influence of pressure up to 20 kbar on the elastic properties of rubidium chloride and iodite", *J. Exp. Theor. Phys.*, **50**, 1173–1182 (in Russian); English transl: **23**, 777–784.

Weaver J. S., Chipman D. W. and Takahashi T. (1979), "Comparison between thermochemical and phase stability data for the quartz–coesite–stishovite transformations", *Am. Min.*, **64**, 604–614.

7. Hypothetical Phase Transformations in the Earth's Mantle

7.1 INTRODUCTION

According to models of the Earth's interior structure considered in the previous chapters the seismic discontinuities in the Earth's mantle are most often recognized as having the nature of phase discontinuities. In this chapter phase transformations of those minerals which most probably occur in the mantle are discussed. The most important phase transformations which selected minerals undergo, under the conditions of p and T in the mantle, are listed in Table 7.1.1. The table contains both the minerals which most probably do exist in the mantle and the minerals whose presence there is doubtful, and also some minerals which are not present in the mantle at all. The latter are of interest either because they are composed of elements typical for the mantle (e.g., low-pressure polymorphs of crystalline silica) or because their studies are of methodological significance (e.g., germanates). Studies of germanates or silicate–germanate solid solutions can be carried out at lower pressures than those needed for studies of pure silicates. Because of the resemblance of crystal structures, the results obtained for germanates can occasionally be extrapolated towards higher pressures and applied to silicates.

Besides phase transformations of the low-pressure phase → high-pressure phase type, reactions of synthesis or disproportionation are also given in Table 7.1.1. These reactions have been included in phase transformations because they are accompanied by distinct changes in density and by thermal effects (exo- or endothermic).

Some of the phase transformations listed in Table 7.1.1 can also be found in Fig. 7.1.1, where the mineral composition of the Earth's mantle down to a depth of about 1500 km is shown. This diagram, made by Ringwood (1975, 1976), is based on his pyrolite model, and its most important feature is the correlation of the physical properties of the mantle (discontinuities of elastic body wave velocities, i.e., the discontinuities in density and in elastic constants) with the hypothetical phase transformations of minerals. It is assumed that the chemical composition of the mantle does not change at least down to a depth of 1500 km, while mineral composition and the phase state of minerals do change.

TABLE 7.1.1
Data related to some phase transformations important for the main transitions in the Earth's mantle

Minerals or chemical compounds, their structures and type of the phase transition	Temperature T [K]	Pressure p [GPa]	Density change $\Delta\varrho/\varrho$ [%] or change of molecular volume ΔV [cm^3]	Author (year)
SiO_2 quartz → coesite	1300	3.12	11.5%	Boyd and England (1960)
SiO_2 coesite → stishovite	1900±100	10.6±0.5	47%	Stishov (1963)
Fe_2SiO_4 olivine → spinel	1300	9.8	47%	Akimoto (1972)
	900	3.8±0.3	10.5%	Ringwood (1966)
	1300	4.9	10.4%	Ringwood (1970)
	1800	6.2	—	Navrotsky (1977)
Mg_2SiO_4 olivine → spinel	900	13.0±0.2	9%	Ringwood (1966)
	1300	12.5	10.6%	Ringwood (1970)
	1800	13.7	—	Navrotsky (1977)
$(Mg, Fe)_2SiO_4$ olivine → spinel	1300	10	10%	Ringwood and Major (1970)
Mg_2SiO_4 (spinel) → $2MgO + SiO_2$ (periclase + stishovite)	—	(680 km)	11.6%	Ringwood (1966)
	1800	23.6	—	Navrotsky (1977)
Synthesis of γMg_2SiO_4 (γ spinel)	1270	25	—	Ito *et al.* (1974)
$2FeSiO_3$ (clinopyroxene) → Fe_2SiO_4 (spinel) + SiO_2 (stishovite)	1300	10	15%	Ringwood (1970), Akimoto (1972)
$2CaMgSi_2O_6$ (clinopyroxene) → Mg_2SiO_4 (spinel) + SiO_2 (stishovite) + $2CaSiO_3$ (garnet)	—	12	—	Ringwood (1966)
$2MgSiO_3$ (clinopyroxene) → Mg_2SiO_4 (β spinel) + SiO_2 (stishovite)	1300÷1700	17÷18	15.2%	Liu (1976b)
$MgGeO_3$ orthopyroxene → ilmenite	970±10	2.8±0.3	15%	Ringwood (1966)

TABLE 7.1.1 (continued)

$MgSiO_3$ clinopyroxene → ilmenite	1300	20÷30	19%	Ringwood	(1970)
$Mg_2SiO_4 + SiO_2 \rightarrow 2\,MgSiO_3$ γ spinel +stishovite → ilmenite	1300÷1700	>20	2%	Liu	(1976b)
60% mol. $MgSiO_3$+40% mol. $Mg_3Al_2Si_3O_{12}$ $[Mg_3(MgSi)Si_3O_{12}]_{60}[Mg_3Al_2Si_3O_{12}]_{40}$ transformation of the pyroxene → garnet type	1300	9÷10	10%	Ringwood	(1970)
eclogite with the following composition: (SiO_2 46.2%, Al_2O_3 14.5%, Fe_2O_3 0.5%, FeO 11.8%, MnO 0.3%, MgO 12.5%, CaO 13.0%, Na_2O 0.8%, Cr_2O_3 0.2%) → garnetite	1300	10÷12	10%	Ringwood	(1970)
$CaGeO_3$ garnet → perovskite	1300	10	−2.53 cm^3	Ringwood	(1970)
$Mg_3Al_2Si_3O_{12}$ garnet → perovskite	1100	30	−1.73 cm^3	Liu	(1974)
$[Mg\,SiO_3]_{80}[Mg_3Al_2Si_3O_{12}]_{20}$ garnet → ilmenite	1300÷1700	22	—	Liu	(1977b)
$[MgSiO_3]_{60}[Mg_3Al_2Si_3O_{12}]_{40}$ garnet → ilmenite	1300÷1700	23	—	Liu	(1977b)
$Mg_3Al_2Si_3O_{12}$ garnet → ilmenite	1300÷1700	24÷25	8%	Liu	(1977b)
$ScAlO_3$ ilmenite → perovskite	1300	12	—	Reid and Ringwood	(1975)
$MgSiO_3$ ilmenite → perovskite	1300÷1700	25	7%	Liu	(1976b, 1977a, 1977b)
95% wt $MgSiO_3$ 5% wt Al_2O_3 ilmenite → perovskite	1300÷1700	26	—	Liu	(1977b)
90% wt $MgSiO_3$ 10% wt Al_2O_3 ilmenite → perovskite	1300÷1700	27÷28	—	Liu	(1977b)
75% wt $MgSiO_3$ 25% wt Al_2O_3 ilmenite → perovskite	1300÷1700	⩾ 30	—	Liu	(1977b)
Synthesis of $MgSiO_3$ with the perovskite structure	1270	28	—	Ito and Matsui	(1978)
$NaAlSi_3O_8 \rightarrow NaAlSi_2O_6 + SiO_2$ albite → jadeite +quartz	1400	2.8	12%	Birch and Le Comte	(1960)
$NaAlSi_2O_6 \rightarrow NaAlSiO_4 + SiO_2$ jadeite →calcium ferrite structure + stishovite	—	20	20%	Reid *et al.*	(1967)
	1270	⩾ 28	20%	Liu	(1978)

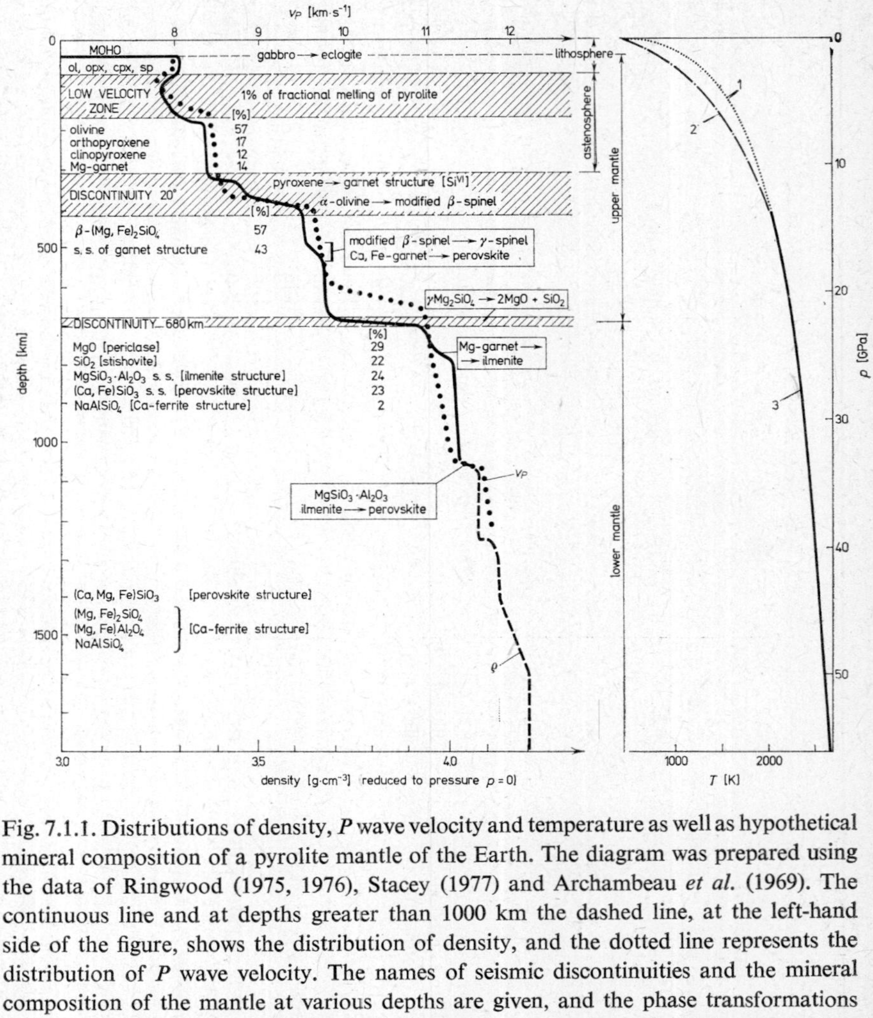

Fig. 7.1.1. Distributions of density, P wave velocity and temperature as well as hypothetical mineral composition of a pyrolite mantle of the Earth. The diagram was prepared using the data of Ringwood (1975, 1976), Stacey (1977) and Archambeau *et al.* (1969). The continuous line and at depths greater than 1000 km the dashed line, at the left-hand side of the figure, shows the distribution of density, and the dotted line represents the distribution of P wave velocity. The names of seismic discontinuities and the mineral composition of the mantle at various depths are given, and the phase transformations most probably corresponding to stepwise density changes are described. The temperature distribution is presented at the right-hand side of the figure: 1—oceanic geotherm, 2—continental geotherm, 3—mantle geotherm at depths greater than 400 km.

It should be noted that the relative density changes during the processes listed in Table 7.1.1 are of the order of 10 per cent. The actual density jumps in the mantle (Fig. 7.1.1), corresponding to particular transformations, are smaller. This can easily be explained by the fact that the medium is not composed entirely of minerals undergoing transformation at a given depth, but a certain part of it consists of minerals which, under given conditions, have the character of a neutral matrix (see Section 6.6.5).

7.2 TRANSFORMATIONS OF CRYSTALLINE SILICA POLYMORPHS

Silica is the main rock forming oxide in the Earth's crust and mantle. Under relatively low pressure SiO_2 occurs in the form of the following polymorphs: quartz α, quartz β, tridymite and cristobalite, whose stability depends mainly on temperature (Fig. 7.2.1). In these varieties Si occurs in coordination 4 with respect to oxygen in spatially linked $[SiO_4]^{4-}$ tetrahedrons. Under the temperature and pressure conditions of the lower lithosphere quartz, which is common in the Earth's crust, is transformed into monoclinic coesite. The quartz → coesite transformation can be

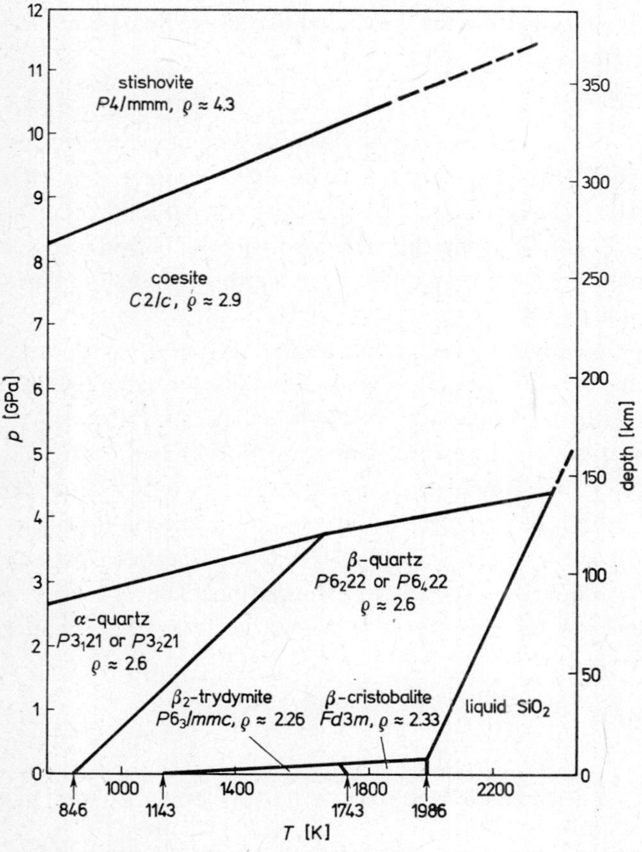

Fig. 7.2.1. Phase diagram of the main silica polymorphs, according to Yoder (1950), Boyd and England (1960), Stishov (1963), Kern and Weisbrod (1967) and Akimoto (1972). The space groups, zero-pressure densities ϱ [$g \cdot cm^{-3}$] and the temperature of transformations under atmospheric pressure of low-pressure polymorphs, in the α-quartz → β-quartz → β_2-trydymite → β-cristobalite series, are given.

described by the equation (p in GPa, T in kelvin)

$$p = 1.42+1.27\times 10^{-3}T-0.69\times 10^{-6}T^2+0.78\times 10^{-9}T^3 \qquad (7.2.1)$$

given by Leliwa-Kopystyński and Bakun-Czubarow (1980). This equation is the result of the integration of the Clausius–Clapeyron equation in a form accounting for the variability of the material parameters. At a depth $z \approx 300$ km in the upper mantle the transition of coesite into tetragonal stishovite with the structure of rutile takes place. This transformation, accompanied by a relative density increase by 47 per cent, consists in the change of the coordination of silicon with respect to oxygen from tetrahedral (coordination number CN = 4) into octahedral (CN = 6). Akimoto (1972) gives the following equation for the coesite–stishovite phase boundary (p in GPa, T in kelvin):

$$p = 5.96+0.0028T \qquad (7.2.2)$$

The high sensitivity of the coesite → stishovite transformation to pressure changes makes it possible to use this transition as a barometer in high-temperature mineralogical experiments. In a pyrolite model of the Earth's mantle stishovite originating from the decomposition of silicate spinels is one of the four main rock forming minerals at depths from 680 to about 1000 km (Ringwood, 1975).

Data relating to phase transformations of the crystalline silica polymorphs obtained in various laboratories display substantial deviations (Weaver *et al.*, 1979). In particular, the phase boundaries for the quartz–coesite and coesite–stishovite transformations are not known with sufficient accuracy, and so the coefficients in eqns. (7.2.1) and (7.2.2) can therefore undergo substantial changes when new data become available. Weaver *et al.* (1979) list a dozen or so linear relations between pressure and temperature in a quartz → coesite transformation. They propose as the most proper relation for this transformation the following equation (p in GPa, T in kelvin):

$$p = 1.47-0.0013T \qquad (7.2.3)$$

7.3 THE GABBRO → ECLOGITE TRANSITION AND THE NATURE OF THE MOHOROVIČIĆ DISCONTINUITY

An attractive hypothesis ascribing to the Mohorovičić discontinuity the nature of a phase transition of the gabbro–eclogite type was a stimulus for numerous experimental investigations, such as the already classical works of Ringwood and Green (1966) and Green and Ringwood (1967).

These authors studied phase transformations occurring in dry basalts with various chemical composition at temperatures from 1300 K to the solidus temperature of rocks, under pressure increasing from 0.1 MPa to 3 GPa. Between the pyroxene–plagioclase paragenesis characteristic for gabbros or pyroxene granulites and the eclogite paragenesis composed mainly of garnet and clinopyroxene rich in sodium they found a transition zone of garnet granulite, in which garnet coexists with pyroxene and plagioclase (Fig. 7.3.1). Under increasing pressure in the zone of pyroxene granulites in silica saturated rocks, reactions occur between pyroxenes and plagioclases, resulting in the formation of alumina-rich ortho- and clinopyroxenes, coexisting with quartz. In silica undersaturated rocks a product of the reactions between olivine and plagioclase is the crystalli-

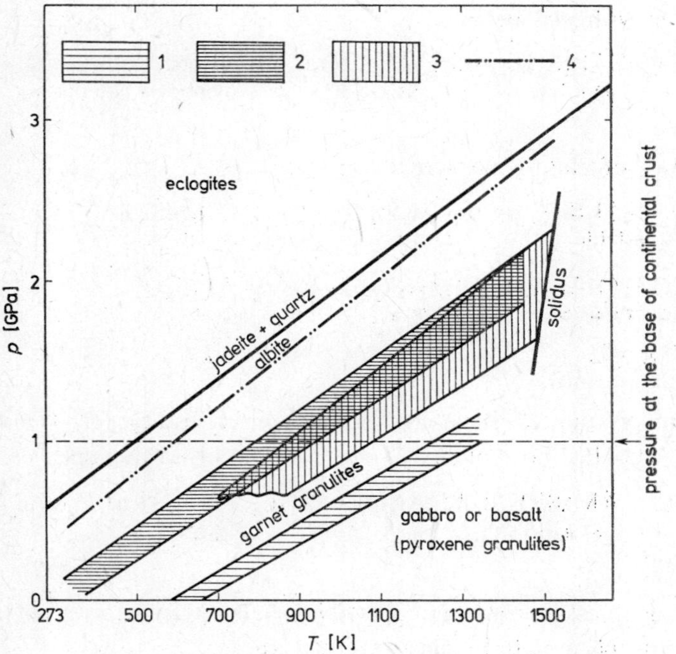

Fig. 7.3.1. Composite phase diagram for the gabbro → eclogite transformation. According to Ito and Kennedy (1970), the eclogitization process of high-alumina olivine tholeiite is accompanied by a rapid density increase of rocks in the following, relatively narrow zones: 1—the zone of initial crystallization of garnet, characterized by a density increase from 3.14 to 3.28 $g \cdot cm^{-3}$; 2—the zone of rapid growth of garnet, in which the density increases from 3.36 to 3.46 $g \cdot cm^{-3}$; 3—the zone of garnet granulite, in which the eclogitization process of quartz tholeiite is continuous within a rather broad interval of pressure (Ringwood and Green, 1966; Green and Ringwood, 1967); 4—the line of total disappearance of plagioclase, according to Ito and Kennedy (1970).

zation of aluminous pyroxenes or an assemblage of alumina-free pyroxenes with aluminous spinel. When the pressure reaches a certain characteristic value, different for basalts with different chemical composition, garnets begin to crystallize along the boundary of the stability field for garnet granulites. In this zone under increasing pressure there is a continuous increase of garnet content and the amount of magnesium in it, and also a gradual disappearance of plagioclase poorer in calcium.

According to Green and Ringwood (1967), the eclogitization process of basaltic rocks, i.e., the transformations occurring in a zone of garnet granulites, can be described by the following simplified reactions between the minerals, in which magnesium cations Mg^{2+} can be replaced by iron cations Fe^{2+} in varying degrees.

Garnets in silica saturated rocks crystallize as a result of either reactions of orthopyroxene with plagioclase:

$$\underset{\text{enstatite}}{4MgSiO_3} + \underset{\text{anorthite}}{CaAl_2Si_2O_8} \rightarrow \underset{\text{garnet}}{Mg_3Al_2Si_3O_{12}} + \underset{\text{diopside}}{CaMgSi_2O_6} + \underset{\text{quartz}}{SiO_2} \tag{7.3.1}$$

or the breakdown of aluminous pyroxenes:

$$\underset{\text{aluminous enstatite}}{mMgSiO_3 \cdot MgAl_2SiO_6} \rightarrow \underset{\text{garnet}}{Mg_3Al_2Si_3O_{12}} + \underset{\text{enstatite}}{(m-2)MgSiO_3} \tag{7.3.2}$$

$$\underset{\text{aluminous diopside}}{mCaMgSi_2O_6 \cdot CaAl_2SiO_6} + \underset{\text{enstatite}}{2MgSiO_3} \rightarrow \underset{\text{garnet}}{CaMg_2Al_2Si_3O_{12}} + \underset{\text{diopside}}{mCaMgSi_2O_6} \tag{7.3.3}$$

In the silica undersaturated rocks the aluminium necessary for garnet crystallization is supplied not only by plagioclases but also by spinels:

$$\underset{\text{anorthite}}{CaAl_2Si_2O_8} + \underset{\text{enstatite}}{4MgSiO_3} + \underset{\text{spinel}}{MgAl_2O_4} \rightarrow \underset{\text{garnet}}{2Ca_{0.5}Mg_{2.5}Al_2Si_3O_{12}} \tag{7.3.4}$$

Under increasing pressure sodium plagioclase albite transforms into a clinopyroxene structure both in silica saturated rocks:

$$\underset{\text{albite}}{NaAlSi_3O_8} + \underset{\text{diopside}}{mCaMgSi_2O_6} \rightarrow \underset{\text{omphacite}}{mCaMgSi_2O_6 \cdot NaAlSi_2O_6} + \underset{\text{quartz}}{SiO_2} \tag{7.3.5}$$

and in silica undersaturated systems:

$$\underset{\text{albite}}{NaAlSi_3O_8} + \underset{\text{enstatite}}{2MgSiO_3} + \underset{\text{spinel}}{MgAl_2O_4} \rightarrow \underset{\text{jadeite}}{NaAlSi_2O_6} + \underset{\text{garnet}}{Mg_3Al_2Si_3O_{12}} \tag{7.3.6}$$

Because of the variety of reactions occurring during the gabbro → eclogite transformations, even relatively small changes in the chemistry of basalts cause significant displacements of the phase equilibrium curves. An increase in the silica saturation degree of basalts results in an increase in the pressure of the onset crystallization of garnets, i.e., of a transition of the gabbro → garnet granulite type. In silica saturated rocks the pressure of a plagioclase disappearance, i.e., of the garnet granulite → eclogite transition, is also considerably higher than in silica undersaturated rocks, in which the transition of albite into the pyroxene structure is combined with the displacement of aluminium from spinel to garnet (7.3.6), which occurs under considerably lower pressure than reactions in which quartz is released, e.g., the breakdown: albite → jadeite + quartz. The pressure of plagioclase disappearance depends also upon the Na_2O content in basalts, and its value decreases with the decrease of sodium content in the rock. The FeO/MgO ratio in basalts distinctly affects the gabbro → garnet granulite phase boundary. An increase of the FeO/MgO ratio results in a decrease of the pressure of the onset crystallization of garnets.

The succession of the main phase transitions in the gabbro → garnet granulite → eclogite transformation series is constant for rocks with different chemical compositions, and thus both for basalts and for intermediate and acid rocks. A typical paragenesis of intermediate and acid rocks, under the conditions of eclogite stability, is garnet + omphacite + quartz + + alkali feldspar, in which albite in a solid solution is stabilized by potassium feldspar, stable under high pressure up to about 10 GPa.

The straight-line segments of the phase equilibrium curves, obtained in basalts during experiments at high temperatures, have been extrapolated towards low temperatures. It has been assumed that the slope of the extrapolated segments is the same as the slope of the phase equilibrium lines relating to eclogitization in simple mineral systems. The following average values of the slopes of the phase equilibrium curves have been adopted: for a gabbro → garnet granulite transformation $dp/dT = 1.5$ MPa $\cdot$ K^{-1}, and for a garnet granulite → eclogite transformation $dp/dT = 2.0$ MPa $\cdot$ K^{-1} (Fig. 7.3.1).

According to Green and Ringwood (1967) the position of the garnet granulite–eclogite phase boundary indicates that eclogites form a stable mineral assemblage under the conditions of the continental Earth's crust, i.e., at a relatively low geothermal gradient. On the other hand, it is not possible to relate the nature of the isochemical gabbro → eclogite transformation to the relatively sharp Mohorovičić discontinuity, since the width of the intermediate zone of garnet granulite assemblage stability is generally large and the changes occurring there are (as a function of pressure) of a continuous nature. Thus, according to the above-mentioned authors, the Mohorovičić discontinuity is most probably caused by a change in chemical composition on the boundary between the intermediate and basic rocks of the lower crust and the ultrabasic rocks of the uppermost mantle.

Phase transformations of the eclogitization process were also studied experimentally by Ito and Kennedy (1970), who found that during the eclogitization of the olivine tholeüte basalt there occurs twice a stepwise increase in rock density. The first stepwise increase in density corresponds to the gabbro → garnet granulite transformation, i.e., to the onset of crystallization of garnets, which in a narrow range of pressures coexist with olivine, orthopyroxene, clinopyroxene and plagioclase (Fig. 7.3.1). The second stepwise increase in denisity is a result of a sudden increase of the garnet phase at the expence of plagioclase during the transition from garnet granulites to eclogites containing sodium-rich plagioclase. With a further increase in pressure, there is a complete disappearance of plagioclase. The rock becomes common eclogite, consisting chiefly of two minerals, namely of garnet and of jadeite-rich clinopyroxene, called omphacite.

Ito and Kennedy, taking into account two well-known changes of P wave velocity v_P, from 6.7 to 7.5 km · s^{-1} and from 7.5 to 8.2 km · s^{-1}, on the boundaries of the intermediate layer between the crust and the upper mantle, proposed the hypothesis that the layer characterized by the velocity $v_P = 7.5$ km · s^{-1} is composed of garnet granulites, while the velocity step to 8.2 km · s^{-1} corresponding to the Mohorovičić discontinuity can be correlated with the garnet granulite → eclogite phase transformation.

The important role played by eclogites in the Earth's upper mantle is confirmed by some results of geophysical studies. The distribution of density in a number of models, consistent with the travel times of compressional and shear waves, with the mass and the moment of inertia of the Earth, with the surface-wave phase velocities and with the eigenperiods of the Earth's free oscillations, was estimated by Press (1970, 1972). A common feature of these models is the presence of rocks with a density of 3.5 g · cm^{-3} at depths from 80 to 150 km. Rock density in large parts of the upper mantle is therefore, according to Press, closer to the density of eclogites than to that of peridotites.

Studies of seismic wave velocities in different eclogites, common bimineral as well as transition eclogites, containing apart from garnet and clinopyroxene also olivine and orthopyroxene or plagioclase, and also in eclogites altered in various degrees, have shown that the ratio $R = v_P/v_S$

in eclogites generally decreases with the increase of density (Anderson, 1977—see Fig. 7.3.2), which can be correlated with the degree of retrogressive alteration of rocks and with changes in their mineral composition. From the variability of R as a function of S wave velocity v_S for olivine and orthopyroxene it follows that neither simplified olivine-orthopyroxene pyrolite nor residual dunites or harzburgites, remaining after the extraction of basalts from the upper mantle peridotites, can be considered the main components of the lower lithosphere, which is generally characterized by higher values of R. High density common eclogites, therefore, can be significant components of the lower lithosphere. Eclogites can be found in three different geological settings (Smulikowski, 1964; Coleman *et al.*, 1965):

(1) in the form of enclaves in kimberlites and alkali basalts and as layers in ultrabasic rocks,

(2) as lenses in gneisses and migmatites,

(3) among metamorphic rocks in ophiolitic complexes.

These groups of eclogites are characterized by different chemical composition of rocks and minerals. The chemical composition of eclogites, as we go from the first to the third group, changes from the chemism of

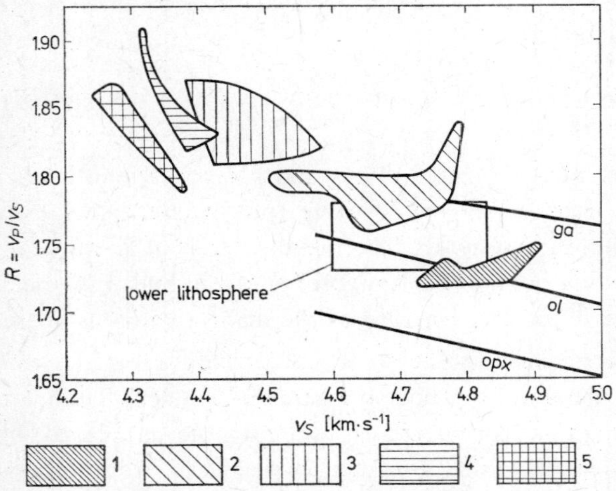

Fig. 7.3.2. Parameter $R = v_P/v_S$ as a function of S wave velocities v_S for various types of eclogites, and for orthopyroxene, olivine and garnet, and also for the lower lithosphere, according to D. Anderson (1977); 1—bimineral eclogites with predominance of garnet over clinopyroxene, $\varrho = 3.50 \div 3.64$ g · cm^{-3}; 2—bimineral eclogites with predominance of clinopyroxene over garnet, $\varrho = 3.39 \div 3.47$ g · cm^{-3}; 3—eclogites with orthopyroxene and olivine, $\varrho = 3.23 \div 3.57$ g · cm^{-3}; 4—retrogressively altered chlorite-bearing eclogites, $\varrho = 3.32 \div 3.44$ g · cm^{-3}; 5—plagioclase-bearing transition eclogites, $\varrho = 3.43$ g · cm^{-3}. (Reproduced, with permission, from *The Annual Review of Earth and Planetary Sciences*, Vol. 5, © 1977 by Annual Reviews Inc.).

olivine basalts to that of tholeiites, the content of magnesium in garnets decreasing, and that of jadeite in pyroxenes increasing. Most probably only eclogites of the first group originated in the upper mantle, while eclogites of the second and third groups have been formed in the Earth's crust. Numerous geological observations indicate, however, that in general small and seldom found eclogitic bodies, unlike basalts, gabbros and garnet granulites, are unstable within the Earth's crust. The formation of eclogites in the crust is often explained by tectonic overpressure.

The results of geological observations of eclogite occurrences, the physical properties of rocks in the lower lithosphere and experiments of Ito and Kennedy on the gabbro → garnet granulite → eclogite transformation of rocks with the chemical composition of olivine tholeiites imply that the hypothesis according to which in large parts of the continental lithosphere, especially in folded areas of old geosynclines, the Mohorovičić discontinuity is a garnet granulite → eclogite transformation is highly probable. Thus the Mohorovičić discontinuity may have the nature both of a chemical composition discontinuity in the oceanic lithosphere and of a phase transformation in large continental areas. It can also have a twofold nature due to the superimposition of phase transitions on even small changes in chemical composition.

7.4 PHASE TRANSITIONS IN PYROLITE OF THE LOWER LITHOSPHERE AND ASTHENOSPHERE

The primitive nondifferentiated matter of the Earth's upper mantle may appear in the form of four varieties of pyrolite (plagioclase, spinel, pyroxene, or garnet pyrolite) or ampholite. The main minerals of pyrolite are olivine and pyroxenes, while ampholite is composed of olivine and amphibole. The varieties of pyrolite differ in the nature of the main aluminous phase.

The stability of particular pyrolite varieties depends upon the p–T conditions and on the partial water vapour pressure in the system. The main role in phase transformations occurring in dry pyrolite is played by aluminium, which under increasing pressure is displaced from plagioclases to spinels or pyroxenes and then to garnets. The phase transformations between different types pyrolite are described below according to the papers of Boyd and MacGregor (1964), Ringwood *et al.* (1964), MacGregor (1967, 1974), O'Hara *et al.* (1971) and Ringwood (1975).

Plagioclase pyrolite is unstable in the upper mantle along normal continental or oceanic geotherms (Fig. 7.4.1). It can probably occur in areas with a particularly high heat flow and a thin Earth's crust, that is under midoceanic ridges or under young island arcs with volcanic activity.

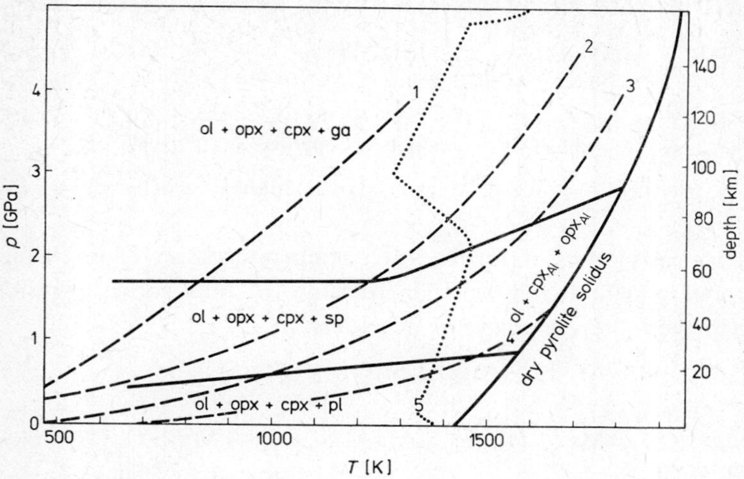

Fig. 7.4.1. Phase diagram for transformations occurring in pyrolite of the uppermost mantle of the Earth, according to Boyd and MacGregor (1964), MacGregor (1967), O'Hara *et al.* (1971) and Ringwood (1975); 1—geotherm of Precambrian shields, 2—oceanic geotherm, 3—geotherm of island arcs with volcanic activity, 4—geotherm of midoceanic ridges, 5—solidus of pyrolite containing 0.1 per cent of H_2O. The deep depression in the water-containing pyrolite solidus, at depths from about 75 to 150 km, corresponds to the low seismic velocity zone. The area of ampholite stability is characterized by temperatures lower than the solidus temperatures of wet pyrolite and by pressures $\leqslant$ 2.5 GPa.

Plagioclase pyrolite under increasing pressure is transformed at low temperatures into spinel pyrolite, or at the highest temperatures, close to solidus temperatures, into pyroxene pyrolite devoid of a separate aluminous phase. These transformations are described by the following simplified reactions:

$$\underset{\text{olivine}}{2Mg_2SiO_4} + \underset{\text{anorthite}}{CaAl_2Si_2O_8} \rightarrow \underset{\text{clinopyroxene}}{CaMgSi_2O_6} + \underset{\text{orthopyroxene}}{2MgSiO_3} + \underset{\text{spinel}}{MgAl_2O_4} \quad (7.4.1)$$

$$\underset{\text{albite}}{NaAlSi_3O_8} + \underset{\text{olivine}}{Mg_2SiO_4} \rightarrow \underset{\text{aluminous clinopyroxene}}{NaAlSi_2O_6} + \underset{\text{orthopyroxene}}{2MgSiO_3} \quad (7.4.2)$$

$$\underset{\text{anorthite}}{CaAl_2Si_2O_8} + \underset{\text{olivine}}{Mg_2SiO_4} \rightarrow \underset{\text{aluminous clinopyroxene}}{CaAl_2SiO_6} + \underset{\text{orthopyroxene}}{2MgSiO_3} \quad (7.4.3)$$

Spinel pyrolite is stable down to a depth of about 60 km in the oceanic litosphere, where its stability area appears to be much greater than in the continental lithosphere. The solubility of Al_2O_3 in pyroxenes increases with temperature, and a gradual transition from spinel to pyroxene pyrolite takes place. This process can be described by the following reactions:

$$\underset{\text{orthopyroxene}}{2MgSiO_3} + \underset{\text{spinel}}{MgAl_2O_4} \rightarrow \underset{\text{aluminous orthopyroxene}}{MgAlSiO_6} + \underset{\text{olivine}}{Mg_2SiO_4} \tag{7.4.4}$$

$$\underset{\text{clinopyroxene}}{CaMgSi_2O_6} + \underset{\text{spinel}}{MgAl_2O_4} \rightarrow \underset{\text{aluminous clinopyroxene}}{CaAl_2SiO_6} + \underset{\text{olivine}}{Mg_2SiO_4} \tag{7.4.5}$$

Pyroxene pyrolite is stable in areas with the highest geothermal gradient.

Under increasing pressure in the lower temperature zone of the spinel pyrolite stability field this pyrolite is transformed into garnet pyrolite, according to the following simplified reactions:

$$\underset{\text{orthopyroxene}}{4MgSiO_3} + \underset{\text{spinel}}{MgAl_2O_4} \rightarrow \underset{\text{garnet}}{Mg_3Al_2Si_3O_{12}} + \underset{\text{olivine}}{Mg_2SiO_4} \tag{7.4.6}$$

$$\underset{\text{clinopyroxene}}{2CaMgSi_2O_6} + \underset{\text{spinel}}{MgAl_2O_4} \rightarrow \underset{\text{garnet}}{Ca_2MgAl_2Si_3O_{12}} + \underset{\text{olivine}}{Mg_2SiO_4} \tag{7.4.7}$$

which can also be presented in the form of the following comprehensive, more complex reaction:

$$\underset{\text{clinopyroxene}}{CaMgSi_2O_6} + \underset{\text{orthopyroxene}}{10MgSiO_3} + \underset{\text{spinel}}{3MgAl_2O_4} \rightarrow \underset{\text{garnet}}{CaMg_8Al_6Si_9O_{36}} + \underset{\text{olivine}}{3Mg_2SiO_4} \tag{7.4.8}$$

The essence of these reactions is the increase of the ion packing density in minerals. The last reaction can be illustrated by the following equation of the change of coordination of magnesium by oxygen:

$$\underset{\text{pyroxene}}{5[Mg]^{VI}} + \underset{\text{spinel}}{3[Mg]^{IV}} \rightarrow \underset{\text{garnet}}{8[Mg]^{VIII}} \tag{7.4.8a}$$

Below this equation we indicate the minerals into whose structure enter the cations with a variable coordination number. The coordination numbers are denoted by Roman numerals.

At higher temperatures, as the pressure increases, the transformations of the spinel pyrolite → garnetiferous pyrolite type are characterized by the participation of aluminium-rich pyroxenes:

$$\underset{\text{aluminous clinopyroxene}}{CaMgAl_2Si_2O_9} + \underset{\text{aluminous orthopyroxene}}{Mg_8Al_2Si_8O_{27}} + \underset{\text{spinel}}{MgAl_2O_4} \rightarrow \underset{\text{garnet}}{CaMg_8Al_6Si_9O_{36}} + \underset{\text{olivine}}{Mg_2SiO_4} \tag{7.4.9}$$

and the coordination number of both magnesium and aluminium by oxygen increases

$$\underset{\text{pyroxene}}{7[Mg]^{VI}} + \underset{\text{spinel}}{[Mg]^{IV}} + \underset{\text{spinel}}{2[Al]^{IV}} \rightarrow \underset{\text{garnet}}{8[Mg]^{VIII}} + \underset{\text{garnet}}{2[Al]^{VI}} \tag{7.4.9a}$$

Pyroxene pyrolite, stable at the highest temperatures, is transformed, under either increasing pressure or decreasing temperature, into garnet pyrolite according to the following simplified reaction:

$$\underset{\text{aluminous orthopyroxene}}{m\mathrm{MgSiO_3}\cdot n\mathrm{MgAl_2SiO_6}} \rightarrow \underset{\text{garnet}}{\mathrm{Mg_3Al_2Si_3O_{12}}} + \underset{\text{aluminous orthopyroxene}}{(m-2)\mathrm{MgSiO_3}\cdot(n-1)\mathrm{MgAl_2SiO_6}} \qquad (7.4.10)$$

which in view of the similarity of the principal chemical formulas for garnets and pyroxenes, can be interpreted as the decrease of solubility of garnet in pyroxene. The pressure of the spinel pyrolite → garnet pyrolite transformation increases with the increasing Cr/Al ratio in rocks (MacGregor, 1970). This phenomenon explains the extension towards high pressures of the stability field of residual spinel peridotites rich in Cr_2O_3, which remain after the extraction of basalts by partial melting of primitive pyrolite.

Garnet pyrolite is the most stable type of pyrolite under the p, T conditions dominating in the lower lithosphere and asthenosphere. Its stability area extends down to a depth of about 350 km. The equilibration conditions of various types of pyrolite can be determined by means of geological thermometers and barometers based on the phase correspondence principles (Perchuk, 1970, Perchuk and Ryabchikov, 1976; Mueller and Saxena, 1977). In particular, the conditions of phase equilibrium in garnet pyrolite have been estimated by means of methods based on the changes of the mutual solubility of pyroxenes and on the solubility of alumina in pyroxenes coexisting with garnet as functions of temperature and pressure (MacGregor, 1974; Mercier and Carter, 1975; Bakun-Czubarow, 1975, 1980).

The stability of ampholite in the upper mantle depends on the stability of amphiboles. Amphiboles are stable down to a depth of about 90 km and at temperatures up to about 1400 K. Under a pressure of about $2.5 \div 3.0$ GPa amphiboles are decomposed into an eclogitic mineral assemblage (clinopyroxene+garnet) and H_2O. The stability field of ampholite covers a considerable part of the stability areas of anhydrous low-pressure varieties of pyrolite, i.e., of plagioclase and spinel pyrolite (Fig. 7.4.1). The total crystallization of pyrolite in the form of ampholite can take place only in a medium rich in water, i.e., containing $0.3 \div 0.4$ weight per cent of H_2O. At the average content of water in pyrolite, which is about 0.1 per cent, an intermediate assemblage—olivine+amphibole+pyroxene—is formed, with greater extension in the upper mantle than that of ampholite, which is probably formed only in areas enriched in water, such as the uppermost parts of the upper mantle and the lower lithosphere at the plate boundaries above the Benioff zone. The poorest in amphiboles is the lower lithosphere under the Precambrian shields and under the oceanic crust, deficient in water and significantly differentiated due to the repeated extracting of the basalt magmas.

The low seismic velocity zone, extending at depths from 75 to 150 km in the mantle (Fig. 7.1.1 and Fig. 7.4.1), is probably formed as a result of amphibole breakdown and a sharp increase in water vapour pressure, which in turn causes a decrease in solidus temperature and slight, about 1 per cent, melting of pyrolite.

7.5 HIGH-PRESSURE PHASE TRANSFORMATIONS IN A_2BO_4 COMPOUNDS — Mg AND Fe ORTHOSILICATES

Ferrous–magnesium orthosilicates as the main components of the Earth's upper mantle are a subject of numerous experimental studies. In studies of the phase transformation of orthorhombic orthosilicate (olivine) into cubic orthosilicate with a spinel structure (olivine → spinel) the following three discoveries were of great importance: the discovery of modified spinel (β phase, β spinel), the construction of the phase diagram for a Fe_2SiO_4–Mg_2SiO_4 system (Ringwood and Major, 1970; Akimoto, 1972), and the synthesis of Mg_2SiO_4 with the true spinel structure (γ phase) (Ito *et al.*, 1974).

A short structural characteristic of polymorphs of the crystalline orthosilicates of divalent metals, based on the works of Kamb (1968), Moore and Smith (1970), Baur (1972), Yagi *et al.* (1974), Morimoto *et al.* (1974), Akimoto *et al.* (1976) and Sung and Burns (1978), will be given below.

Crystalline ferrous–magnesium orthosilicates can occur in the Earth's mantle in three polymorphs—α, β and γ. The lattice of orthorhombic olivine, i.e., of the α phase of Me_2SiO_4, where Me^{2+} denotes the divalent Mg^{2+} and Fe^{2+} cations, is composed of oxygen anions O^{2-} arranged according to a pattern close to the hexagonal closest packed structure (hcp). Half of the octahedral sites existing among oxygen ions are occupied by cations of divalent metal Me^{2+}, while one eighth of the tetrahedral sites are filled with silicon cations Si^{4+}.

The β phase of Me_2SiO_4, called the modified spinel, crystallizes in an orthorhombic system. In the lattice of the β phase oxygen ions form the face centred cubic structure (fcc), i.e., the cubic closest packed structure (ccp). In this lattice Si^{4+} cations occupy tetrahedral positions, and Me^{2+} octahedral ones. Tetrahedra $[SiO_4]^{4-}$ combine in twos through a common oxygen atom, forming $[Si_2O_7]^{6-}$ groups, while one oxygen atom remains not directly bound with silicon.

The γ phase of Me_2SiO_4, called true spinel, crystallizes in a cubic system. Its lattice is also based on the cubic closest packed structure of oxygen ions, Si^{4+} cations are packed in tetrahedral sites, and Me^{2+} in octahedral ones. In the structure of the true spinel the tetrahedrons $[SiO_4]^{4-}$ are isolated from one another.

The structure of the three polymorphs of ferrous–magnesium orthosilicate can be presented by means of a set of octahedra and tetrahedra situated between the layers of close packed oxygen ions. In all the polymorphs two structural motives can be distinguished: (1) chains of octahedra sharing edges with neighbouring octahedra and (2) chains in which octahedra and tetrahedra share corners (Fig. 7.5.1). In the α phase structure we find

(a) (b)

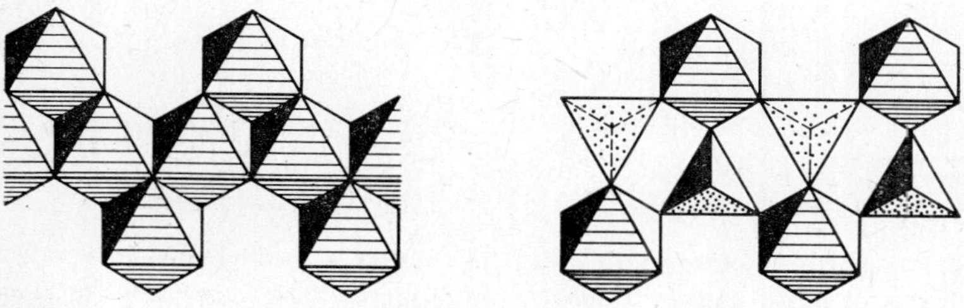

Fig. 7.5.1. The key structural units of the magnesium and iron orthosilicate polymorphs: (*a*) the chain of octahedra sharing edges with neighbouring octahedra, (*b*) the chain in which octahedra and tetrahedra share corners (according to Morimoto *et al.*, 1974).

identical layers composed of chains of octahedra of the first structural motive, bound with one another by means of tetrahedra sharing corners with octahedra. The β phase structure can be represented by identical layers composed of double octahedral chains, alternating with double octahedral-tetrahedral chains. In the γ phase structure two different layers occur alternately: (a) a layer composed exclusively of chains of octahedra sharing edges with neighbouring octahedra, and (b) a layer composed exclusively of chains of octahedra sharing corners with tetrahedra.

The $\alpha \rightarrow \beta \rightarrow \gamma$ phase transformations occur under increasing pressure. The appearance of more stable polymorphs under high pressures is not the result of a coordination number increase of Si^{4+} and Me^{2+} cations by oxygen, but it probably results from a decrease, in the structures of the phases from α to γ, of the number of common edges and corners, whose presence diminishes the stability of crystalline phases.

The phase diagrams for the system of Fe and Mg orthosilicates at temperatures of 1100 and 1300 K and under pressures between 4.0 and 14.0 GPa, according to Akimoto (1972), are shown in Fig. 7.5.2.

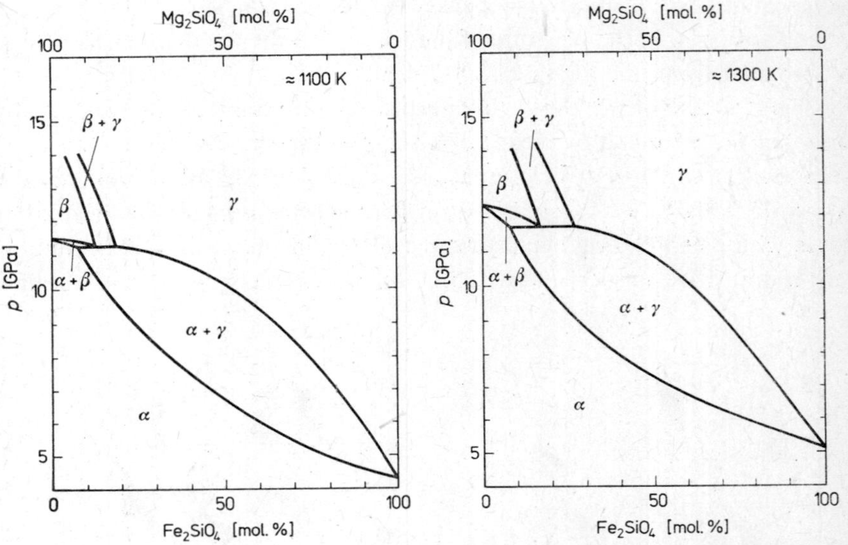

Fig. 7.5.2. Phase diagram of the Mg_2SiO_4–Fe_2SiO_4 system, according to Akimoto (1972). The isotherms of phase transformations at temperatures 1100 K and 1300 K, under pressures between 4 and 14 GPa, are shown. (α—olivine, β—modified spinel, γ—true spinel)

For orthosilicates with the composition of olivine dominating in the upper mantle—$(Mg_{0.9}Fe_{0.1})_2SiO_4$, an important role is played by transformation into the β phase (modified spinel) and then into the γ phase (true spinel). The stability field of the β phase is very sensitive to temperature changes: with increasing temperature it extends and shifts towards higher pressures. Further expansion of the stability field of the β phase at higher temperatures is highly probable. Akimoto (1972) did not obtain at a temperature of 1300 K a homogeneous γ phase for the substratum with the composition $(Mg_{0.9}Fe_{0.1})_2SiO_4$ even under a pressure of 14.0 GPa, but obtained a mixture of phases β and γ, while at a temperature of 1100 K the homogeneous γ phase was synthetized under a pressure not smaller than 14.0 GPa.

The dependence of the β phase stability upon temperature has been taken into account by Sung and Burns (1976) in their considerations of the kinetics of olivine → spinel transformation. They believe that there is a significant difference between the phase transformations of the main component of the upper mantle—orthosilicate with the $(Mg_{0.9}Fe_{0.1})_2SiO_4$ composition—under oceanic and continental lithospheric plates and the transitions occurring in slabs descending in subduction zones. The schematic phase diagrams in both these cases are shown in Figs. 7.5.3 *a*, *b*. This

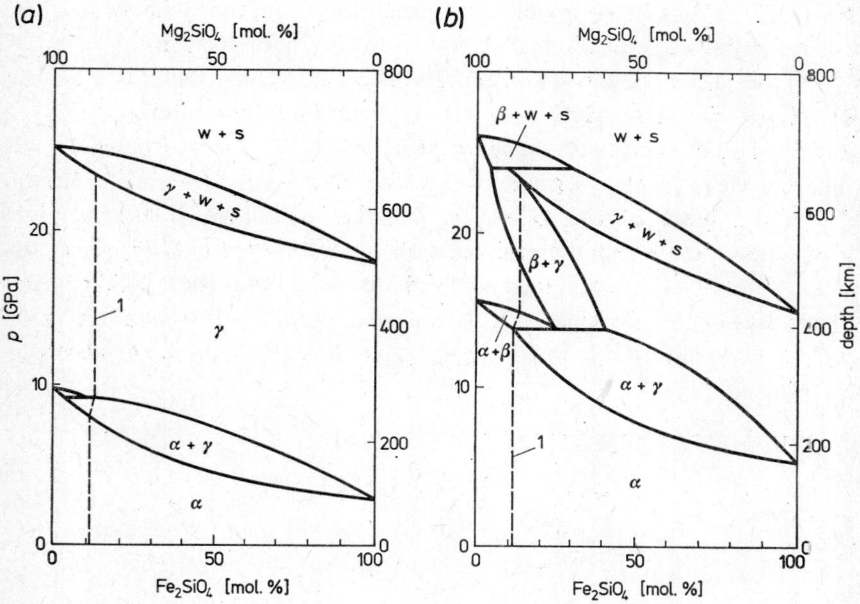

Fig. 7.5.3. Phase transformations of the magnesium and iron orthosilicates in the Earth's upper mantle, according to Sung and Burns (1976); (*a*) in descending lithospheric plates, in subduction zones, (*b*) under stable oceanic and continental lithospheric plates, the temperatures at corresponding depths are higher here than in the case (*a*). The dashed line marks the average composition of olivines in the mantle. The pressure-depth relationship slightly different from that commonly assumed. (α—olivine, β—modified spinel, γ—proper spinel, w—magnesiowustite, s—stishovite)

figure can be regarded as the Akimoto's diagram (Fig. 7.5.2) extended towards high pressures and supplemented by geophysical interpretation (temperature variation). The vertical dashed lines in Fig. 7.5.3 mark the average composition of orthosilicates in the mantle.

According to Sung and Burns the following transformations occur in descending slabs:

$$\alpha \rightarrow \alpha+\gamma \rightarrow \gamma \qquad (7.5.1)$$

while under oceanic and continental lithospheric plates the transition series is most probably much more diversified:

$$\alpha \to \alpha+\gamma \to \beta+\alpha \to \beta \to \beta+\gamma \to \gamma \qquad (7.5.2)$$
$$(\alpha \gg \gamma) \quad (\beta \gg \alpha)$$

Inequalities of the $\alpha \gg \gamma$ and $\beta \gg \alpha$ type mean a clear quantitative dominance of the corresponding phase. Formula (7.5.2) is a good description of the olivine → spinel transformations, most probably occurring in motionless areas of the mantle, at depths of about 400 km.

At a depth of about 680 km spinel is probably decomposed into a mixture of oxides—periclase MgO and wustite FeO with the structure of NaCl, and stishovite SiO_2 with a rutile structure (Fig. 7.5.3). There are two different views on the thermal effect of the spinel → oxides transformation. The first is based on the conviction that the breakdown of silicate spinels into oxides is an endothermic reaction, and the gradient dp/dT of the phase equilibrium curve is negative (Ahrens and Syono, 1967; D. Anderson, 1967; Basset and Ming, 1972; Kumazawa *et al.*, 1974; Schubert *et al.*, 1975). The phase diagram for Fe_2SiO_4 is shown in Fig. 7.5.4, according

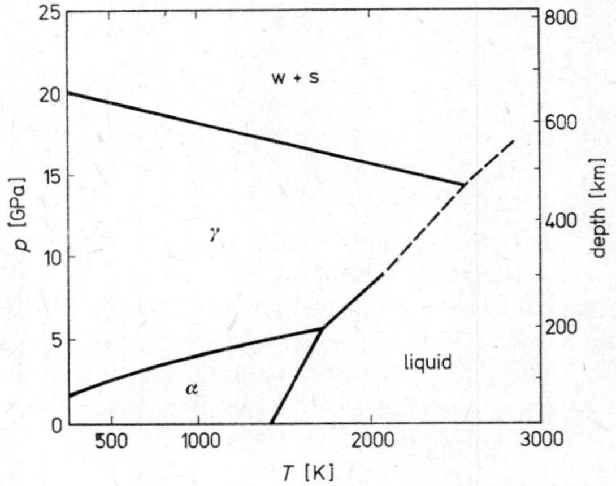

Fig. 7.5.4. Phase transformations of ferrous orthosilicate (fayalite), according to Basset and Ming (1972). The negative slope of the curve of the spinel → mixed oxides disproportionation indicates the endothermic thermal effect of transformations. (α—fayalite, γ—true spinel, w—wustite, s—stishovite)

to Basset and Ming (1972). The other, different view on the thermal effect of the spinel → oxides transformation is represented by Minear and Toksöz (1970 a, b), Ringwood (1973, 1975, 1976), Jackson *et al.* (1974), Liu (1975), Navrotsky and Kasper (1976) and Navrotsky (1977).

Ascertaining the character of the thermal effect of the spinel → oxides transformation would be of essential geophysical significance for the dynamics of descending lithospheric slabs. In the case of an endothermic character of the reaction, bringing about density increase, the phase equilibrium surface in the descending slab is convex downwards (Fig. 7.5.5—the spinel

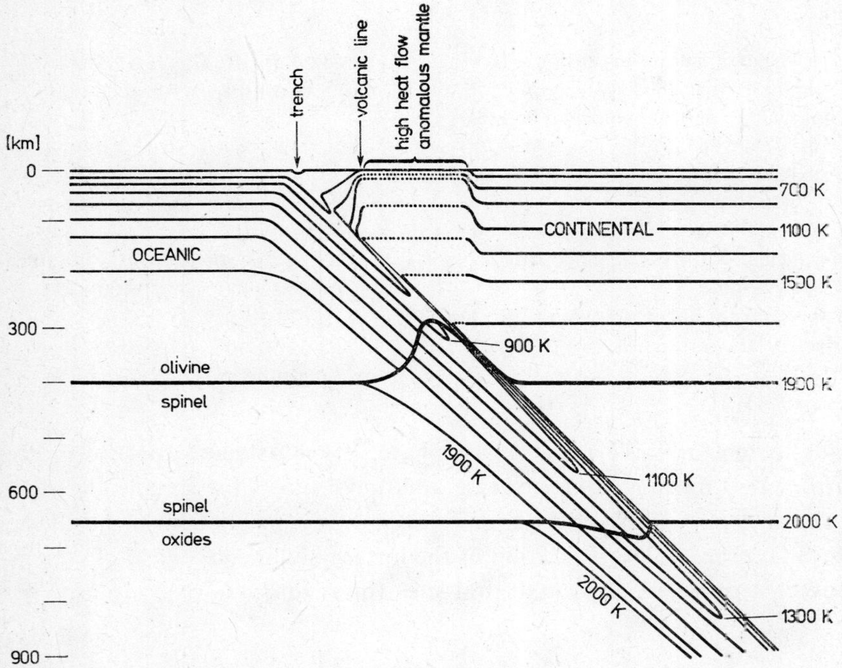

Fig. 7.5.5. Isotherms and phase separation surfaces in the neighbourhood of a descending lithospheric plate. The calculations were carried out using the data listed in Table 7.5.1 and under an assumption that the velocity of downgoing slab is 8 cm · year^{-1}. (According to C. Schubert, D. A. Yuen and D. L. Turcotte, *Geophys. J. R. Astron. Soc.*, **42**, 728 (1975), copyrighted by the American Geophysical Union).

→ oxides transformation at a depth of about 680 km). In an exothermic reaction, on the other hand, causing an increase in density, the phase separation surface is convex upwards (Fig. 7.5.5—the olivine → spinel transformation at a depth of about 400 km). The convexity of the equilibrium surface downwards can be interpreted as a hydrostatic barrier preventing the descent of lithospheric slabs: the medium above the equilibrium surface, still in the state of the less dense low-pressure phase, is forced into a denser environment composed of the high-pressure phase. And conversely, the convexity of the phase separation surface upwards

TABLE 7.5.1
Parameters of the olivine → spinel and spinel → oxides phase transformations (according to Schubert *et al.*, 1975)

Parameter	Transformation	
	olivine → spinel	spinel → oxides
Coefficient of thermal expansion α [K^{-1}]	$(3 \div 3.5) \times 10^{-5}$	$(2 \div 2.5) \times 10^{-5}$
Relative change in density $\Delta\varrho/\varrho$	$0.08 \div 0.1$	$0.08 \div 0.1$
Slope of the phase equilibrium boundary dp/dT [$MPa \cdot K^{-1}$]	$+3.0 \div +6.2$	$-2.0 \div -1.0$
Absolute temperature of the phase equilibrium T [K]	$1800 \div 1900$	$2300 \div 2400$
Coefficient of kinematic viscosity ν [$cm^2 \cdot s^{-1}$]	$10^{22} \div 10^{24}$	$10^{23} \div 10^{26}$
Coefficient of the heat conductivity $\varkappa$ [$cm^2 \cdot s^{-1}$]	$(10^{-2} \div 2) \times 10^{-2}$	$(10^{-2} \div 2) \times 10^{-2}$
Depth D [km]	$300 \div 400$	$600 \div 700$
Geothermic gradient at a depth D [$K \cdot km^{-1}$]	$1 \div 5$	$1 \div 5$
Heat of transformation $c_{1,2}$ [$J \cdot kg^{-1}$]	$+1.67 \times 10^5$ (egzotherm.)	-0.75×10^5 (endotherm.)

leads to the formation of a denser phase, above its standard level in the surrounding medium, thus creating additional forces for sinking the slab.

In silicate spinels with the composition of, say, magnesium orthosilicate Mg_2SiO_4 there occurs a reaction of the intracrystalline exchange of cations between tetrahedral and octahedral sites (inversion):

$$Si^{4+}_{tet} + Mg^{2+}_{oct} \rightarrow Si^{4+}_{oct} + Mg^{2+}_{tet} \tag{7.5.3}$$

Jackson *et al.* (1974) believe that the inversion degree of spinels has a significant influence on the thermal effect of their decomposition into oxides. The slope of the phase equilibrium curve for silicate spinels should not be smaller than 1.0 $MPa \cdot K^{-1}$, and may also assume a small positive value. This idea was expounded by Navrotsky and Kasper (1976) and Navrotsky (1977). Their results are given below.

The inversion degree η is defined by the formula

$$\eta = \frac{n}{N} \tag{7.5.4}$$

where n is the number of octahedral sites occupied by Si^{4+} cations, equal to the number of tetrahedral sites occupied by cations of bivalent metals Me^{2+}, and N is the number of tetrahedral sites occupied by cations, equal to the number of molecules of silicate spinel and also to the number of Si^{4+} cations. When $\eta = 0$, all Me^{2+} ions are situated in octahedra and all Si^{4+} ions are in tetrahedra. Since cations occupy twice as many octahedral as tetrahedral sites in the spinel structure, the constant of equilibrium of intracrystalline exchange reaction (7.5.3) can be described by the formula

$$K = \left(\frac{n}{2N}\frac{n}{N}\right)\left(\frac{2N-n}{2N}\frac{N-n}{N}\right)^{-1} = \frac{\eta^2}{(2-\eta)(1-\eta)} \tag{7.5.5}$$

in which $2N$ is the number of octahedral sites occupied by cations. From eqn. (7.5.5) and the Van't Hoff equation

$$K = \exp\left(-\frac{\Delta H_{int}}{RT}\right) \tag{7.5.6}$$

where ΔH_{int} is the enthalpy of interexchange of cation sites and $R = 8.317$ $J \cdot mol^{-1} \cdot K^{-1}$ is the gas constant, the dependence of the inversion degree upon temperature $\eta(T)$ has been determined. The interexchange enthalpy ΔH_{int} can be found when the inversion degree of spinels in the equilibrium state at a given temperature is known. The total entropy of the spinel phase is equal to the sum

$$S_{spinel}(\eta) = S_{spinel}(\eta = 0) + S_{conf}(\eta) \tag{7.5.7}$$

where $S_{spinel}(\eta = 0)$ is the entropy of normal noninversed spinel, i.e., spinel situated at sufficiently low temperatures. The entropy of the spinel configuration with the inversion degree η can be found from the formula

$$S_{conf} = k \ln\left[\binom{2N}{n} \times \binom{N}{N-n}\right] \tag{7.5.8}$$

where $k = 1.381 \times 10^{-23}$ $J \cdot K^{-1}$ is the Boltzmann constant and the expression in brackets represents the probability that the system of N molecules Me_2SiO_4 is in such a state that n ions Si^{4+} occupy octahedral sites and N—n ions Si^{4+} occupy tetrahedral sites. Using the Stirling formula for factorials and definition (7.5.4), we find that the entropy of configuration for one mole ($N = 6.22 \times 10^{23}$ molecules of Me_2SiO_4) is

$$S_{conf} = R[2\ln 2 - 2\eta \ln\eta - (1-\eta)\ln(1-\eta) - (2-\eta)\ln(2-\eta)] \tag{7.5.9}$$

Dependence (7.5.9) is shown in Fig. 7.5.6.

The enthalpy of spinel is

$$H_{spinel}(\eta) = H_{spinel}(\eta = 0) + \eta\Delta H_{int} \tag{7.5.10}$$

The formula for the interchange enthalpy ΔH_{int} can be derived from eqns. (7.5.5) and (7.5.6)

$$\Delta H_{int} = RT[\ln(2-\eta) + \ln(1-\eta) - 2\ln\eta] \tag{7.5.11}$$

TABLE 7.5.2
Parameters of the olivine → spinel and spinel → oxides phase transformations for magnesium orthosilicate Mg_2SiO_4 (according to Navrotsky, 1977)

Parameter	Transformation					
	olivine → spinel			spinel → oxides		
Temperature of the phase equilibrium T [K]	1273	1773	2373	1273	1773	2373
The degree of the spinel inversion η	0.00	0.05	0.11	0.00	0.05	0.11
ΔH [$J \cdot kg^{-1}$]	2.61×10^5	2.97×10^5	3.38×10^5	5.48×10^5	5.13×10^5	4.71×10^5
ΔS [$J \cdot kg^{-1} \cdot K^{-1}$]	−89.3	−63.7	−43.8	+23.8	−1.8	−22.0
Pressure of the phase equilibrium p [GPa]	12.5	13.7	14.8	23.7	23.6	23.9
The slope of the Clapeyron curve dp/dT [$MPa \cdot K^{-1}$]	+2.94	+2.10	+1.44	−1.07	+0.08	+1.00
Heat of transformation $c_{1,2}$ [$J \cdot kg^{-1}$]	$+11.4\times10^4$	$+11.3\times10^4$	$+10.4\times10^4$	-3.0×10^4	$+0.3\times10^4$	$+5.2\times10^4$

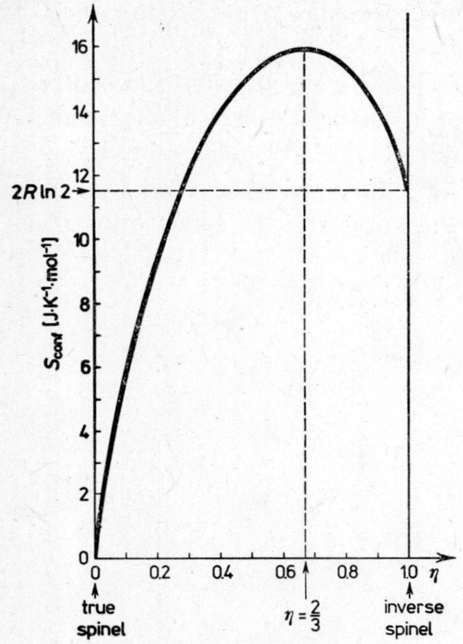

Fig. 7.5.6. Configuration entropy S_{conf} as a function of the degree of spinel inversion η, see eqn. (7.5.9).

From eqns. (7.5.9) and (7.5.11) it follows that always $0 < \eta < 1$, i.e., that a certain inversion of cation sites always exists.

With the use of the transformation parameters known from experiments and relating to sufficiently low temperatures at which inversion is nearly impossible, i.e., $\eta \approx 0$, the transformation parameters at higher temperatures, at which $\eta > 0$, have been found (Table 7.5.2). The theoretical phase diagram for Mg_2SiO_4, calculated with a correction for spinel inversion, is shown in Fig. 7.5.7. The influence of spinel inversion on the shape of the phase boundary of the spinel → oxides disproportionation is of special interest here. Even a small degree of spinel inversion (from 1 to 10 per cent of Si^{4+} in octahedral sites) changes the slope of the phase equilibrium curve from negative at temperatures smaller than 1300 K for true spinel, through zero at a temperature of about 1800 K, to positive at temperatures between 1800 and 2400 K. The decomposition of spinel into oxides is thus, according to Navrotsky (1977), an exothermic transformation under the conditions characteristic for the Earth's mantle, and the corresponding discontinuity does not form a mechanical barrier for sinking lithospheric slabs: the convexity of the phase separation surface is directed upwards and not downwards as in the Schubert diagram (Fig. 7.5.5).

A significant influence on the olivine → spinel transformation pressure is exerted by the stabilization effect in the crystal field of the cations of transition metals, mainly of Fe^{2+}. Akimoto *et al.* (1976) have shown that the cations of transition metals, such as Fe^{2+}, Co^{2+}, Ni^{2+}, have an excess crystal field stabilization energy in the octahedral sites in spinel over olivine.

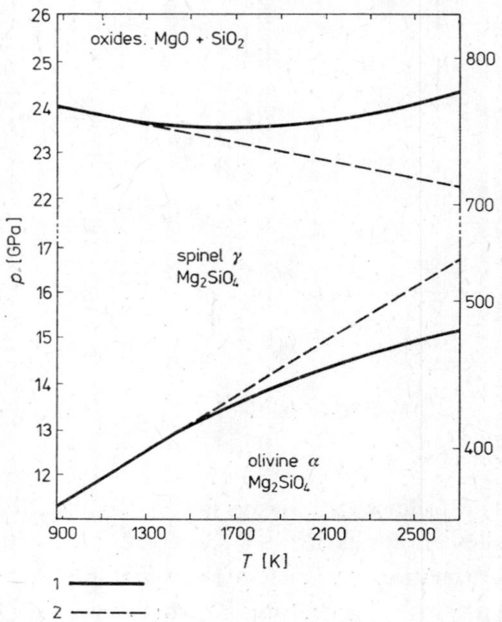

Fig. 7.5.7. The olivine → spinel and spinel → oxides phase transformations for magnesium orthosilicate, according to Navrotsky (1977); 1—the Clapeyron's curve taking into account the degree of spinel inversion as a function of temperature, 2—the Clapeyron's curve neglecting the degree of cation ordering in spinel structure. The consideration of the inversion degree of spinel widens the stability area of spinel structure and changes the slope of the Clapeyron's curve of spinel decomposition into oxides from negative to positive, at temperatures greater than 1800 K.

This energy causes a significant decrease in the pressure of the olivine–spinel phase transformation. For ferrous orthosilicate Fe_2SiO_4 the pressure in question reaches about 10 GPa at a temperature of 1300 K. Burns and Sung (1978) estimated that the olivine–spinel phase boundary is displaced upwards by about 15 km as a result of a transformation pressure decrease caused by the presence in the upper mantle olivines of the Fe^{2+} cation, stabilized in the crystal field.

From experiments on phase transformations in forsterites (Liu, 1976a) it follows that the decomposition of true spinel into a mixture of oxides is improbable under the conditions of the Earth's mantle, since direct transition of true spinel γMg_2SiO_4 into the mixture of periclase and magnesium metasilicate with an ilmenite structure has been observed at a temperature of 1300 K and under pressure of 20 GPa.

7.6 HIGH-PRESSURE PHASE TRANSFORMATIONS IN ABO_3 COMPOUNDS — Mg AND Fe METASILICATES

Metasilicates (pyroxenes) occupy quantitatively the second place, after olivine, in the mineral composition of the upper mantle down to a depth of about 300 km. Accordingly, phase transformations of ferrous–magnesium pyroxenes have been studied in numerous experiments. Ringwood (1970) considers the following three types of magnesium metasilicate transformations:

(1) disproportionation into modified β spinel and stishovite,
(2) transformation leading to an ilmenite type structure,
(3) transformation of the pyroxene → garnet type.

The disproportionation of ferrous metasilicate into a mixture of spinel and stishovite occurs at a temperature of 1300 K and under pressure of about 10 GPa, i.e., under pressure considerably higher than that of the transformation of fayalite (ferrous orthosilicate) into a structure of modified β spinel ($4.5 \div 5.5$ GPa and $1100 \div 1500$ K). According to Akimoto and Syono (1970), and Akimoto (1972), the high pressure of the pyroxene disproportionation into β spinel and stishovite is conditioned by the relatively high pressure of the coesite → stishovite phase transformation (Fig. 7.2.1). In a series of $(Mg,Fe)SiO_3$ pyroxene solid solutions an increasing share of magnesium is accompanied, at a constant temperature, by an increasing pressure of disproportionation of pyroxene into the mixture of β spinel and stishovite (Fig. 7.6.1). Ringwood (1970) estimated that the pressure of disproportionation of the pure magnesium member—clinoenstatite—is between 20 and 30 GPa. In comparison with olivines, ferrous–magnesium pyroxenes are stable up to considerably higher pressures than those which induce the olivine → β spinel phase transformation, under the assumption that the value of the Mg/Fe ratio is the same both in pyroxenes and in olivines.

The alternative phase transformation of $MgSiO_3$ pyroxene with regard to the disproportionation into spinel and stishovite is its transition into the ilmenite type structure with the closest hexagonal packing of oxygen

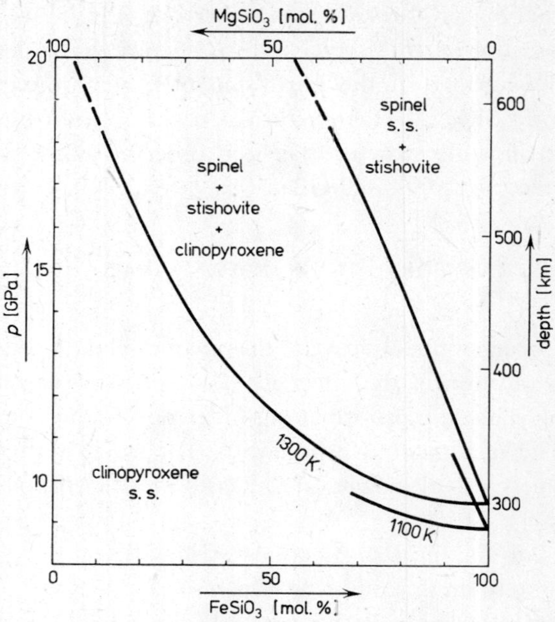

Fig. 7.6.1. Phase diagram of a $MgSiO_3$–$FeSiO_3$ system at temperatures of 1100 and 1300 K, according to Akimoto (1972).

atoms and with Mg and Si atoms occupying the octahedral sites. This transformation was first found in germanates (Table 7.1.1) and later studied in solid solutions $MgSiO_3$–$MgGeO_3$. The transformation pressure of pure $MgSiO_3$ was estimated by extrapolation of phase boundaries to be between 20 and 30 GPa at a temperature of about 1300 K.

The pyroxene → garnet type transformation is the most important high-pressure transformation in metasilicates which can occur in the pyrolite mantle of the Earth (Ringwood, 1975). Most probably this transformation occurs at depths of 300 ÷ 350 km. The essence of this transition is a partial change (up to 1/4) of silicon atom coordination from tetrahedral into octahedral. As a result, complex granet is formed. From a general crystallochemical formula of garnets $A_3^{VIII}B_2^{VI}C_3^{IV}O_{12}$ (where A^{VIII} is the symbol of divalent cations in coordination 8 by oxygen, B^{VI} is the symbol of trivalent cations in eightfold coordination (CN = 6), and C^{IV} is the symbol of tetravalent cations in fourfold coordination (CN = 4)), the formula of an extreme high-pressure member of the solid solutions series of complex garnets can be represented in the form $A_3^{VIII}(AC)^{VI}C_3^{IV}O_{12}$. This form indicates that the tetravalent silicon cation alternately with a divalent cation can occupy the octahedral sites, occupied in common rock-forming garnets of lower pressures by Al^{3+}, Fe^{3+} and Cr^{3+} cations. Moreover, high-pressure complex garnets display in general a high degree of cation ordering in octahedral coordination.

Among the numerous experimental studies devoted to pyroxene → garnet transformation particular attention should be paid to an experiment in the crystallization of garnets from glass with the composition of 90 weight per cent $MgSiO_3$ and 10 weight per cent Al_2O_3, corresponding to the mixture of 60 molecular per cent of enstatite and 40 molecular per cent of pyrope, at a temperature of 1300 K under increasing pressure (Ringwood, 1970; Fig. 7.6.2., Table 7.1.1). In that experiment the crystallization of pyrope and clinoenstatite up to pressure of about 9 GPa was ob-

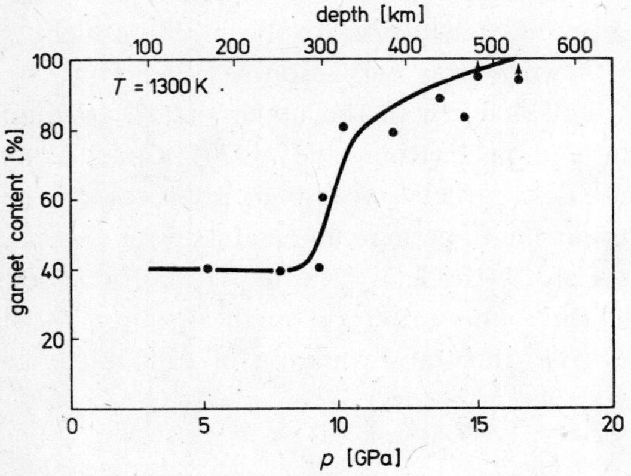

Fig. 7.6.2. Participation of garnets crystallizing from glass with the composition 90% of $MgSiO_3$ and 10% of Al_2O_3 (by weight) as a function of pressure, at temperature of about 1300 K, according to Ringwood (1970).

served. Under pressures of between 9 and 10 GPa a sudden increase of the garnet phase occurred, while under a pressure of 15 GPa only garnets with the $[Mg_3MgSiSi_3O_{12}]_{60}$ $[Mg_3Al_2Si_3O_{12}]_{40}$ composition were crystallized. The crystallization in the garnet structure substratum poorer in Al_2O_3, corresponding to the mixture of 70 molecular per cent of enstatite and 30 molecular per cent of pyrope, occurs under a pressure of 20 GPa, while the transformation of aluminium-free clinoenstatite into garnet should occur under pressure considerably exceeding 20 GPa. Since, however, all models of the upper mantle assume a considerable amount of Al_2O_3 in it, and also the presence of Cr_2O_3 and Fe_2O_3, the pyroxene → garnet transformation most probably occur under a pressure of about 10 GPa in natural conditions, and thus at a shallower level of the upper mantle than the transition of olivine into modified β spinel. Similar experiments have also been carried out with the use of glass of calcium, ferrous and calcium–magnesium metasilicates. Their entire crystallization in a garnet structure was obtained under pressures of between 10 and 15 GPa.

Of great importance for the transition occurring in natural mineral parageneses were the experimental studies of transformations occurring in a substratum with the composition of alkali-poor olivine tholeiite. This substratum was crystallized under a pressure of 2 GPa in the form of eclogite paragenesis (50 per cent of garnet + 50 per cent of pyroxene), while under pressures between 10 and 15 GPa and at a temperature of about 1300 K pyroxene disappeared and a monomineral garnet phase was crystallized in which silicon atoms are partly in octahedral coordination; thus the eclogite → garnetite transformation occurred, which is a specific case of the pyroxene → garnet transformation.

At depths of between 300 and 400 km in the upper mantle calcium pyroxenes are transformed into garnets (Ringwood, 1970). Conclusions relating to transformations of calcic garnets under high pressures can be drawn from transformations occurring in germanium analogues of garnets, as well as in solid solutions $CaSiO_3$–$CaGeO_3$ (Table 7.1.1). Ringwood (1975) maintains that at a depth of about 500 km in the upper mantle, under a pressure of about 16 GPa, the transformation of calcium–ferric garnets into a perovskite type structure occurs.

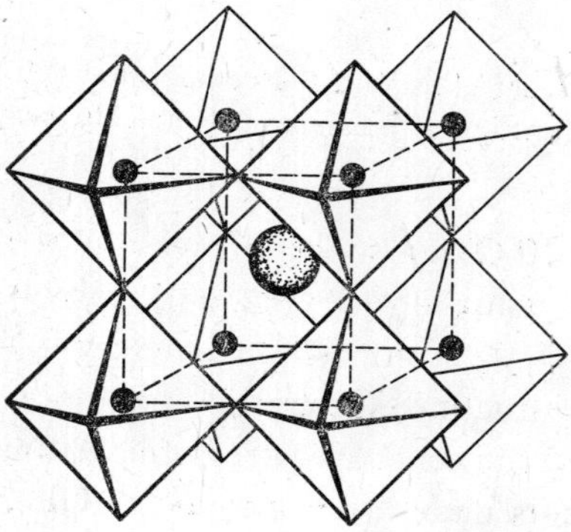

Fig. 7.6.3. Simplified structure of perovskite $CaTiO_3$, according to Náray-Szabó (1943). Octahedrons $[TiO_6]$ are marked by a solid line while the cube $[CaTi_8]$ is marked by a dashed line.

In a perovskite structure (Fig. 7.6.3) calcium and oxygen ions with large ionic radii form a lattice of the closest cubic packing, while the octahedral sites between them are occupied by titanium atoms. Calcium cations remain surrounded by 12 oxygen anions $[CaO_{12}]$. Under normal conditions Mg cannot enter a perovskite structure, as its ionic radius is too small as compared to the oxygen anion to form a lattice of the closest cubic packing with it. The magnesium garnet–pyrope transformation into a perovskite structure is possible only when the pressure is considerably higher than that needed for calcium garnet. The experimental transition of pyrope into a perovskite structure was obtained by Liu (1974) under a pressure of 30 GPa (Table 7.1.1). In his later experiments, carried out at temperatures from 1300 to 1700 K, Liu (1977b) could not obtain a direct transformation of pyrope into a perovskite structure, but he obtained the following series of phase transformations:

$$\text{pyrope} \xrightarrow[24\div25\ \text{GPa}]{} \text{ilmenite} \xrightarrow[30\ \text{GPa}]{} \text{perovskite} \qquad (7.6.1)$$

The transformation of magnesium garnet into an ilmenite structure occurs according to Ringwood (1975) in the upper parts of the lower mantle under a pressure of about 23 GPa. This transformation was initially studied in a two-component system of pyrope and its germanium analogue $Mg_3Al_2Si_3O_{12}$–$Mg_3Al_2Ge_3O_{12}$ (Ringwood, 1970). The existence of a continuous series of solid solutions of germanium garnets with pyrope was found, in the interval from 0 to 50 molecular per cent of $Mg_3Al_2Si_3O_{12}$, in which at a constant temperature the pressure of the garnet → ilmenite transformation increased with the share of the pyrope molecule. Ilmenite displays higher density than garnet of the same chemical composition since most of the silicon cations in garnet are situated in tetrahedra $[SiO_4]$, while in an ilmenite structure all the cations are of octahedral coordination. Liu (1977b) synthesized the solid solutions $MgSiO_3 \cdot xAl_2O_3$ with an ilmenite structure, where $0 \leqslant x \leqslant 25$ weight per cent. The ilmenite was formed as a result of the high-pressure transition of garnet from all substrata containing alumina (Al_2O_3). The transformation pressure increased with the share of the pyrope molecule from 22 GPa for 20 molecular per cent of pyrope to 25 GPa for 100 molecular per cent of pyrope.

Reid and Ringwood (1975) synthesized $ScAlO_3$ with the structure of orthorhombic perovskite at a temperature of 1300 K and under pressure of 12 GPA, in which Sc is in coordination 8 and aluminium is in coordination 6 by oxygen. Since, on the one hand, the ionic radius of Sc^{3+} is similar to those of Mg^{2+} and Fe^{2+}, and, on the other hand, the ionic radius of Al^{3+} is similar to that of Si^{4+}, $ScAlO_3$ is a close analogue of ferrous–magnesium metasilicates, which in the lower mantle should appear in a perovskite structure (Reid and Ringwood, 1975).

Liu (1976b) studied the consecutive phase transformations of $MgSiO_3$ experimentally at temperatures from 1300 to 1700 K and under pressures from 15 to 30 GPa, using new static experimental techniques, i.e., a diamond high-pressure chamber with laser heating. Glass with the composition of magnesium metasilicate was the initial material. From this substratum under a pressure of 15 GPa clinoenstatite was crystallized, which under a pressure of $17 \div 18$ GPa was broken down into a mixture of β spinel and stishovite. This transition was accompanied by a decrease of mole volume by 13.2 per cent. In the mixture with stishovite, under a pressure of 20 ± 1 GPa, β spinel was transformed into true γ spinel. A small decrease of mole volume by 1.7 per cent was observed. Magnesium metasilicate with an ilmenite structure was formed in all the experiments in a wide range of pressures changing from 17.5 to 27.5 GPa. The volume effect of the Mg_2SiO_4 (γ spinel)$+SiO_2$ (stishovite) $\rightarrow 2MgSiO_3$ (ilmenite) transition is -2.0 per cent. Under pressures of between 24 and 30 GPa mixed oxides (periclase+stishovite) were not obtained experimentally. Most probably therefore the $MgSiO_3$ polymorph with an ilmenite structure is directly transformed into a perovskite structure under pressure of about 25 GPa. The ilmenite → perovskite transition is accompanied by a decrease in mole volume by 6.6 per cent. The lack of mixed oxides in an experimental series of phase transformations of $MgSiO_3$ does not exclude, however, the possibility of the occurrence of their mixture as one of the links of a chain of transitions which natural ferrous–magnesium metasilicates undergo under the conditions dominating in the Earth's mantle. Recently Liu (1977a) has shown by means of an analysis of mole volume changes for the phase transitions of ilmenite → perovskite and mixed oxides → perovskite that oxides with the structure of NaCl, rutile and corundum do not form a stable mineral paragenesis under the conditions of the lower mantle. $MgSiO_3$ in the structure of orthorhombic perovskite has recently been synthesized by Ito and Matsui (1978) at a temperature of 1270 K and under a pressure of 28 GPa.

The series of high-pressure phase transformations in $MgSiO_3$ has, according to Liu (1977a), the following form:

$$\begin{array}{ccccc} 2MgSiO_3 & \xrightarrow[15.2\%]{} & Mg_2SiO_4 + SiO_2 & \xrightarrow[1.8\%]{} & Mg_2SiO_4 + SiO_2 \\ \text{clinoenstatite} & & \beta\ \text{phase} + \text{stishovite} & & \gamma\ \text{spinel} + \text{stishovite} \\ \varrho = 3.19 & & \bar{\varrho} = 3.67 & & \bar{\varrho} = 3.74 \end{array}$$

$$\begin{array}{ccccc} & \xrightarrow[2\%]{} & 2MgSiO_3 & \xrightarrow[7\%]{} & 2MgSiO_2 \\ & & \text{ilmenite} & & \text{perovskite} \\ & & \varrho = 3.81 & & \varrho = 4.08 \end{array} \qquad (7.6.2)$$

Under the arrows appear the values of $\Delta\varrho/\varrho$ percentage characterizing the density increase during each transformation. $MgSiO_3$ with a perovskite structure, displaying the closest packing of all the known magnesium silicates, is probably the main mineral of the Earth's lower mantle. The transition of metasilicates with a ilmenite structure into a perovskite structure (Fig. 7.6.3) occurs in the lower mantle at depths greater than 1000 km. The pressure of the ilmenite → perovskite transformation for solid solutions with the $MgSiO_3 \cdot xAl_2O_3$ composition increases with the content of Al_2O_3 from 25 GPa for 5 weight per cent of Al_2O_3 to over 30 GPa for 25 weight per cent of Al_2O_3 (Liu, 1977b).

7.7 PHASE TRANSFORMATIONS IN SODIUM ALUMINOSILICATES

Alkali metals in the Earth's crust are mainly concentrated in feldspars and other aluminosilicates, which are unstable under high pressures. Sodium in the upper mantle occurs in clinopyroxenes of the jadeite type. The disproportionation of albite into jadeite and quartz has been observed experimentally at a temperature of 1400 K and under a pressure of 2.8 GPa (Birch and Le Comte, 1960). The high-pressure disproportionation of jadeite has been investigated experimentally on its germanium analogue $NaAlGe_2O_6$ (Table 7.1.1). Jadeite is stable within a broad interval of pressure. According to Reid *et al.* (1967), when pressure is greater than 20 GPa, jadeite probably undergoes a transformation according to the formula

$$\underset{\text{jadeite}}{NaAlSi_2O_6} \rightarrow \underset{\text{calcium ferrite structure}}{NaAlSiO_4} + \underset{\text{stishovite}}{SiO_2} \tag{7.7.1}$$

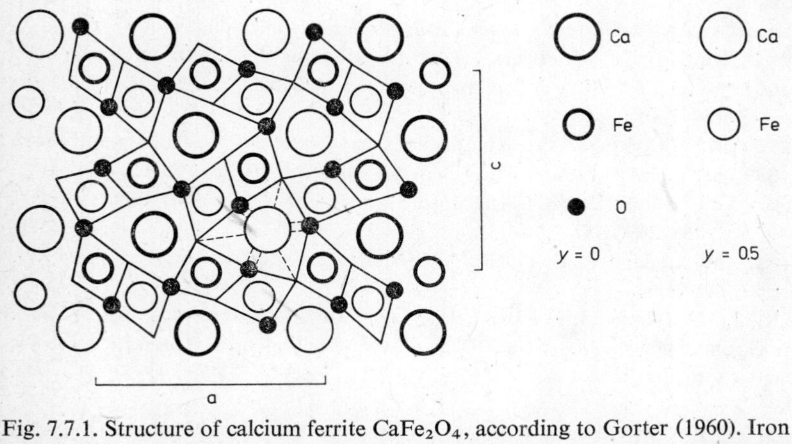

Fig. 7.7.1. Structure of calcium ferrite $CaFe_2O_4$, according to Gorter (1960). Iron occurs in octahedra $[FeO_6]$ while calcium occurs in "tunnels", with coordination number 8 or 9 with respect to oxygen.

This transformation was obtained experimentally by Liu (1978), at a temperature of 1270 K and under a pressure of 28 GPa. In $NaAlSiO_4$ with the structure of calcium ferrite (Fig. 7.7.1) the coordination number (CN) is 8 or 9 for Na atoms, and 6 for Al and Si. Minerals with the structure of calcium ferrite are significant components of the Earth's lower mantle, as their atoms are very closely packed (Reid *et al.*, 1967; Ringwood, 1975; Liu, 1978).

REFERENCES

Ahrens T. J. and Syono Y. (1967), "Calculated mineral reactions in Earth's mantle", *J. Geophys. Res.*, **72**, 4181–4188.

Akimoto S. I. (1972), "The system MgO–FeO–SiO_2 at high pressures and temperatures—phase equilibria and elastic properties", in: *The Upper Mantle*, A. R. Ritsema (ed.), Elsevier, Amsterdam; *Tectonophysics*, **13**, 161–187.

Akimoto S. I., Matsui Y. and Syono Y. (1976), "High-pressure crystal chemistry of orthosilicates and the formation of the mantle transition zone", in: *The Physics and Chemistry of Minerals and Rocks*, R. G. J. Strens (ed.), J. Wiley and Sons, London–New York–Sydney–Toronto, 327–363.

Akimoto S. I. and Syono Y. (1970), "High-pressure decomposition of the system $FeSiO_3$–$MgSiO_3$", *Phys. Earth Planet. Inter.*, **3**, 186–188.

Anderson D. L. (1967), "Phase changes in the upper mantle", *Science*, **157**, 1165–1173.

Anderson D. L. (1977), "Composition of the mantle and core", *Ann. Rev. Earth Planet. Sci.*, **5**, 179–202.

Archambeau C. B., Flint E. A. and Lambert D. G. (1969), "Fine structure of the upper mantle", *J. Geophys. Res.*, **74**, 5825–5865.

Bakun-Czubarow N. (1975), "Geochemistry of ultramafic rocks applied to geothermometry", *Publ. Inst. Geophys. Pol. Acad. Sci.*, **8** , 127–140.

Bakun-Czubarow N. (1980), "Pyroxene geothermometry applied to the Sowie Góry ultramafic rocks", *Publ. Inst. Geophys. Pol. Acad. Sci.*, A–8, **130**, 57–66.

Basset W. A. and Ming L. C. (1972), "Disproportionation of Fe_2SiO_4 to $2FeC+SiO_2$ at pressures up to 250 kb and temperatures up to 3000°C", *Phys. Earth Planet. Inter.*, **6**, 154–160.

Baur W. H. (1972), "Computer–simulated crystal structures of observed and hypothetical, Mg_2SiO_4 polymorphs of low and high density", *Am. Min.*, **57**, 709–731.

Birch F. and Le Comte P. (1960), "Temperature–pressure plane for albite composition", *Am. J. Sci.*, **258**, 209–217.

Boyd F. R. and England J. L. (1960), "The quartz–coesite transition", *J. Geophys. Res.*; **65**, 749–756.

Boyd F. R. and MacGregor I. D. (1964), *Ultramafic Rocks*, Carnegie Inst., **63**, 152–156.

Burns R. G. and Sung C. M. (1978), "The effect of crystal field stabilization on the olivine → spinel transition in the system Mg_2SiO_4–Fe_2SiO_4", *Phys. Chem. Minerals*, **2**, 349–364.

Coleman R. G., Lee D. E., Beatty L. B. and Brannock W. W. (1965), "Eclogites and eclogites; their differences and similarities", *Geol. Soc. Am. Bull.*, **76**, 483–508.

Gorter E. W. (1960), "Some structural relationships of ternary transition metal oxides", in: *Proc. 17th Int. Congress for Pure and Applied Chemistry, Munich, 1959*, **1**, Inorganic Chemistry, 303–328.

Green D. H. and Ringwood A. E. (1967): "An experimental investigation of the gabbro to eclogite transition and its petrological application", *Geochim. Cosmochim. Acta*, **31**, 767–833.

Ito K. and Kennedy G. C. (1970), "An experimental study of the basalt–garnet granulite–eclogite transition", Publ. No. 884. Institute of Geophysics and Planetary Physics, University of California.

Ito E. and Matsui Y. (1978), "Synthesis and crystal-chemical characterization of $MgSiO_3$ perovskite", *Earth Planet. Sci. Lett.*, **38**, 443–450.

Ito E., Matsui Y., Suito K. and Kawai N. (1974), "Synthesis of Mg_2SiO_4", *Phys. Earth Planet. Inter.*, **8**, 342–344.

Jackson I. M. S., Liebermann R. C. and Ringwood A. E. (1974), "Disproportionation of spinels to mixed oxides: significance of cation configuration and implications for the mantle", *Earth Planet. Sci. Lett.*, **24**, 203–208.

Kamb B. (1968), "Structural basis of the olivine-spinel stability", *Am. Min.*, **53**, 1439–1455.

Kern R. and Weisbrod A. (1967), *Thermodynamics for Geologists*, Freeman Cooper, San Francisco, California.

Kumazawa M., Sawamoto H., Ohtani E. and Masaki K. (1974), "Postspinel phase of forsterite and evolution of the Earth's mantle", Nature, **247**, 356–358.

Leliwa-Kopystyński J. and Bakun-Czubarow N. (1980), "The effect of material parameters on the shape of the phase separation surfaces within the Earth's mantle", *Phys. Earth Planet. Inter.*, **22**, 244–254.

Liu L. G. (1974), "Silicate perovskite from phase transformations of pyrope garnet at high pressure and temperature", *Geophys. Res. Lett.* **1**, 6, 277–280.

Liu L. G. (1975), "Disproportionation of Ni_2SiO_4 to stishovite plus bunsenite at high pressures and temperatures", *Earth Planet. Sci. Lett.*, **24**, 357–362.

Liu L. G. (1976a), "The post-spinel phase of forsterite", *Nature*, **262**, 770–772.

Liu L. G. (1976b), "The high-pressure phases of $MgSiO_3$", *Earth Planet. Sci. Lett.* **31**, 200–208.

Liu L. G. (1977a), "Post-ilmenite phases of silicates and germanates", *Earth Planet. Sci. Lett.*, **35**, 161–168.

Liu L. G. (1977b), "Ilmenite-type solid solutions between $MgSiO_3$ and Al_2O_3 and some structural systematics among ilmenite compounds", *Geochim. Cosmochim. Acta*, **41**, 1355–1361.

Liu L. G. (1978), "High-pressure phase transformations of albite, jadeite and nepheline" *Earth Planet. Sci. Lett.*, **37**, 438–444.

MacGregor I. D. (1967), "Mineralogy of model mantle compositions", in: *Ultramafic and Related Rocks*, P. J. Wyllie (ed.), John Wiley and Sons, New York–London–Sydney, 382–393.

MacGregor I. D. (1970), "The effect of CaO, Cr_2O_3, Fe_2O_3 and Al_2O_3 on the stability of spinel and garnet peridotites", *Phys. Earth Planet. Inter.*, **3**, 372–377.

MacGregor I. D. (1974), "The system $MgO–Al_2O_3–SiO_2$: Solubility of Al_2O_3 in enstatite for spinel and garnet peridotite compositions", *Am. Min.*, **59**, 110–119.

Mercier J. C. and Carter N. L., (1975) "Pyroxene geotherms", *J. Geophys. Res.*, **80**, 3349–3362.

Minear J. W. and Toksöz M. N. (1970a) "Thermal regime of downgoing slab", *Tectonophysics*, **10**, 367–390.

Minear J. W. and Toksöz M. N. (1970b), "Thermal regime of downgoing slab and new global tectonics", *J. Geophys. Res.*, **75**, 1397–1419.

Moore P. B. and Smith J. V. (1970), "Crystal structure of β Mg_2SiO_4—crystal-chemical and geophysical implications", *Phys. Earth Planet. Inter.*, **3**, 166–177.

Morimoto N., Tokonami M., Watanabe M. and Koto K. (1974), "Crystal structures of three polymorphs of Co_2SiO_4", *Am. Min.*, **59**, 475–485.

Mueller R. F. and Saxena S. K. (1977), *Chemical Petrology with Applications to the Terrestrial Planets and Meteorites*, Springer-Verlag, New York–Heidelberg–Berlin.

Náray-Szabó St. V. (1943), "Der Strukturtyp des Perovskits $CaTiO_3$", *Naturwiss.*, **31**, 202–203.

Navrotsky A. (1977), "Calculation of effect of cation disorder on silicate spinel phase boundaries", *Earth Planet. Sci. Lett.*, **33**, 437–442.

Navrotsky A. and Kasper R. S. (1976), "Spinel disproportionation at pressure: calorimetric determination of enthalpy of formation of Mg_2SnO_4 and Co_2SnO_4 and some implications for silicates", *Earth Planet. Sci. Lett.*, **31**, 247–254.

O'Hara M. J., Richardson S. W. and Wilson G. (1971), "Garnet–peridotite stability and occurence in crust and mantle", *Contrib. Mineral. Petrol.*, **32**, 48–68.

Perchuk L. L. (1970), *Equilibria of Rockforming Minerals*, Nauka, Moskva (in Russian).

Perchuk L. L. and Ryabchikov I. D. (1976), *Phase Correspondence in Mineral Systems*, Nedra, Moskva (in Russian).

Press F. (1970), "Earth models consistent with geophysical data", *Phys. Earth Planet. Inter.*, **3**, 3–22.

Press G. (1972), "The Earth's interior as inferred from a family of models", in: *The Nature of the Solid Earth*, E. C. Robertson (ed.), McGraw-Hill, New York, 147–171.

Reid A. F. and Ringwood A. E. (1975), "High-pressure modification of $ScAlO_3$ and some geophysical implications", *J. Geophys. Res.*, **80**, 3363–3370.

Reid A. F., Wadsley A. D. and Ringwood A. E. (1967), "High-pressure $NaAlGeO_4$ a calcium ferrite isomorph and model structure for silicates at depth in the Earth's mantle", *Acta Cryst.*, **23**, 736–739.

Ringwood A. E. (1966), "Mineralogy of the mantle", in: *Advances in Earth Science*, P. M. Hurley (ed.), MIT Press, Boston, 357–399.

Ringwood A. E. (1970), "Phase transformations and the constitution of the mantle", *Phys. Earth Planet. Inter.*, **3**, 109–155.

Ringwood A. E. (1973), "Phase transformations and their bearing on the dynamics of the mantle", *Fortschr. Min.*, **50**, 113–139.

Ringwood A. E. (1975), *Composition and Petrology of the Earth's Mantle*, McGraw-Hill, New York.

Ringwood A. E. (1976), "Phase transformations in descending plates and implications for mantle dynamics", *Tectonophysics*, **32**, 129–143.

Ringwood A. E. and Green D. H. (1966), "An experimental investigation of the gabbro-eclogite transformation and some geophysical implicatons", *Tectonophysics*, **3**, 383–427.

Ringwood A. E., MacGregor I. D. and Boyd F. R. (1964), *Petrological Constitution of the Upper Mantle*, Carnegie Inst., **63**, 147–152.

Ringwood A. E. and A. Major (1970), "The system Mg_2SiO_4–Fe_2SiO_4 at high pressures and temperatures", *Phys. Earth Planet. Inter.*, **3**, 89–108.

Schubert C., Yuen D. A. and Turcotte D. L. (1975), "Role of phase transitions in a dynamic mantle". *Geophys. J. R. Astron. Soc.*, **42**, 705–735.

Smulikowski K. (1964), "An attempt at eclogite classification", *Bull. Acad. Pol. Sci., Sér. Sci. Géol. Géogr.*, **XII**, 1.

Stacey F. D. (1977), "A thermal model of the Earth", *Phys. Earth Planet. Inter.*, **15**, 341–348.

Stishov S. M. (1963), "On the equilibrium line between coesite and rutile-like silica polymorph", *DAN SSSR*, **148**, 1186–1188 (in Russian).

Sung C. M. and Burns R. G. (1976), "Kinetics of the olivine–spinel transition: implication to deep-focus earthquake genesis", *Earth Planet. Sci Lett.*, **32**, 165–170.

Sung C. M. and Burns R. G. (1978), "Crystal structural features of the olivine → spinel transitions", *Phys. Chem. Minerals*, **2**, 177–197.

Weaver J. S., Chipman D. W. and Takahashi T. (1979), "Comparison between thermochemical and phase stability data for the quartz–coesite–stishovite transformations", *Am. Min.*, **64**, 604–614.

Yagi T., Marumo F. and Akimoto S. J. (1974), "Crystal structures of spinel polymorphs of Fe_2SiO_4 and Ni_2SiO_4", *Am. Min.*, **59**, 486–490.

Yoder H. S. (1950), "High-low quartz inversion up to 10 000 bars", *Am. Geophys. Union Trans.*, **31**, 827–835.

8. Petrology and Elements of Geochemistry

8.1 INTRODUCTION

The chemical composition of the Earth can be estimated under the assumption that there is a unity of matter in the solar system and in the universe. Meteorites are an invaluable source of information about the composition of the Earth's interior, inaccessible to direct studies. The composition of primitive matter forming the Earth before its zonal differentiation was probably very similar to that of unique carbonaceous chondrites, which are extremely poor in the metallic phase but rich in water, carbon and carbon compounds. The composition of carbonaceous chondrites does not differ essentially either from the average composition of all chon-

TABLE 8.1.1
Average composition of meteorites and the whole Earth (in weight per cent)

Element	Average composition of chondrites (after Ronov and Yaroshevsky, 1972)	Average composition of meteorites (after Heide, 1957)	The Earth (after Mason, 1966)	The Earth (after Polański and Smulikowski, 1969)
Fe	25.0	30.7	35	40.91
O	35.0	31.6	30	25.92
Si	18.0	15.8	15	13.00
Mg	14.0	12.2	13	10.00
Ni	1.35	1.45	2.4	2.78
S	2.0	2.7	1.9	2.74
Ca	1.4	1.5	1.1	1.25
Al	1.3	1.3	1.1	1.10
Na	0.7	0.6	0.57	0.49
Cr	0.25	0.28	0.26	0.22
Co	0.08	0.11	0.13	0.21
P	0.05	0.10	0.10	0.06
K	0.08	0.15	0.07	0.13
Ti	0.05	0.14	0.05	0.11
Mn	0.20	0.24	0.22	0.18

drites or from the average composition of all types of meteorites, both metallic and stone ones (Table 8.1.1). The Earth's primitive matter was differentiated into a metallic core and a silicate mantle, from which the Earth's crust was subsequently separated by partial melting. Any model of the global chemical composition of the Earth must take into account the Earth's subdivision into three main geospheres as well as the possibility of the extraction of magma from the mantle by partial melting. The chemical composition of the entire Earth was estimated by Mason (1966), who assumed that the chemistry of the Earth's core is that of the meteorite metallic phase, the Earth's mantle is of peridotite composition, and the weight ratio of metals to sulphides and silicates is 3.6 : 1 : 7.9. The material forming the crust, atmosphere, and hydrosphere, being quantitatively subordinate, was neglected. Mason's model was corrected by Polański and Smulikowski (1969) who assumed that the silicate composition of the mantle corresponds to the average composition of the silicate phase of meteorites (Table 8.1.1).

The main views on the composition of consecutive geospheres and on selected processes of the exchange of matter between the geospheres are considered below, with special regard to magmatic processes which control the differentiation of the primitive matter of the Earth's mantle.

8.2 THE EARTH'S CORE

The Earth's core is separated from the outer geospheres by a sharp geophysical discontinuity at a depth of 2900 km, where an abrupt change of density from 5.5 to 10 $g \cdot cm^{-3}$ takes place. The core contains 1/3 of the mass of the Earth, occupying merely 1/7 of its volume. The pressure at the core centre is about 0.36 TPa and the temperature about 4300 K, while at the boundary between the core and the lower mantle the pressure is about 0.14 TPa and the temperature is about 3000 K (Brett, 1976; Stacey, 1977). The solid inner core with a radius of 1250 km is separated from the liquid outer core by a transition zone with a thickness varying in different models from 0 to 950 km.

The boundary between the lower mantle and the Earth's core is recognized as a discontinuity in chemical composition and consistency state between the liquid metallic outer core and the lower mantle composed mainly of silicon, magnesium and iron oxides. There are also hypotheses, however, that this discontinuity is of a phase transformation nature. The hypothesis of Ramsey (1949) is particulary interesting. He believes that the boundary between the core and the mantle corresponds to structural changes of matter while its chemical composition remains unchanged, and that the main components of the mantle are transformed into the metallic state under high pressure present at a depth of 2900 km. Ramsey's hypothesis, however, is disproved by the results of high-pressure experiments with shock waves, by the results of Birch's studies (1963) of the relations between hydrodynamic sound velocity and density for various metals, olivines and materials forming the Earth's core and mantle, and by the relations between the elastic properties and the density of various materials under increasing pressure and during isochemical phase transitions, obtained experimentally and theoretically (Ringwood, 1979).

In shock wave experiments, magnesium orthosilicate—Mg_2SiO_4 and crystalline silica—SiO_2 were exposed to pressure reaching 0.5 TPa, which is more than three times as high as the pressure existing at the boundary between the core and the mantle, but they were not transformed into a metallic state.

From Birch's studies (see Fig. 1.10.1) it follows that the mean atomic weight of the core matter should be close to the atomic weight of V and Cr on the one hand and to Fe and Co on the other. From this fact and from cosmochemical data it follows in turn that iron is the main component of the Earth's core.

From studies of elastic properties as a function of density for various materials it has been found that the bulk modulus increases abruptly with density, and this can be described by the formula

$$K \propto \varrho^n \tag{8.2.1}$$

where most often $3 < n < 4$. The bulk moduli of the lower mantle and outer core are approximately the same, while the densities differ considerably. This contradicts the hypothesis that isochemical phase transitions can occur at the boundary between these two geospheres. That would necessitate a value of the bulk modulus in the outer core several times as high as that in the lower mantle.

A rather interesting hypothesis, though not having many supporters now, was proposed by Bullen (1973), who considered density distributions in the interior of the Earth and in planetary interiors. According to that hypothesis the outer cores of planets of the type of the Earth are composed of Fe_2O, unstable under normal conditions, while their inner cores are composed of metallic iron. High-pressure experiments, however, have not confirmed so far the existence of Fe_2O, and this is the main drawback of Bullen's hypothesis.

Ringwood (1979) has drawn attention to the possibility of the presence of FeO in considerable amounts in the outer core. Analysing metallurgical data related to the Fe–O system, he found that the solubility of FeO in molten iron increases abruptly with temperature, and at a temperature corresponding to that at the boundary between the mantle and the core it reaches 40 molecular per cent. An increase in pressure is also favourable for the increase of FeO solubility in melted Fe. Considering the density distribution, Ringwood estimated that the contribution of FeO in the core is 44 ± 16 weight per cent, which computed in terms of oxygen is 10 ± 4 weight per cent.

The hypothesis that metallic iron is the main component of the Earth's core is generally accepted. From seismic wave velocities, hydrodynamic sound velocities, and density distributions it follows that the outer core should contain up to 20 weight per cent of elements lighter than iron, while the inner core should contain up to 50 weight per cent of nickel. The acceptance of the cosmic ratio of Fe/Ni for the Earth leads to the conclusion that nickel should constitute in the entire core 10 weight per cent of iron. In such a model the mean atomic mass of the core matter is 51.9 $g \cdot mole^{-1}$ (Anderson, 1977).

The density of the liquid outer core is about 10% higher than that of liquid iron under the same pressure and temperature, while the density of the lower mantle at the core boundary is higher by about 10% than that corresponding to its hypothetical mineral composition. This difference is simply explained by the hypothesis of Alder (1972), according to which magnesium and silicon in periclase and stishovite can be replaced by iron, while the released magnesium and silicon can be dissolved in liquid iron. Furthermore, iron and its oxide can form solid solutions with magnesium and silicon.

An interesting hypothesis concerning the chemical composition of the outer core, based on analogies with chondrites, was recently proposed by Herndon (1980). He noted a consistent differentiation of the chemical composition of chondrites according to their oxygen content. Carbonaceous chondrites, rich in oxygen, are composed of silicates of magnesium and iron and sulphides of iron and nickel, but they are devoid of the metallic phase. Ordinary chondrites, which are poorer in oxygen, are composed of magnesium and iron silicates, iron sulphide, and the iron–nickel metallic phase. Finally, enstatite chondrites, which are the poorest in oxygen, contain metasilicate of magnesium which is devoid of iron, sulphides of calcium, magnesium, manganese and iron, the metallic phase in the form of a Fe–Ni–Si alloy, and nickel silicide Ni_2Si. According to Herndon, the primitive matter of the Earth's interior was poor in oxygen and therefore silicon was easily dissolvable in liquid Fe–Ni. The silicide of nickel was in turn crystallized in a liquid Fe–Ni–Si alloy, and then accumulated gravitationally in the Earth's centre, forming its inner core.

The nature of elements lighter than iron, which are possible components of the core, needs a more extensive description. These elements should fulfil the following requirements:

(1) forming alloys with iron,

(2) high abundance in the universe,

(3) greater tendency to appear in the metallic phase than in silicate or oxide phases.

These conditions are fulfilled by H, C, N, O, Mg, Si and S. From these elements, H, C and N should be eliminated since they form additive solutions with iron, which means that they fill the gaps between the atoms of iron, and thus their presence would increase the density of the core. Oxygen can also form additive solutions with iron, though its contribution to the composition of the core in the form of FeO dissolved in liquid iron is highly probable. It was shown in laboratory experiments by Ahrens *et al.* (1978) that oxygen in this form can enter into the composition of the core to the amount of 12 weight per cent, which agrees well with the estimation of Ringwood (1979). The concentration of magnesium in the core is improbable since this element shows a much stronger geochemical affinity to oxygen than silicon, and its accumulation in the core would have been preceded by a much more intensive concentration of silicon. Of the remaining two light elements, i.e., sulphur and silicon, sulphur, by analogy to chondrites, appears to be the most probable component of the core, diminishing its density. According to Ahrens *et al.* (1978) and Ahrens (1979), the share of sulphur in the core material is from 9 to 12 weight per cent. According to Anderson (1977), the elastic properties of the core correspond to a composition in which the amount of sulphur constitutes 11 weight per cent of the amount of the alloy iron and nickel appearing in cosmic proportions. In such chemical composition of the core the value of the ratio Fe/S lies in the interval of its usual variability in chondrites.

The possibility of potassium concentration in the Earth's core should be underlined. Under reduction conditions potassium reveals chalcophile tendencies to form chemical compounds with sulphur. In experimental studies of the partitioning of potassium between silicates and the Fe–FeS alloy it was shown that the sulphide-metallic phase can accumulate an amount of potassium sufficient to release by the radioactive decay of ^{40}K the energy necessary to generate convection in the Earth's core (Jacobs, 1975).

8.3 THE EARTH'S MANTLE

The Earth's mantle is the largest geosphere. It occupies about 4/5 of the volume of the globe and comprises 68% of its mass. The mantle extends from the Mohorovičić discontinuity down to 2900 km, that is to the boundary of the outer core. This geosphere is distinctly dual. Its upper part, contained between the Mohorovčić discontinuity and the discontinuity at a depth of 680 km, is called the upper mantle, while its inner part down to a depth of 2900 km is called the lower mantle. Ringwood (1975, 1979) distinguishes a transition zone 600 km thick between the upper and the lower mantle at depths from 400 to 1000 km. It is the zone of main phase transformations.

A hypothetical chemical and mineral composition of the Earth's mantle, especially of its upper part, can be estimated from:

(1) the balance of matter between the upper mantle and the Earth's crust,

(2) extraterrestrial analogies,

(3) the composition of rocks and minerals accessible to direct studies and most probably originating from the upper mantle,

(4) data of experimental petrology,

(5) seismic data.

The balance of matter in external geospheres was made by Ringwood (1966) under the assumption that the composition of the primitive mantle corresponds to the chemical composition of the differentiated upper mantle, taken jointly with the Earth's crust which was extracted from the mantle. The average composition of the primitive undifferentiated mantle, from which basaltic magmas could be extracted, would correspond to a mixture of one part of tholeiitic basalt and three parts of residual peridotite, remaining after the extraction of basaltic magma. This hypothetical material of the Earth's upper mantle, an alumina-bearing perdotite, has been called *pyrolite* because of its most probable mineral composition—pyroxene+olivine. The chemical composition of pyrolite (Table 8.3.1) found in this way most probably shows overestimated quantities of Ti, Na, K and P. The principle that the upper mantle peridotites and basaltic magmas are complementary was also used, in an extended form, by Ringwood (1975, 1979) (Table 8.3.1) in his successive attempts to estimate the chemical composition of pyrolite. He studied the chemical equilibrium between various types of basalts and residual peridotites using the results of experiments and taking into account the partitioning of trace elements, mainly of rare earth elements, between complementary rocks. The chemical composition of basaltic magmas depends on the degree of partial melting of

TABLE 8.3.1
Composition of the Earth's upper mantle (in weight per cent)

Oxides	Pyrolite (after Ringwood)			Chondritic model of the Earth's primitive mantle (Ringwood, 1979)	Ophiolitic complex of Vourinos, Greece (Ringwood, 1975)	Earth's upper mantle (Middlemost, 1975)	Earth's upper mantle (Harris *et al.*, 1967)	Earth's primitive upper mantle (Maaløe and Aoki, 1977)	Xenoliths of garnetiferous lherzolites from kimberlites (Carswell, 1980)
	1966	1975	1979						
SiO_2	45.16	45.1	45.1	48.2	45.6	44.43	44.2	44.71	45.89
TiO_2	0.71	0.2	0.2	0.15	0.2	0.28	0.1	0.16	0.09
Al_2O_3	3.54	4.6	3.3	3.5	3.9	3.31	2.7	2.46	1.57
Cr_2O_3	0.43	0.3	0.4	0.7	0.4	0.34	0.30	0.42	0.32
Fe_2O_3	0.46	0.3				1.13	1.1		
FeO	8.04	7.6	8.0*	8.1*	5.2*	7.85	7.3	8.15*	6.91*
MnO	0.14	0.1	0.15	0.5	0.1	0.14	0.15	0.18	0.11
NiO	0.20	0.2			0.3	0.27	0.20	0.26	0.29
MgO	37.47	38.1	38.1	34.0	37.7	38.96	41.30	41.00	43.46
CaO	3.08	3.1	3.1	3.3	3.0	2.74	2.40	2.42	1.16
Na_2O	0.57	0.4	0.4	1.6	0.5	0.39	0.25	0.29	0.16
K_2O	0.13	0.02	0.03	0.15	0.1	0.08	0.015	0.09	0.12
P_2O_5	0.06	0.02			0.03	0.08		0.06	0.04

* Total iron as FeO.

the pyrolite mantle. As a result of slight, $1\div5$ per cent melting of the mantle matter nephelinite and basanite magmas are generated; alkali basalts are the result of $5\div10$ per cent melting of pyrolite; a $15\div25$ per cent melting provides tholeiites; and the most intensive, $30\div60$ per cent melting generates picritic and komatiitic magmas. Garnetiferous or spinel lherzolites are complementary peridotites to nephelinites, basanites and alkali basalts, harzburgites form residua after the extraction of tholeiitic magmas, and dunites are residua after the extraction of picrites and komatiites. The chemical composition of pyrolite was estimated by combining the chemical composition of various basalt types and complementary peridotites in appropriate proportions. The composition changes slightly with regard to the content of the main and the compatible elements, while the content of the incompatible elements varies significantly. Depending on p and T conditions and on the partial pressure of water, pyrolite can occur in five mineral varieties:

(1) plagioclase pyrolite: olivine+orthopyroxene+clinopyroxene+plagioclase,

(2) spinel pyrolite: olivine+orthopyroxene+clinopyroxene+spinel,

(3) pyroxene pyrolite: olivine+alumina-bearing orthopyroxene++alumina-bearing clinopyroxene,

(4) garnet pyrolite: olivine+orthopyroxene+clinopyroxene+garnet,

(5) ampholite: olivine+amphibole.

The stability of the different varieties of pyrolite and the phase transformations between them were described in Chapter 7.

Extraterrestrial analogies of the Earth's mantle, with regard to the chemical composition, are stone meteorites called *chondrites*. In a chondritic model, according to geophysical and geochemical data, the composition of the whole Earth corresponds to carbonaceous chondrites. The mantle composition can be found from this model by eliminating the metallic iron concentrated in the core and by subtracting the reduced part of ferrous oxide in order to obtain the value of the ratio $FeO/(FeO+MgO) = 0.12$, characteristic for the Earth's mantle (Ringwood, 1966; 1975; 1979). In the chemical composition of the primitive mantle estimated in this way the content of most of the major elements is not significantly different from their concentration in pyrolite (Table 8.3.1). Relative concentrations of the components most difficult to volatilize under cosmochemical conditions, Al_2O_3 and CaO, in relation to MgO and SiO_2, do not differ significantly in the two models either. Pyrolite is poorer in relative content in relation to MgO, SiO_2, Cr_2O_3, MnO, Na_2O, and K_2O in comparison with the primitive mantle in a chondritic model. Under cosmochemical conditions these components are characterized by greater volatility than that of MgO, the volatility increasing in the above series. They may have already differentiated in relation to silicates of magnesium in the accretion process of globe matter.

The values of the ratio Fe^{3+}/Fe^{2+} in the rocks and minerals of the Earth's mantle is between 0.05 and 0.1, which is at least one order of magnitude higher than the characteristic value of silicate and oxide minerals coexisting in equilibrium with metallic iron. Moreover, H_2O and CO_2 dominate over H_2 and CO in fluids arising from the mantle, and this corresponds to a relatively high degree of oxidation of the mantle minerals which most probably were not in the state of equilibrium with metallic iron during the formation of the Earth's core.

The ultramafic rocks of the upper mantle accessible to direct studies are not in general representative of its primitive matter, as they originate from materials poor, in different degrees, in easily melting components. Among these rocks several groups can be distinguished:

1. Peridotites of the alpine type, occurring in linear folded zones, where they most probably penetrated from the mantle during the most intensive folding of deep geosynclines.

2. Ophiolite complexes which are the remnants of ancient oceanic lithosphere and consist of tectonized peridotites, assemblages of layered basic and ultrabasic rocks, and basic and intermediate volcanic rocks. The average composition of the rocks of the whole ophiolite complex is close to that of the primitive matter of the mantle (e.g., the Vourinos ophiolitic complex in Greece, Table 8.3.1).

3. Tectonic xenoliths of garnetiferous peridotites in old crystalline granulite complexes.

4. Ultramafic enclaves in kimberlites and ultramafic parageneses, occurring in the form of inclusions in natural diamonds.

5. Ultramafic enclaves in alkali olivine basalts, among which xenoliths of spinel lherzolites are predominant. Lherzolite enclaves are in chemical contrast with the surrounding alkali basalts. Furthermore, they show traces of mechanic deformations and parallel textures, and therefore probably could not have originated by fractional crystallization from basaltic magma. These xenoliths are probably fragments of the products of an early differentiation of the mantle matter, i.e., fragments of residual rocks which remained after the first basaltic magmas were extracted from the mantle.

Among the methods of estimation of the mantle composition based on ultramafic rocks accessible to direct investigations the methods used by

Harris *et al.* (1967, 1972) and by Maaløe and Aoki (1977) are of special interest. The first group of authors estimated the upper mantle composition using those ultramafic rocks, accessible to direct studies and originating from the mantle, which most probably were not poor in easily melting components. Maaløe and Aoki, on the other hand, assumed that spinel lherzolites originating in the mantle, which are characterized by a chondritic distribution of the rare earth elements, represent the primitive mantle of the Earth. They estimated statistically the chemical composition of the lower lithosphere. The composition of the ultrabasic upper mantle estimated by different methods is contained in a narrow range of variation of the main rock-forming oxides. Small systematic differences appear only between the pyrolite model and models based on the data obtained for mantle ultrabasic rocks accessible to direct investigations. The upper mantle composition in ultrabasic rock models (Harris *et al.*, 1967, 1972; Maaløe and Aoki, 1977; Middlemost, 1975) is in comparison with the pyrolite model less rich in SiO_2, CaO and Na_2O, but more rich in MgO, and this could imply that ultrabasites accessible to direct investigation have undergone an initial partial melting process, and thus have features of residual material as compared with the primitive mantle of the Earth.

Yoder (1976), who studied the sources and melting conditions of basaltic magmas, concluded that the main components of the upper mantle are most probably formed by garnetiferous peridotites accessible to direct investigation, which have stable mineral paragenesis under high pressures and a chemical composition close to the composition of chondrites. They are characterized, furthermore, by the densities and velocities of seismic waves typical for the uppermost mantle of the Earth.

Some aspects of the application of high-pressure experimental petrology to the estimation of the composition and state of the mantle material were considered in Chapter 7.

The upper mantle of the Earth is distinctly differentiated along its vertical cross-section, both chemically and in its mineral composition. The upper parts of the upper mantle are composed mainly of residual peridotites (dunites, harzburgites and some lherzolites) forming a layer called the peridotitic layer, while the deeper parts of the mantle approach nondifferentiated pyrolite as to their composition. Residual peridotites are significantly impoverished as regards easily melting components and incompatible elements. On the other hand, they are relatively rich in some transition elements. The tendency of Ni^{2+} and Cr^{3+} to concentrate in residual ultrabasic rock in the mantle can be explained by the crystal field theory (Burns, 1973). These cations are characterized by a high site preference energy in octahedral coordination. Consequently, during the partial melting process they are not passed to the melt, in which a partial drop of their coordination number to 4 would inevitably occur.

Regional differences in the upper mantle are also observed. For example, from analyses of ultramafic enclaves in kimberlites it follows that the upper mantle under the south African shield is richer in iron than under the Siberian shield.

Moreover, from an analysis of lherzolite enclaves of the upper mantle origin, made by Carswell (1980), it follows that of all the areas with kimberlite and alkali–basalt magmatic activity the Earth's upper mantle beneath Southern Africa and Yakutia is the most impoverished as regards elements easily extracted into basaltic magmas. This observation is in good agreement with the residual character of the chemical composition of garnetiferous lherzolites forming xenoliths in kimberlites (Table 8.3.1). The average chemical composition of these lherzolites is distinctly impoverished with regard to Al_2O_3, CaO and Na_2O in comparison with the upper mantle composition estimated by other methods.

The primitive mantle was most probably chemically homogeneous. However, there are differences of opinion as to whether the chemical composition of the contemporary upper mantle differs significantly from that of the lower mantle, that is whether the evolution of the mantle during 3 or 4 milliard years could lead to a significant differentiation in its Fe/Mg ratio. The value of the Fe/Mg ratio is approximately constant in the whole mantle according to Ahrens (1972) and Ringwood (1975), while according to Kumazawa *et al.* (1974) the discontinuity at a depth of 680 km dividing the upper mantle from the lower mantle is of a double character. On the one hand this discontinuity corresponds to a spinel → oxides phase transformation (see Chapter 7), and on the other hand it is a discontinuity of chemical composition caused by migration of iron to the lower mantle and of magnesium in the opposite direction. Ringwood (1979) believes that the whole of the Earth's mantle down to a depth of 2700 km, considered on a global scale, is chemically uniform, and its state and transformations can be explained by the pyrolite model.

An interesting hypothesis concerning the irreversible differentiation of the upper mantle material has recently been proposed by Anderson (1979 a,b). He believes that the top part of the Earth's upper mantle down to a depth of 220 km is composed of alumina-poor garnetiferous peridotites. Beneath this area there extends another zone, reaching depths of 670 km, of alumina-rich eclogites and garnetites. The model of Don Anderson is in good agreement with the distributions of density and seismic velocities as well as with the heat flow and geochemical data observed in the mantle.

According to Ringwood (1975) the evolution of the mantle, that is the extraction of basalt magmas and the growth of the Earth's crust on the one hand, and the sinking of the lithospheric plates into the mantle in the subduction zones on the other hand, has caused significant and irreversible differentiations of the mantle material.

8.4 IGNEOUS PROCESSES AND ROCKS

In the history of the Earth igneous rocks were the primary components of the Earth's crust. The criteria of their classification are based on their texture and fabric on the one hand, and on the main characteristics of the substance which forms them on the other hand. According to the conditions and depths of solidification of magma within the Earth's crust we divide igneous rocks into two groups:

1. Plutonic rocks, which crystallized slowly at great depths, under high pressures and at great pressures of volatiles. They are as a rule coarse-grained rocks.

2. Volcanic rocks, which solidified from the magma which poured out into the Earth's surface and rapidly cooled under atmospheric pressure. They are fine-grained rocks, often partially glassy.

Two intermediate types of secondary importance are often distinguished between plutonic and volcanic rocks. They are subvolcanic rocks, which crystallized from the magma just under the Earth's surface, and hypabyssal rocks, which solidified at intermediate depths.

The classification of igneous rocks, following their substantial characteristics, can be based on either their chemical or their mineral composition. The oldest chemical classification is the commonly used, though not very accurate, division of rocks into four groups, differing in the absolute content of SiO_2. The content of SiO_2 in these groups varies within the following intervals (Hyndman, 1972):

acid rocks	above	66	weight per cent of SiO_2
intermediate rocks		$52 \div 66$	weight per cent of SiO_2
basic rocks		$45 \div 52$	weight per cent of SiO_2
ultrabasic rocks	below	45	weight per cent of SiO_2

A discussion initiated by Streckeisen (1967) led to a unification of the most comprehensive and precise classification of igneous rocks, based on their mineral composition. The use of this classification is recommended by the Subcommission on the Systematics of Igneous Rocks of the Interna-

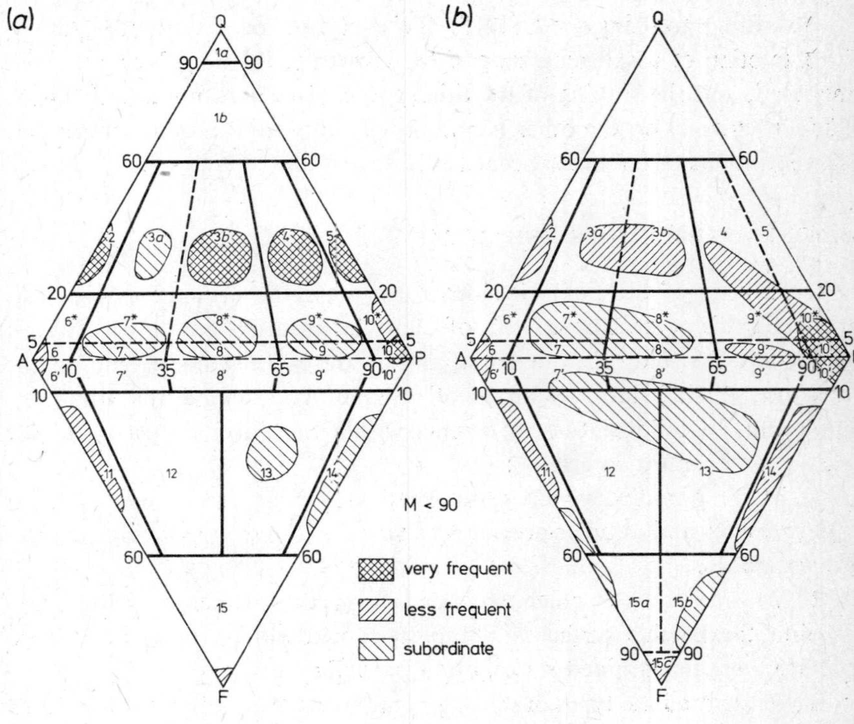

Fig. 8.4.1. General classification and nomenclature of igneous rocks, based on their modes, according to the recommendations of the IUGS Subcommission on the Systematics of Igneous Rocks (1973, 1978); (*a*) plutonic rocks: 1*a*—quartzolite, 1*b*—quartz rich granitoids, 2—alkali feldspar granite, 3*a*+3*b*—granite, 4—granodiorite, 5—tonalite, 6*—quartz alkali-feldspar syenite, 6—alkali-feldspar syenite, 6′—foid-bearing alkali-feldspar syenite, 7*—quartz syenite, 7—syenite, 7′—foid-bearing syenite, 8*—quartz monzonite, 8—monzonite, 8′—foid-bearing monzonite, 9*—quartz monzodiorite or quartz monzogabbro, 9—monzodiorite or monzogabbro, 9′—foid-bearing monzodiorite or monzogabbro, 10*—quartz diorite or quartz gabbro, 10—diorite or gabbro, 10′—foid-bearing diorite or gabbro, 11—foid syenite, 12—foid plagiosyenite or monzosyenite, 13—foid monzodiorite or foid monzogabbro, 14—foid diorite or foid gabbro, 15—foidolites; (*b*) volcanic rocks: 2—alkali-feldspar rhyolite, 3*a*+3*b*—rhyolite, 4++5—dacite, 6*—quartz alkali-feldspar trachyte, 6—alkali-feldspar trachyte, 6′—foid-bearing alkali-feldspar trachyte, 7*—quartz trachyte, 7—trachyte, 7′—foid-bearing trachyte, 8*—quartz latite, 8—latite, 8′—foid-bearing latite, (9*+9+9′)+(10*+10++10′)—andesite and basalt, 11—phonolite, 12—tephritic phonolite, 13—phonolitic tephrite and phonolitic basanite, 14—tephrite and basanite, 15*a*—phonolitic foidite, 15*b*—tephritic foidite, 15*c*—foidite. Parameters of the graphical scheme QAPF: Q—quartz, A—alkali feldspar, P—plagioclase, F—feldspathoid = foid.

tional Union of Geological Sciences (1973, 1978). The rock systematics under consideration, shown schematically in Fig. 8.4.1, is based on the following mineral groups, expressed in volume per cent:

Q—minerals of silica (quartz, tridymite, cristobalite),

A—alkali feldspars (orthoclase, microcline, sanidine, perthite, anorthoclase, albite An_{0-5}),

P—plagioclases (An_{5-100}, scapolite),

F—feldspathoids (nepheline, leucite, sodalite, nosean, hauyne, analcime, cancrinite and others),

M—mafic minerals and others (micas, amphiboles, pyroxenes, olivines, opaque and accessory minerals, epidotes, garnets, melilites, monticellites, primary carbonates and others).

This classification comprises rocks containing below 90 volume per cent of mafic minerals and is based on the mutual volume relations between light minerals (Q, A, P, F). For some rocks, such as gabbro and basalt, additional classification diagrams are needed. Separate triangular diagrams are also used for the classification of ultramafic rocks, i.e., rocks containing above 90% of minerals from group M. The classification of ultramafic rocks, which are probably the most widespread in the Earth's upper mantle, is shown in Fig. 8.4.2. Table 8.4.1, which gives the average chemical composition of common igneous rocks according to Le Maître (1976), Nickolds (1954) and Daly (1933), is a supplement to Fig. 8.4.1.

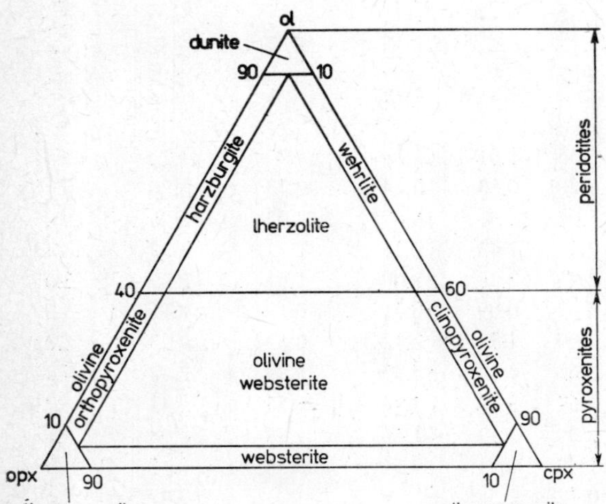

Fig. 8.4.2. Classification and nomenclature of ultramafic rocks which are the main components of the upper mantle.

TABLE 8.4.1
Average chemical composition of the main types of igneous rocks; after Le Maître denoted by D (in brackets is given number of analyses)

Plutonic rocks / Oxides	Alkali feldspar granite N (48)	Granite LM (2485)	Granodiorite LM (885)	Tonalite LM (97)	Alkali feldspar syenite N (25)	Syenite LM (517)	Monzonite N (336)
SiO_2	73.86	71.30	66.09	61.52	61.86	58.58	62.60
TiO_2	0.20	0.31	0.54	0.73	0.58	0.84	0.78
Al_2O_3	13.75	14.32	15.73	16.48	16.91	16.64	15.65
Fe_2O_3	0.78	1.21	1.38	1.83	2.32	3.04	1.92
FeO	1.13	1.64	2.73	3.82	2.63	3.13	3.08
MnO	0.05	0.05	0.08	0.08	0.11	0.13	0.10
MgO	0.26	0.71	1.74	2.80	0.96	1.87	2.02
CaO	0.72	1.84	3.83	5.42	2.54	3.53	4.17
Na_2O	3.51	3.68	3.75	3.63	5.46	5.24	3.73
K_2O	5.13	4.07	2.73	2.07	5.91	4.95	4.06
H_2O^+	0.47	0.64	0.85	1.04	0.53	0.99	0.90
H_2O^-*		0.13	0.19	0.20		0.23	0.19
P_2O_5	0.14	0.12	0.18	0.25	0.19	0.29	0.25
CO_2		0.05	0.08	0.14		0.28	0.08

Volcanic rocks / Oxides	Alkali rhyolite N (21)	Rhyolite LM (670)	Rhyodacite LM (100)	Dacite LM (651)	Quartz andesite N (50)	Alkaline trachyte N (15)	Trachyte LM (534)	Latite LM (188)
SiO_2	74.57	72.82	65.55	65.01	63.58	61.95	61.21	61.25
TiO_2	0.17	0.28	0.60	0.58	0.64	0.73	0.70	0.81
Al_2O_3	12.58	13.27	15.04	15.91	16.67	18.03	16.96	16.01
Fe_2O_3	1.30	1.48	2.13	2.43	2.24	2.33	2.99	3.28
FeO	1.02	1.11	2.03	2.30	3.00	1.51	2.29	2.07
MnO	0.05	0.06	0.09	0.09	0.11	0.13	0.15	0.09
MgO	0.11	0.39	2.09	1.78	2.12	0.63	0.93	2.22
CaO	0.61	1.14	3.62	4.32	5.53	1.89	2.34	4.34
Na_2O	4.13	3.55	3.67	3.79	3.98	6.55	5.47	3.71
K_2O	4.73	4.30	3.00	2.17	1.40	5.53	4.98	3.87
H_2O^+	0.66	1.10	1.09	0.91	0.56	0.54	1.15	1.09
H_2O^-*		0.31	0.42	0.28			0.47	0.57
P_2O_5	0.07	0.07	0.25	0.15	0.17	0.18	0.21	0.33
CO_2		0.08	0.21	0.06			0.09	0.19

* H_2O^- is the moisture, H_2O^+ is the constitutional water.

(1976) denoted by LM, after Nickolds (1954) denoted by N and after Daly (1933)

Monzodiorite N (56)	Diorite N (50)	Gabbro LM (1451)	Nepheline (feldspathoid) syenite LM (115)	Essexite N (15)	Theralite D (6)	Ijolith N(11)	Peridotite LM (37)
54.66	51.86	50.14	54.99	46.88	45.61	42.58	42.26
1.09	1.50	1.12	0.60	2.81	1.96	1.41	0.63
16.98	16.40	15.48	20.96	17.07	14.35	18.46	4.23
3.26	2.73	3.01	2.25	3.62	6.17	4.01	3.61
5.38	6.97	7.62	2.05	5.94	4.03	4.19	6.58
0.14	0.18	0.12	0.15	0.16	0.19	0.20	0.41
3.95	6.12	7.59	0.77	4.85	6.05	3.22	31.24
6.99	8.40	9.58	2.31	9.49	9.49	11.38	5.05
3.76	3.36	2.39	8.23	5.09	5.12	9.55	0.49
2.76	1.33	0.93	5.58	2.64	3.69	2.55	0.34
0.60	0.80	0.75	1.30	0.97	2.60	0.55	3.91
		0.11	0.17				0.31
0.43	0.35	0.24	0.13	0.48	0.74	1.52	0.10
		0.07	0.20			0.38	0.30

Latite andesite N (38)	Andesite LM (2600)	Tholeiitic basalt LM (202)	Alkaline olivine basalt N (96)	Phonolite LM (340)	Tephrite LM (85)	Nephelinite LM (176)	Olivine leucitite N (11)	Olivine melilitite N (10)
56.00	57.94	49.58	45.78	56.19	47.80	40.60	43.64	37.08
1.29	0.87	1.98	2.63	0.62	1.76	2.66	2.54	3.31
16.81	17.02	14.79	14.64	19.04	17.00	14.33	10.82	8.08
3.74	3.27	3.38	3.16	2.79	4.12	5.48	5.11	5.12
4.36	4.04	8.03	8.73	2.03	5.22	6.17	5.89	7.23
0.13	0.14	0.18	0.20	0.17	0.15	0.26	0.15	0.18
3.39	3.33	7.30	9.39	1.07	4.70	6.39	13.86	16.19
6.87	6.79	10.36	10.74	2.72	9.18	11.89	10.66	16.30
3.56	3.48	2.37	2.63	7.79	3.69	4.79	2.16	2.30
2.60	1.62	0.43	0.95	5.24	4.49	3.46	4.09	1.36
0.92	0.83	0.91	0.76	1.57	1.03	1.65	0.72	1.89
	0.34	0.50		0.37	0.22	0.54		
0.33	0.21	0.24	0.39	0.18	0.63	1.07	0.63	0.96
	0.05	0.03		0.08	0.02	0.60		

For comparison, the average chemical composition of selected ultramafic rocks, i.e., plutonic peridotites and volcanic melilitites, is also listed. In the upper part of the Earth's crust, accessible to direct investigations, two groups of igneous rocks are dominant: plutonic—granites and granodiorites rich in silica, and volcanic—tholeiitic basalts significantly poorer in silica (Fig. 8.4.1, Wedepohl, 1969).

Basalts, though considerably differentiated in terms of their mineral composition and geological setting, and also their connections with other rocks in petrogenetic series, in general do not differ significantly in the concentrations of the two main rock-forming oxides, SiO_2 and Al_2O_3. Basalts are products of the solidification of magmas extracted from the upper mantle by partial melting, thus providing an excellent source of information about the state and composition of the mantle material. Basalt magmas can be divided into two main groups with contrasting tendencies to differentiate under relatively low pressures (Fig. 8.4.3).

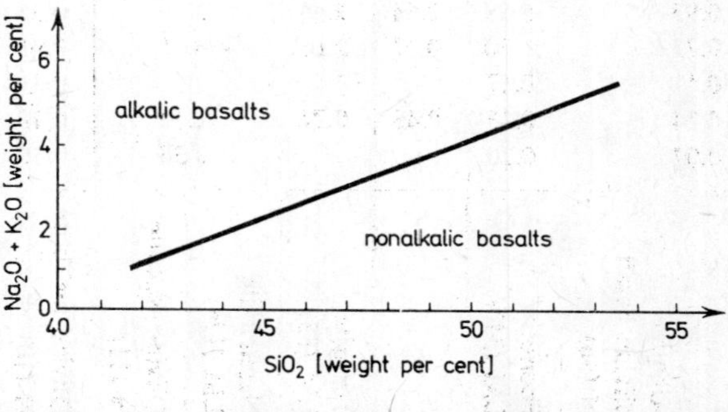

Fig. 8.4.3. Alkali content as a function of silica content in alkalic and nonalkalic basalts (tholeiites), according to MacDonald and Katsura (1964).

They are the magmas of nonalkalic basalts, often called tholeiites, whose differentiation progresses in the direction of silica saturation and oversaturation of the melt, and the magmas of alkalic basalts, differentiating in the direction of an increasing degree of silica undersaturation. The systematics of basalts is based on their normative mineral composition, calculated from their chemical composition. For basalt classification we use Yoder's and Tilley's "basalt tetrahedron" (Yoder and Tilley, 1962) with quartz, olivine, clinopyroxene–diopside and nepheline in the corners, and placioclase and orthopyroxene–hypersthene on the suitable edges of the tetrahedron (Fig. 8.4.4). The cpx–pl–ol plane divides the tetrahedron into a zone of nonalkalic basalts extending towards the quartz corner, and a zone of alkalic basalts extending towards the nepheline corner. The nonalkalic basalts can either be saturated with silica (olivine tholeiites) or oversaturated with silica (quartz tholeiites). The boundary between them is formed by the cpx–opx–pl plane, which is the surface of maximum silica saturation of magma. As to their chemistry, alkalic basalts with the same concentration of SiO_2 as nonalkalic basalts are richer in Na_2O+K_2O (Fig. 8.4.3) and poorer in CaO.

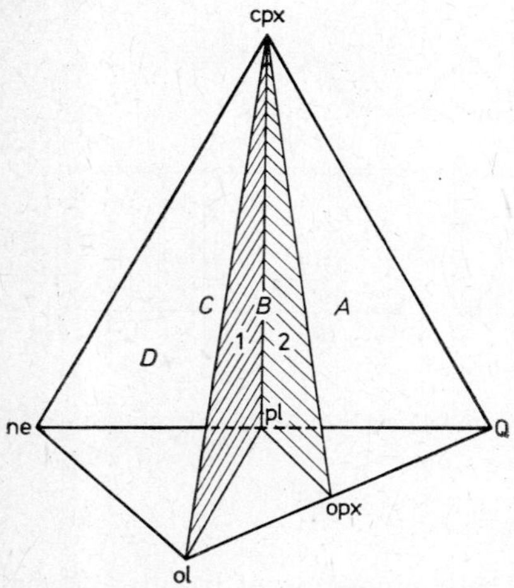

Fig. 8.4.4. Normative basalt tetrahedron applied to simplified classification of basalts, according to Yoder and Tilley (1962); 1—cpx–pl–ol, the plane of minimum silica undersaturation of basalts; 2—cpx–pl–opx, the surface of maximum silica saturation of basalts; A—the field of quartz tholeiites; B—the field of olivine tholeiites, in which at the surface cpx–pl–ol olivine basalts are present, containing 0÷3 per cent of the normative opx; C—the field of alkalic olivine basalts containing less than 5 per cent of the normative ne; D—the field of olivine basanites, in which at the surface cpx–ol–ne olivine nephelinites are located.

Yoder (1976) noticed periodicity in the intensity of magmatic activity on the scale of the whole globe. In the Earth's history it is possible to distinguish at least seven main stages of basalt magma generation. This was proved by Gastil (1960), who constructed a histogram of the absolute age of minerals, which may be compared with a diagram of the volumes of continental tholeiitic basalts of different ages, estimated for the whole Earth by cartographic methods (Fig. 8.4.5). The composition of basalts varied significantly in the Earth's history. The oldest, primitive basaltic magmas were tholeiitic magmas, while alkali basalts appeared within the last milliard years, and an especially intensive increase in their share in the group of basaltic magmas took place within the last 250 million years.

The origin of basaltic magmas has been studied in a number of experiments, among which the most comprehensive were those carried out by Yoder and Tilley (1962), Green and Ringwood (1967), O'Hara (1968),

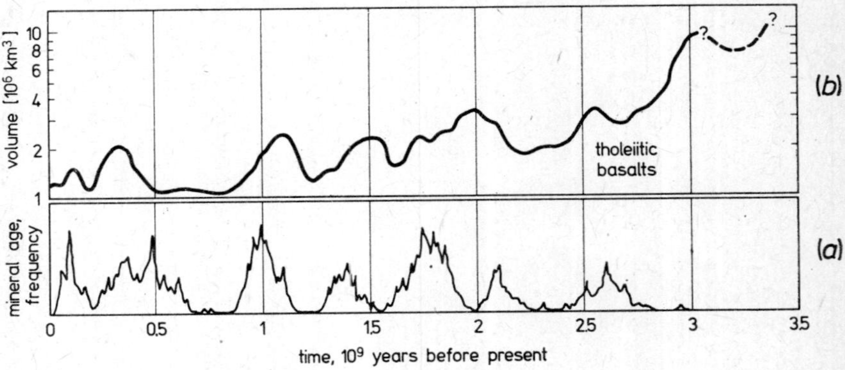

Fig. 8.4.5. Periodicity of magmatic activity in the Earth's history, according to Yoder (1976); (*a*) distribution of the absolute age of minerals, (*b*) volumes of continental tholeiitic basalts.

Green (1969) and Ringwood (1975). The composition of basaltic magmas is affected by three main processes:

1. Partial melting, occurring in a given volume of the mantle material, and the formation of a magma chamber by the segregation of homogeneous magma from residual refractory crystalline phases.

2. Fractional crystallization, occurring at different depths in a magma chamber during its rise towards the surface of the Earth.

3. Reactions of the magma with the surrounding rocks, called wall-rock reactions and other kinds of magma contamination which can change the content of certain elements in the magma to a great degree.

The extraction of basalt magmas from the upper mantle by partial melting is a complex process. In the low-velocity zone, at depths of between 75 and 150 km in the upper mantle, where an approximately 1% partial melting of pyrolite occurs due to the decrease in solidus temperature, diapir is formed under the thermodynamic conditions p and T characteristic for a mantle geotherm, and starts its adiabatic rise towards shallower horizons. During this rise the degree of melting of the diapir material increases, and so does the velocity of its movement upwards. Initially, the magma is in the state of equilibrium with residual solid phases. During the next stage magma segregation from solid phases occurs at shallower horizons and a magma chamber is formed. The composition of primitive basaltic magma depends on the degree of melting of pyrolite, on the thermodynamic conditions p and T and vapour pressure p_{H_2O} under which the magma chamber was formed. For example, at small depths, under a pressure of 0.5÷0 GPa, a magma with the composition of quartz tholeiite will be separated from anhydrous pyrolite melted in a low degree (⩽5%), while under a higher degree of pyrolite melting (> 20%) a magma with the composition of tholeiite picrite is formed. Under a pressure of 0.5÷1 GPa, magmas with the composition of high-alumina olivine tholeiites can separate from anhydrous pyrolite melted in about 20%. In magma chambers formed under a pressure of 1.2÷2 GPa, olivine tholeiite magma is generated in the case of considerable melting of pyrolite (about 25%) or magma with the composition of alkali olivine basalt when the pyrolite is less melted (5÷10%). Under a pressure of 2÷3 GPa, chambers of alkali picrite magma are segregated from dry pyrolite slightly melted, while tholeiite picrite magma originates from pyrolite more intensively melted. The presence of water considerably changes the temperature of the initial melting of pyrolite. Water is also favourable, even when the pyrolite is only slightly melted, for the generation of magmas extremely undersaturated by silica—melilitic, nephelinitic and kimberlitic magmas. Kimberlitic magmas rich in potassium are most probably separated from slightly melted (< 1%) pyrolite rich in volatile components (H_2O and CO_2) at depths of 150 km in continental areas with low values of heat flow.

High-pressure experiments concerning the melting of natural basalts (Green and Ringwood, 1967; Green, 1969) permit the construction of a model of basalt magma differentiation due to the fractional crystallization (Fig. 8.4.6). Common low-pressure differentiation of basaltic magmas occurring at depths down to 15 km under a pressure ⩽ 0.5 GPa leads either to the enrichment of the tholeiitic magma in silica through the crystallization of olivine and afterwards of plagioclase or clinopyroxene, and thus towards quartz tholeiites, or among alkali magmas, to the increase of their undersaturation by silica. At a depth of about 30 km the differentiation of tholeiite magmas progresses by the crystallization of olivine and enstatite towards high-alumina basalts. At depths of between 35 and 70 km tholeiites are differentiated towards alkali basalts and olivine basanites by fractional crystallization of aluminous enstatite, which is replaced by aluminous clinopyroxene as the temperature decreases. At depths of between 70 and 100 km, under increasing p_{H_2O}, fractional crystallization of aluminous enstatite, and afterwards also of garnet and clinopyroxene, leads to the

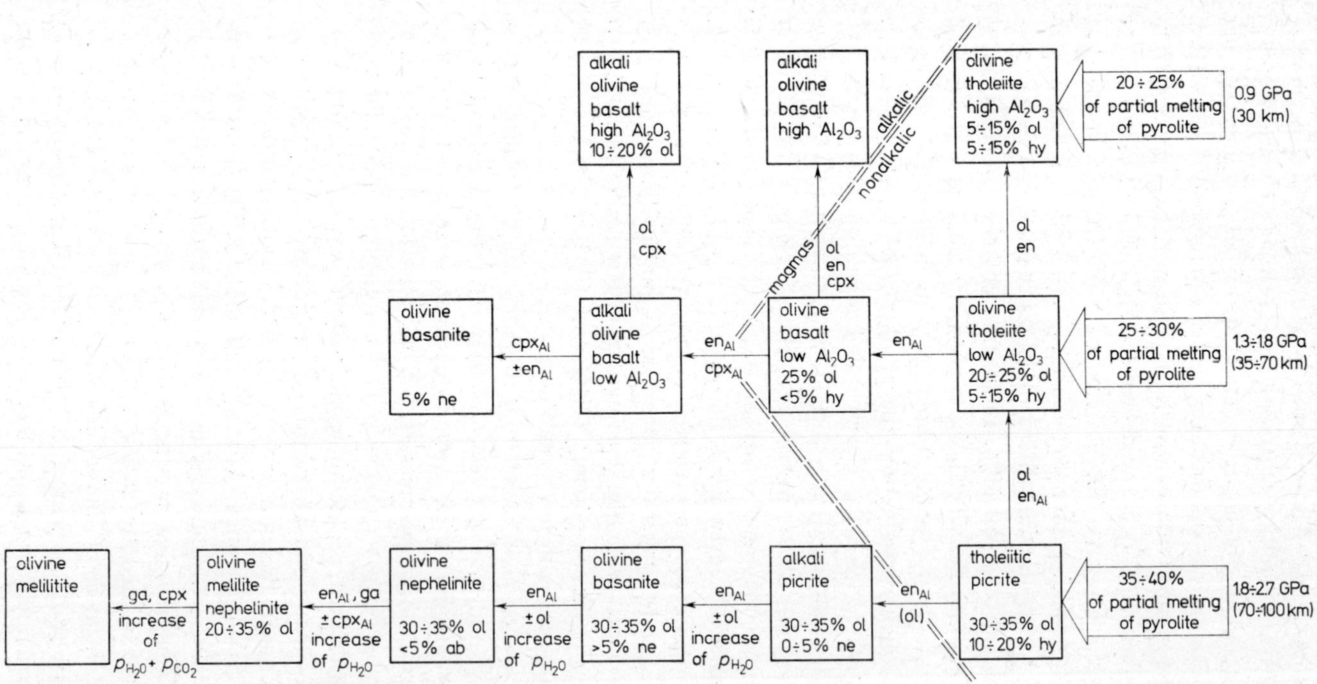

Fig. 8.4.6. Medium- and high-pressure differentiation of basaltic magmas resulting from fractional crystallization; the simplified model according to Green and Ringwood (1967) and Green (1969). The arrows with the mineral names show the trend of differentiation resulting from the crystallization of these minerals from magmas. The advance of fractional crystallization causes an increase in the alkali content and the saturation degree of magmas by silica. High-pressure differential series of basalts cross the thermal barrier marked by the double dashed line, existing under low pressures between the nonalkalic and alkalic basaltic magmas.

differentiation of tholeiite magmas towards alkali magmas with a high degree of silica undersaturation, with the composition of nephelinites and melilitites.

Acid magmas are only in a small part direct derivatives of basalt magmas. They are formed mainly by anatectic melting of the rocks of the Earth's crust, or by the melting of the rocks of the lithosphere in subduction areas, as is the case with intermediate andesite magmas.

The role played by different processes, such as the partial melting of the mantle material, fractional crystallization and wall–rock reactions, in the origin of basalt magmas can be assessed from, say, geochemical analyses of minor and trace elements. These elements can be divided into two groups according to their capacity for entering into crystalline lattices of olivines and pyroxenes, and also of garnets and spinels:

1. Compatible elements (CE): Mn, Cr, Ni, Co, Cu, Ga, V and Y.

2. Incompatible elements (IE): P, K, Ti, Ba, Cs, Rb, Sr, U, Th, Zr, Hf and the rare earth elements of the cerium group.

In the Earth's mantle IE are concentrated in accessory minerals—apatite, geikielite ilmenite, phlogopite and others. They pass into the melt even when the pyrolite is melted in a minimal degree, and they are highly concentrated in magmas extremely silica undersaturated. With the increasing degree of melting IE become considerably diluted. Significant enrichment of alkali basalts in IE can sometimes be explained by reactions with the surrounding rocks. However, in many cases a high concentration of IE indicates an extremely low degree of melting of the mantle material. The concentrations of CE together with their distribution coefficients for minerals remaining in an equilibrium state with the magma permit us to choose, between partial melting and fractional crystallization, the more probable process responsible for the generation of the basaltic rocks under consideration (Gast, 1968). Analyses of Ni and Cr concentrations in basanites and nephelinites indicate that these rocks were most probably formed from the mantle pyrolite by partial melting rather than by fractional crystallization in the parental magma with an olivine tholeiite composition.

Subalkalic basalt magmas form two differentiation series: tholeiitic and calc-alkalic. The basalt magmas of the tholeiitic series contain normative hypersthene and olivine or quartz. With the development of magma differentiation consecutive links of the petrogenetic tholeiitic series appear: tholeiite basalts—andesites—dacites—rhyolites. These rocks generally lack hornblende and biotite. The basaltic magmas of the calc–alkalic series do not differ from the tholeiitic series either in their normative mineral composition or in the composition of the petrogenetic series. The calc-alkalic basalt magmas, unlike the tholeiitic series magmas, are already considerably impoverished with regard to iron and relatively intensively enriched in SiO_2 in the earliest stage of differentiation. This results from the early crystallization of magnetite caused by high partial pressure of oxygen in the magma (Fig. 8.4.7). Moreover, the rocks formed during the final stages of differentiation of the calc–alcalic series magma contain hornblende

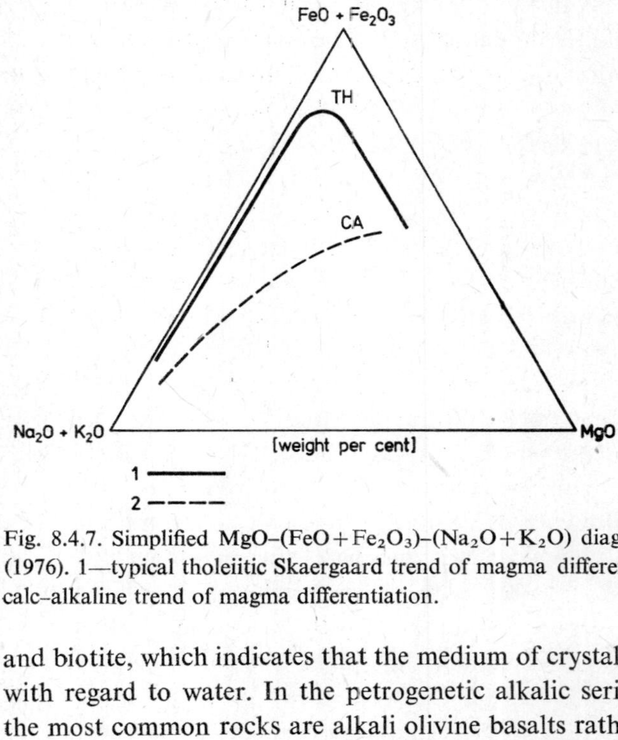

Fig. 8.4.7. Simplified MgO–(FeO + Fe_2O_3)–(Na_2O + K_2O) diagram, according to Ernst (1976). 1—typical tholeiitic Skaergaard trend of magma differentiation; 2—the average calc–alkaline trend of magma differentiation.

and biotite, which indicates that the medium of crystallization was enriched with regard to water. In the petrogenetic alkalic series of basalt magmas the most common rocks are alkali olivine basalts rather poor in normative nepheline (< 5%). The differentiation of alkali magmas usually leads from basalts and basanites towards trachytes and phonolites. The alkalic differentiation series, however, can provide various unusual petrogenetic series, often composed of rocks with a high degree of silica undersaturation. This depends on the conditions of formation of the parental basalt magma chamber and on the conditions of magma differentiation.

There is a distinct relationship between the character of petrogenetic series and their tectonic setting. The rocks of the calc-alkalic series occur exclusively along convergent boundaries of lithospheric plates, i.e., in orogenic zones of island arcs and active continental margins. The mineral parageneses and chemical composition of the rocks of this series indicate vertical movements of the basalt magma, characterized by a relatively high partial pressure of oxygen, from the dry mantle to orogenic environments rich in water. The rocks of the alkalic and tholeiitic series are much more widespread since they appear in all the main tectonic settings, both in intraplate eruptions and on convergent and divergent boundaries of lithospheric plates. Alkalic basalts are characteristically formed through intraplate eruptions, i.e., midoceanic islands and plateau basalts which are the products of linear eruptions on stable continents. Rocks of the tholeiitic series are dominant in midoceanic ridges (abyssal tholeiites) and in some plateau basalts. They can also be found on midoceanic islands and on island arcs in the early stages of their development. The character of magmatism in the same tectonic setting can changed considerably with time; for instance, on island arcs the initially tholeiitic magmatism passes through the calc-alkalic to the alkalic type. During this process the content of IE undergoes considerable differentiation, the content of trace alkalis and alkali earth elements increases, while the distribution of the rare earth elements changes from chondritic, non-differentiated, to strongly differentiated, enriched in elements of the cerium group.

The geochemical discrimination methods of searching for the primary tectonic setting of basalt rocks are now in common use. They were first worked out by Pearce and Cann (1973). They are chiefly based on those trace elements whose concentrations do not change significantly during processes of secondary alterations of the rock (Ti, Zr, Y, Sr an Nb).

8.5 THE EARTH'S CRUST

The Earth's crust contains less than 1% of the Earth's mass. Compared with the other geospheres, the crust is the most easily accessible to direct petrological and geochemical studies. The crust originates from magmas by partial melting in the Earth's primitive mantle, but it displays greater horizontal and vertical differentiation than the mantle, resulting not only from magmatic but also from weathering, sedimentary and metamorphic processes.

The crust is formed by three genetically different groups of rocks:

1. Magmatic rocks, which are primary products of the crystallization of melts of silicates.

2. Sedimentary rocks, which are the products of sedimentation on the Earth's surface of mineral substances originating from disintegration, alteration, and solubilization of primary rocks by their weathering.

3. Metamorphic rocks, which are the products of thermodynamic

transformations at different depths of rocks belonging to the previous two groups.

When the chemistry of various rock types and the proportion between them is known, the average composition of the Earth's crust can be determined. The first estimation of the composition of a sixteen km thick layer of the continental crust was made by Clarke and Washington (1924). An interesting attempt to estimate the composition of the continental crust was made by Taylor (1964), who assumed the proportion of acid to basic rocks to be 1: 1, on the basis of the distribution of the rare earth elements.

The modern estimation of the composition of the present continental and oceanic crust was made by Ronov and Yaroshevsky (1976). They divided the continental crust into Precambrian shields, devoid of sedimentary cover, plates with old crystalline Preriphean basements, and areas of younger Postriphean folded geosynclines. A subcontinental

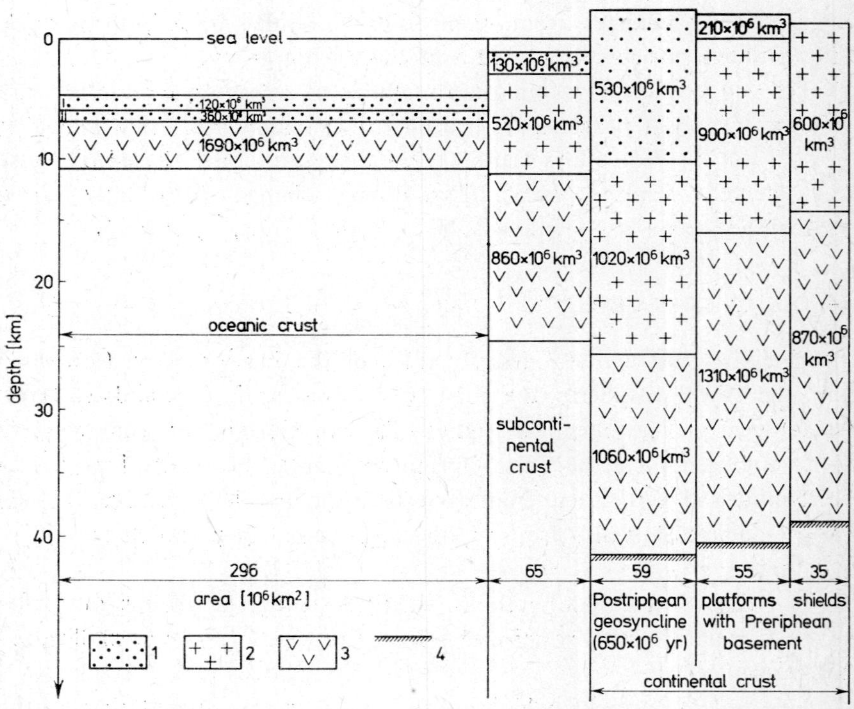

Fig. 8.5.1. Schematic diagram of the Earth's block and zonal structure, according to Ronov and Yaroshevsky (1976). The surface of blocks [km^2] and volumes of individual zones within the blocks [km^3] are given; 1—sedimentary rocks, 2—"granites", 3—"basalts", 4—Mohorovičić discontinuity.

shelf crust was distinguished on the continental margins (Fig. 8.5.1). The volumes of selected crust blocks were calculated from maps of rock thickness down to the Moho discontinuity and from a tectonic map of the world. The volume of sedimentary rocks was estimated from lithological maps. The continued distribution of the traditional "granitic" and "basaltic" layers within the continental crust may be justified by the following geophysical and petrological facts:

1. The velocity v_P of longitudinal seismic waves in acid and intermediate magmatic and metamorphic rocks is considerably lower ($5.0 \div 6.5$ km · s^{-1}) than that in basic magmatic and metamorphic rocks ($6.4 \div 8.0$ km · s^{-1}).

2. Basic rocks of greenstone facies prevail in the lower parts of old archaic shields, which could indicate an increase in rock basicity with depth in the continental crust.

On the basis of geochemical analyses and the elastic properties of rocks in the lower crust the authors agreed to assume that one-third of the continental basaltic layer is composed of rocks of the granitic layer and two-thirds are composed of basic rocks which are similar in chemical composition to the basic rocks of Archaean shields.

Within the oceanic crust, three seismic layers were distinguished on the basis of different velocities of seismic waves:

1. The layer of unconsolidated sediments (I) with an average thickness of 0.4 km and velocity of longitudinal seismic waves $v_P = 1.5 \div 1.8$ km · s^{-1}.

2. The layer composed in one-third of sedimentary rocks and in two-thirds of volcanic rocks (II) with an average thickness of 1.2 km and $v_P = 2.1 \div 5.5$ km · s^{-1}.

3. The layer of true basalts with $v_P = 6.5 \div 7.0$ km · s^{-1}. Sedimentary rocks of the continents are characterized by a distinct preponderance of clays and shales (40 weight per cent) over sandstones, volcanogenic rocks and carbonates. Sedimentary rocks in comparison with rocks of the granitic layer are richer in volatile components and light nonmetals such as H_2O, CO_2, S, C, Cl, F and B; they possess more calcium and display a high K/Na ratio (Table 8.5.1).

The granitic layer is present in the continental and subcontinental crust, while in the oceanic crust it is absent. The layer is composed of granites, granodiorites, granite gneisses and high grade metamorphosed sedimentary and sedimentary-volcanic rocks with the chemical composition of acid and intermediate rocks. The quantity of basic and ultrabasic rocks in this layer reaches 10 volume per cent. The igneous rocks are in general richer in SiO_2 in comparison with rocks which have undergone a weathering-sedimentary-metamorphic geochemical cycle. The average

TABLE 8.5.1

Average chemical composition of the Earth's crust and its individual zones, after Ronov and Yaroshevsky (1976), in weight per cent

Oxides and elements	Continental crust				Subcontinental crust				Oceanic crust				Earth's crust
	sedimentary rocks	"granites"	"basalts"	on the average	sedimentary rocks	"granites"	"basalts"	on the average	sedimentary rocks I	effusive sedimentary rocks II	basalts	on the average	on the average
SiO_2	47.75	63.08	54.84	57.23	47.75	63.08	54.84	56.88	34.62	44.54	49.43	48.17	55.24
TiO_2	0.68	0.54	0.84	0.71	0.68	0.54	0.84	0.73	0.48	1.16	1.49	1.40	0.86
Al_2O_3	12.03	15.38	14.28	14.46	12.03	15.38	14.28	14.43	8.45	13.17	15.50	14.90	14.55
Fe_2O_3	2.52	2.24	2.42	2.36	2.52	2.24	2.42	2.37	4.46	3.13	2.47	2.64	2.42
FeO	2.79	3.60	7.25	5.41	2.79	3.60	7.25	5.64	0.94	5.65	7.97	7.37	5.86
MnO	0.10	0.10	0.17	0.13	0.10	0.10	0.17	0.14	0.83	0.39	0.18	0.25	0.16
MgO	3.30	2.96	6.37	4.77	3.30	2.96	6.37	4.97	2.36	6.06	7.89	7.42	5.37
CaO	13.09	3.79	8.09	6.98	13.09	3.79	8.09	7.14	22.50	14.95	11.23	12.19	8.12
Na_2O	1.55	2.71	2.34	2.40	1.55	2.71	2.34	2.39	2.43	2.54	2.60	2.58	2.44
K_2O	2.00	2.89	1.32	1.98	2.00	2.89	1.32	1.90	1.29	0.59	0.24	0.33	1.61
P_2O_5	0.14	0.16	0.16	0.16	0.14	0.16	0.16	0.16	0.15	0.21	0.23	0.22	0.17
C org.	0.50	0.05	0.02	0.08	0.50	0.05	0.02	0.07	0.65	0.21	—	0.05	0.07
CO_2	9.57	0.81	0.37	1.48	9.57	0.81	0.37	1.37	15.81	5.22	—	1.35	1.44
SO_3	0.58	0.10	0.03	0.12	0.58	0.10	0.03	0.11	—	—	—	—	0.09
S	0.18	0.06	0.08	0.08	0.18	0.06	0.08	0.08	0.03	0.05	0.06	0.06	0.08
Cl	0.31	0.02	0.01	0.04	0.31	0.02	0.01	0.04	—	—	—	—	0.03
F	0.04	0.05	0.02	0.03	0.04	0.05	0.02	0.03	0.05	0.03	0.02	0.02	0.03
H_2O	2.87	1.46	1.40	1.57	2.87	1.46	1.40	1.56	4.95	2.10	0.69	1.05	1.46

chemical composition of the total granitic layer is close to that of granodiorite (Table 8.5.1).

The volume of the continental basaltic layer is about 1.3 times as large as that of the granitic layer. In general, the volume ratio of the granitic to basaltic layers increases in the continental crust from older towards younger tectonic units. The continental basaltic layer is composed mainly of basic metamorphic rocks, while the share of granitoides does not exceed 30 weight per cent. The composition of the whole basaltic layer is thus intermediate between basalts and andesites (Table 8.5.1).

The basaltic layer is composed of primitive nondifferentiated basalts, called oceanic tholeiites, which are characterized by a small amount of incompatible trace and subordinate elements P, K, Rb, Sr, Ba, Zr, Th and U.

In the Earth's history the evolution of the Earth's crust led to a faster and faster growth of the continents at the expense of the oceanic crust. The oldest protocontinents most probably originated in the Archaean, about 3700 million years ago (Tugarinov and Bibikova, 1979). They were characterized by a small thickness (about 10 km) and by a small amount of granitoids (up to about 10%). The growth of the continents was accompanied by an increase in their thickness and by an intensive increase of the amount of acid rocks. At the beginning of the Phanerozoic (about 600 million years ago) the surface of the continental crust, in which the amount of acid rocks was already four times as high as that of alkaline rocks, was equal to 3/4 of its present area. The evolution of the Earth's crust can be explained by the theory of magma melting in two stages. During the first stage basaltic magmas are melted from the Earth's mantle, while during the second stage parent magmas of granite are melted from basalts and other rocks of the Earth's crust. During the Earth's history, with the growing continents, the upper part of the continental crust became considerably enriched in silicon, sodium and potassium, and became impoverished in iron, magnesium and titanium, while the amount of aluminium remained almost unchanged.

The chemical composition of the present Earth's crust corresponds to intermediate rocks (Table 8.5.1) and can be represented approximately as a mixture of the most common two magmatic rocks—granite and basalt, taken with the weight ratio 35 : 65. The principal chemical elements in the Earth's crust are oxygen, composing 47% of the crust weight and 94% of its volume, and silicon, contributing 28% to the weight and less than 1% to the crust volume (Brownlow, 1979). Further elements, in order of decreasing weight contribution to the crust, are Al, Fe, Ca, Mg, Na and

K. The Earth's crust differs from the mantle and primitive chondrite matter in having a higher content of oxygen, silicon, aluminium, calcium, alkali metals and the following trace elements: Be, B, F, Ti, Sr, Ba, Zr, Hf, Nb, Ta, Tl, W, Pb, Bi, Th, U, Y and the rare earth elements. It differs also in having a lower concentration of iron, magnesium, sulphur and the following period IV trace transition metals: Co, Ni and Cr.

During the formation and evolution of the Earth's crust, the upper mantle was also subject to differentiation as a result of extraction of magmas by partial melting. The upper mantle differentiation was considerably more intensive and reached greater depths under the continents than under the oceans.

The crust matter is in constant motion which can be described by a closed geochemical cycle. The magma formed in the upper mantle is

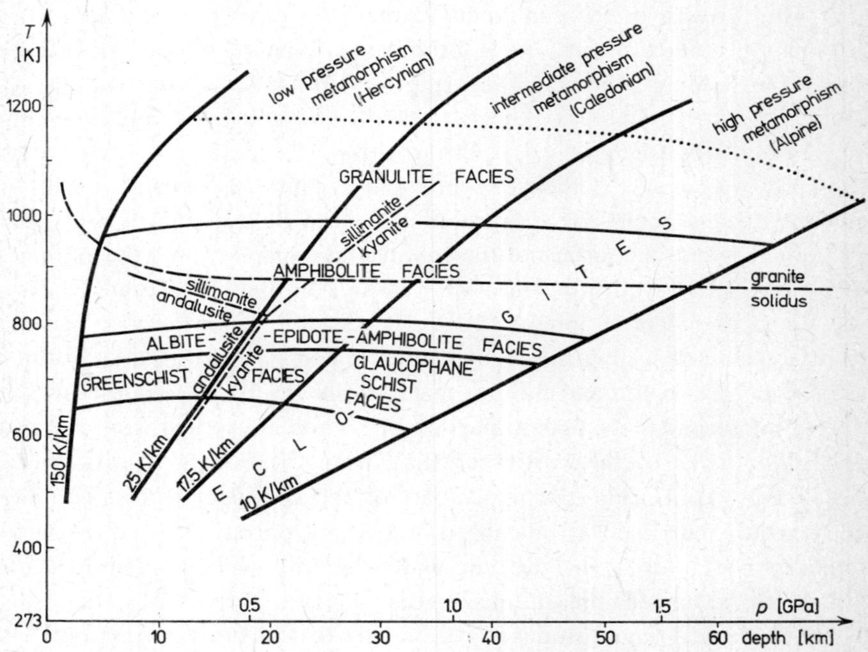

Fig. 8.5.2. Modified diagram of metamorphic facies of the Earth's crust, according to Zwart (1975).

then crystallized in the crust, thus supplying primary magmatic rocks, which are either weathered or, when placed in deeper parts, metamorphosed. The conditions of metamorphic rock formation in the Earth's crust are given in Fig. 8.5.2. In the Earth's history periods of intensification of high-pressure metamorphism of the Alpine type alternate with periods of domination of high-temperature metamorphism of the Hercynian type. The metamorphic rock complexes undergo either weathering or melting and palingenesis resulting in the formation of new magmas. The products of rock weathering undergo transportation, sedimentation and consolidation during the process of diagenesis. Sedimentary rocks formed in this way can reach the area of anatectic melting and palingenesis through a metamorphism zone, thus becoming a source of new magmas. There is also a constant exchange of matter between the upper mantle and the Earth's crust, as not only the mantle continuously enriches the crust in new basaltic magmas, but also the crust matter falls into the upper mantle in the form of whole plates sinking in subduction zones. The Earth's crust also exchanges its matter with the outermost zones of the globe—the atmosphere and hydrosphere.

REFERENCES

Ahrens T. J. (1972), "The state of mantle minerals", in: *The Upper Mantle*, A. R. Ritsema (ed.), Elsevier, Amsterdam.

Ahrens T. J. (1979), "Equations of State of Iron Sulfide and Constraints of the Sulfur Content of the Earth", *J. Geophys. Res.*, **84**, 985–998.

Ahrens T. J., R. Jeanloz and M. R. Sommervile (1978), "Composition of the Earth's Core", Abstracts of the International Conference: *High Pressure Physics and Core and Mantle Dynamics*, Potsdam.

Alder J. (1972), "Core geochemistry", in: *The Encyclopedia of Geochemistry and Environmental Sciences. Encyclopedia of Earth Sciences Series*, R. W. Fairbridge (ed.), Van Nostrand and Reinold, New York, Vol. IV A, 195–196.

Anderson D. L. (1977), "Composition of the mantle and core", *Ann. Rev. Earth Planet. Sci.*, **5**, 179–202.

Anderson D. L. (1979a), "The upper mantle transition region: eclogite?" *Geophys. Res. Lett.*, **6**, 433–436.

Anderson D. L., (1979b), "Chemical stratification of the mantle", *J. Geophys. Res.*, **84**, 6297–6298.

Birch F. (1963), "Some geophysical applications of high pressure research", in: *Solids under Pressure*, W. Paul and D. Warschauer (eds.), McGraw-Hill, New York, 137–162.

Brett R. (1976), "The current status of speculations on the core of the Earth", *Rev. Geophys. Space Phys.*, **14**, 375–383.

Brownlow A. H. (1979), *Geochemistry*, Prentice-Hall, Englewood Cliffs.

Bullen K. E. (1973), "Cores of the terrestrial planets", *Nature*, **243**, 68.

Burns R. G. (1973), "The partitioning of trace transition elements in crystal structures: a provocative review with applications to mantle geochemistry", *Geochim. Cosmochim. Acta*, **37**, 2395–2403.

Carswell D. A. (1980), "Mantle derived lherzolite modules associated with kimberlite, carbonatite and basalt magmatism: A review", *Lithos*, **13**, 121–138.

Clarke F. W. and H. S. Washington (1924), "The composition of the Earth's crust", *US Geol. Surv. Profess. Paper* **127**.

Daly R. A. (1933), *Igneous Rocks and the Depths of the Earth*, McGraw-Hill, New York.

Ernst W. G. (1976), *Petrologic Phase Equilibria*, W. H. Freeman and Company, San Francisco.

Gast P. W. (1968), "Trace element fractionation and the origin of tholeiitic and alkaline magma types" *Geochim. Cosmochim. Acta*, **32**, 1057–1080.

Gastil G. (1960), "The distribution of mineral dates in time and space", *Am. J. Sci.*, **258**, 1–35.

Green D. H. (1969), "The origin of basaltic and nephelinitic magmas in the Earth's mantle", *Tectonophysics*, **7**, 409–422.

Green D. H. and Ringwood A. E. (1967), "The genesis of basaltic magmas", *Contrib. Mineral. Petrol.*, **15**, 103–190.

Harris P. G., Hutchison R. and Paul D. K. (1972), "Plutonic xenoliths and their relation to the upper mantle", *Phil. Trans. R. Soc. Lond.*, **A271**, 313–323.

Harris P. G., Reay A. and White I. G. (1967), "Chemical composition of the upper mantle", *J. Geophys. Res.*, **72**, 6359–6369.

Heide F. (1957), *Kleine Meteoritenkunde*, 2nd ed., Springer-Verlag, Berlin.

Herndon J. M. (1980), "The chemical composition of the interior shells of the Earth", *Proc. R. Soc. Lond.*, **A372**, 149–154.

Hyndman D. W. (1972), *Petrology of Igneous and Metamorphic Rocks*, McGraw-Hill, New York.

IUGS Subcommission on the Systematics of Igneous Rocks (1973), Classification and Nomenclature of Plutonic Rocks, *N. Jb. Min. Mh.*, **4**, 149–164.

IUGS Subcommission on the Systematics of Igneous Rocks (1978), Classification and Nomenclature of Volcanic Rocks, Lamprophyres, Carbonatities and Melilitic Rocks, *N. Jb. Min. Abh.*, **134**, 1–14.

Jacobs J. A. (1975), *The Earth's Core*, Academic Press, London.

Kumazawa M., Sawamoto H., Ohtani E. and Masaki K. (1974), "Postspinel phase of forsterite and evolution of the Earth's mantle", *Nature*, **247**, 356–358.

Le Maître R. W. (1976), "The chemical variability of some common igneous rocks", *J. Petrology*, **17**, 589–637.

Maaløe S. and Aoki K. (1977), "The major element composition of the upper mantle estimates from the composition of lherzolites", *Contrib. Mineral. Petrol.*, **63**, 161–173.

MacDonald G. A. and Katsura T. (1964), "Chemical composition of Hawaiian lavas", *J. Petrology*, **5**; 82–133.

Mason B. (1966), *Principles of Geochemistry*, 3rd ed., J. Wiley and Sons, New York.

Middlemost E. A. K. (1975), "The basalt clan", *Earth Science Rev.*, **11**, 337–364.

Nickolds S. R. (1954), "Average chemical compositions of some igneous rocks", *Bull. Geol. Soc. Am.*, **65**, 1007–1032.

O'Hara M. J. (1968), "The bearing of phase equilibria studies in synthetic and natural systems on the origin of basic and ultrabasic rocks", *Earth–Science Rev.*, **4**, 69–133.

Pearce J. A. and Cann J. R. (1973), "Tectonic setting of basic volcanic rocks determined using trace element analyses", *Earth Planet. Sci. Lett.*, **19**, 290–300.

Polański A. and Smulikowski K. (1969), *Geochemistry*, Wydawnictwa Geologiczne, Warszawa (in Polish).

Ramsey W. H. (1949), "On the nature of the Earth's core", *Mon. Not. R. Astron. Soc. Geophys.*, Suppl. **5**, 409–412.

Ringwood A. E. (1966), "The chemical composition and origin of the Earth", in: *Advances in Earth Sciences*, P. M. Hurley (ed.), MIT Press, Boston, 287–356.

Ringwood A. E. (1975), *Composition and Petrology of the Earth's Mantle*, McGraw-Hill, New York.

Ringwood A. E. (1979), *Origin of the Earth and Moon*, Springer-Verlag, New York.

Ronov A. B. and Yaroshevsky A. A. (1972), "Earth's crust geochemistry", in: *The Encyclopedia of Geochemistry and Environmental Sciences, Encyclopedia of Earth Sciences Series*, R. W. Fairbridge (ed.), Van Nostrand and Reinold, New York, Vol. IV A, 243–254.

Ronov A. B. and Yaroshevsky A. A. (1976), "A new model of the chemical composition of the Earth's crust", *Geokhimiya*, **12**, 1763–1795 (in Russian).

Stacey F. D. (1977), "A thermal model of the Earth", *Phys. Earth Planet. Inter.*, **15**, 341–348.

Streckeisen A. (1967), Classification and Nomenclature of Igneous Rocks, *N. Jb. Min. Abh.*, **107**, 144–240.

Taylor S. R. (1964), "Abundance of chemical elements in the continental crust: a new table", *Geochim. Cosmochim. Acta*, **28**, 1273–1285.

Tugarinov A. I. and Bibikova Ye. V. (1979), "Evolution of the chemical composition of the Earth's crust", in: *Partitioning of the Elements and Isotopes in Geochemical Processes*, Nauka, Moskva, 5–14 (in Russian).

Wedepohl K. H. (1969), "Composition and abundance of common igneous rocks", in: *Handbook of Geochemistry*, K. H. Wedepohl (ed.), Springer-Verlag, Berlin, 227–271.

Yoder H. S., Jr. (1976), *Generation of Basaltic Magma*, National Academy of Sciences, Washington, D. C.

Yoder H. S. Jr. and Tilley C. E. (1962), "Origin of basalt magmas: an experimental study of natural and synthetic rock systems", *J. Petrology*, **3**, 342–532.

Zwart H. J. (1975), "The duality of orogenic belts", in: *Metamorphism and Plate Tectonic Regimes*, W. G. Ernst (ed.), Dowden, Hutchinson and Ross, Stroudsburg, 43–69.

MAIN MINERALS CITED IN CHAPTERS 7 AND 8, THEIR INDICES, CHEMICAL FORMULAS AND SYSTEMS

PART A

Non-silicates

Mineral class	Mineral name	Index	Chemical formula	System
Oxides	periclase	per	MgO	cubic
	wustite	wu	FeO	cubic
	magnesiowustite	mwu	$(Mg,Fe)O$	cubic
	spinel	sp	$(Mg,Fe)(Al,Cr)_2O_4$	cubic
	corundum	cor	Al_2O_3	hexagonal
	ilmenite	il	$FeTiO_3$	trigonal
	geikielite	ge	$MgTiO_3$	trigonal
	perovskite	per	$CaTiO_3$	cubic
	quartz	Q	SiO_2	{ α–trigonal, β–hexagonal
	tridymite	trid	SiO_2	hexagonal
	cristobalite	crist	SiO_2	tetragonal
	coesite	coes	SiO_2	monoclinic
	stishovite	stish	SiO_2	tetragonal
	rutile	ru	TiO_2	tetragonal
Carbonates	calcite	cal	$CaCO_3$	trigonal
	dolomite	dol	$CaMg[CO_3]_2$	trigonal
Phosphates	apatite	ap	$Ca_5[F(PO_4)_3$	hexagonal

PART B
Silicates

Structural group	Mineral name	Index	Chemical formula	System
Ortho-silicates	olivine:	ol	$(Mg,Fe)_2SiO_4$	orthorhombic
	forsterite	fo	Mg_2SiO_4	orthorhombic
	fayalite	fa	Fe_2SiO_4	orthorhombic
	monticellite	mtc	$CaMgSiO_4$	orthorhombic
	garnet:	ga	$X_3^{II}Y_2^{III}Si_3O_{12}$—variable	cubic
	pyrope	py	$Mg_3Al_2Si_3O_{12}$	cubic
	almandine	alm	$Fe_3Al_2Si_3O_{12}$	cubic
	grossular	gross	$Ca_3Al_2Si_3O_{12}$	cubic
	sillimanite	sill	Al_2SiO_5	orthorhombic
	andalusite	and	,,	orthorhombic
	kyanite	ky	,,	triclinic
Diortho-silicates	melilite	mel	$(Ca,Na)_2(Al,Mg)(Si,Al)_2O_7]$	tetragonal
Orthodi-ortho-silicates	epidote	ep	$Ca_2(Al,Fe)_3Si_3O_{12}(OH)$	monoclinic
	zoisite	zo	$Ca_2Al_3[(Si_2O_7)(SiO_4)O(OH)]$	orthohombic or monoclinic
Chain silicates	pyroxenes:	px	variable	
	(1) orthopyroxene	opx	variable	orthorhombic
	enstatite	en	$MgSiO_3$	orthorhombic
	hypersthene	hy	$(Mg,Fe)SiO_3$	orthorhombic
	(2) clinopyroxene	cpx	variable	monoclinic
	diopside	di	$CaMgSi_2O_6$	monoclinic
	clinoenstatite	cen	$MgSiO_3$	monoclinic

PART B (continued)

Double chain silicates	amphiboles:	amph	variable	
	actinolite	act	$Ca_2(Mg,Fe)_5[(OH)Si_4O_{11}]_2$	monoclinic
	tremolite	tr	$Ca_2Mg_5[(OH)Si_4O_{11}]_2$	monoclinic
	hornblende	hbd	$(Ca,Na,K)_{2-3}(Mg,Fe^{2+})_4(Fe^{3+},Al)[(OH)_2Si_6(Si,Al)_2O_{22}]$	monoclinic
Sheet silicates	mica:	mc	variable	
	muscovite	musc	$KAl_2[(OH)_2AlSi_3O_{10}]$	monoclinic
	biotite	bi	$K(Mg, Fe, Al)_3[(OH)_2Al_{1-1.5}Si_{3-2.5}O_{10}]$	monoclinic
	phlogopite	phl	$KMg_3[(OH)_2AlSi_3O_{10}]$	monoclinic
	chlorite	chl	$Al(Mg,Fe)_5[(OH)_8AlSi_3O_{10}]$	monoclinic or triclinic
	serpentine	serp	$Mg_6[(OH)_8Si_4O_{10}]$	monoclinic
Framework silicates	feldspars:	fsp	variable	
	potassium feldspar:	Kfsp	$KAlSi_3O_8$	
	orthoclase		,,	monoclinic
	sanidyne		,,	monoclinic
	microcline		,,	triclinic
	plagioclase:	pl	variable	triclinic
	albite	ab	$NaAlSi_3O_8$	triclinic
	anorthite	an	$CaAl_2Si_2O_8$	triclinic
	feldspathoids:	fdt	variable	
	nepheline	ne	$NaAlSiO_4$	hexagonal
	leucite	lc	$K[AlSi_2O_6]$	tetragonal
	sodalite		$Na_8[(AlSiO_4)_6Cl_2]$	cubic
	nosean		$Na_8[(AlSiO_4)_6(SO_4)]$	cubic
	hauyne		$(Ca, Na)_{8-4}[(AlSiO_4)_6(SO_4)_{2-1}]$	cubic
	analcime		$Na[AlSi_2O_6] \cdot H_2O$	cubic
	cancrinite		$Na_6Ca[(AlSiO_4)_6(CO_3)] \cdot 2H_2O$	hexagonal
	scapolite (solid solution)		$Na_8[(AlSi_3O_8)_6(Cl_2,SO_4,CO_3)]$— —$Ca_8[(Al_2Si_2O_8)_6(Cl_2,SO_4,CO_3)_2]$	tetragonal

PART C

Correlation of silicate structures

	Nesosilicates	Sorosilicates		Inosilicates		Phyllosilicates	Tectosilicates
	orthosilicates	diortho-silicates	orthodiorthosilicates	chain silicates	double chain silicates	sheet silicates	framework silicates
Basic unit formula	$[SiO_4]^{4-}$	$[Si_2O_7]^{6-}$	$[(SiO_4)(Si_2O_7)]^{10-}$	$[SiO_3]^{2-}$	$[Si_4O_{11}]^{6-}$	$[Si_4O_{10}]^{4-}$	$[Al_nSi_{k-n}O_{2k}]^{n-}$

Index